Oil and Gas Production from Carbonate Rocks

Oil and Gas Production FROM Carbonate Rocks

EDITED BY

George V. Chilingar

School of Engineering
University of Southern California
Los Angeles, California

Robert W. Mannon

Consulting Petroleum Engineer
Los Angeles, California

Herman H. Rieke, III

Continental Oil Company
Ponca City, Oklahoma

American Elsevier
Publishing Company, Inc.
NEW YORK

AMERICAN ELSEVIER PUBLISHING COMPANY, INC.
52 Vanderbilt Avenue, New York, N.Y. 10017

ELSEVIER PUBLISHING COMPANY
335 Jan Van Galenstraat, P.O. Box 211
Amsterdam, The Netherlands

International Standard Book Number 0-444-00099-2

Library of Congress Card Number 70-153417

Printed in the United States of America

THIS BOOK IS DEDICATED TO

Edmund C. Babson, Carrol M. Beeson,
and Marshall B. Standing

in recognition for their important contributions to the international oil and gas industry

Contributors

GEORGE V. CHILINGAR	*Department of Petroleum Engineering, University of Southern California, Los Angeles, California (pp.*1–22, 30, 31, 255–308, 340–368)
LEENDERT DE WITTE	*Consultant, San Gabriel, California (pp.* 143–215)
MARTIN FELSENTHAL	*Production Research Division, Continental Oil Company, Ponca City, Oklahoma (pp.* 83–142)
HOWARD H. FERRELL	*Production Research Division, Continental Oil Company, Ponca City, Oklahoma (pp.* 83–142)
ALFRED R. HENDRICKSON	*Dowell of Canada, Inc., Calgary, Alberta, Canada (pp.* 309–339)
RICHARD L. JODRY	*Production Research Department, Sun Oil Company, Richardson, Texas (pp.* 35–82)
GERALD L. LANGNES	*Alaska Division, Exploration and Production, Mobil Oil Corporation, Los Angeles, California (pp.* 255–308, 363–368)
ROBERT W. MANNON	*Consulting Petroleum Engineer, Los Angeles, California (pp.* 1–22, 216–308, 340–368)
HERMAN H. RIEKE, III	*Production Research Division, Continental Oil Company, Ponca City, Oklahoma (pp.* 1–22, 30, 31, 255–308, 340–368)
NICO VAN WINGEN	*Consulting Petroleum Engineer, Adjunct Professor of Petroleum Engineering, University of Southern California, Los Angeles, California (pp.* 23–30)

Contents

Foreword

Many of the problems which arise in the development and production of carbonate reservoirs are uniquely characteristic of these reservoirs; the problems result from the peculiarities of pore structure and the special chemistry of carbonate rocks. Much of the theory and many of the practices acceptable for sandstones apply inexactly or not at all to carbonates. Fluid flow through highly fractured rocks or vugular limestones, for example, is a substantially different process from that through a homogeneous sandstone.

The field and laboratory investigations of the special performance characteristics of carbonate reservoirs have been substantially fewer than their economic value would warrant. Nevertheless, a number of papers have been published which have resulted in a developing technology, specifically applicable to carbonate reservoirs. These papers are scattered, however, throughout the petroleum literature. In this book much of this technology has been collected into a single, comprehensive reference volume.

On some topics specific recommendations are given for treating problems in carbonate reservoirs. In other areas only suggestions of promising approaches are possible. Thus schemes for classifying carbonate reservoir rocks are well developed, but the application of these classifications in predicting reservoir performance needs further investigation. Although recently several studies have been reported on many facets of fluid flow in fractured reservoirs, experimental and theoretical studies of flow in vugular materials are limited. Specific techniques in formation evaluation have been developed for carbonate reservoirs, but the heterogeneity of carbonates makes the interpretations difficult. The particular physical and chemical properties of carbonates have led, however, to the development of stimulation methods which have been remarkably successful for these reservoirs.

One of the more important contributions of this book is to focus attention on carbonate reservoir engineering as a special subject in its own right, meriting and requiring more extensive study in many areas. The material presented should serve as a nucleus around which a more substantial body of technology specifically related to carbonate reservoirs can develop.

LYMAN L. HANDY
Chairman, Division of Petroleum and Chemical Engineering, University of Southern California, Los Angeles, California

Preface

As of January 1, 1968, the estimated proved oil reserves for the world, according to *World Oil*, were 399.5 billion barrels. A large percentage of these reserves resides in carbonate rocks. Although accurate figures are not available, it appears that somewhere between 35 and 50% of the world petroleum reserves are contained in carbonate reservoirs. The enormous reserves and potential of carbonate rocks have resulted in intensified efforts by petroleum engineers and geologists to obtain a clearer insight into the nature of carbonate reservoirs and their performance. This book represents a joint effort by a group of oil industry technologists to focus on some aspects of the broad knowledge of the subject and, where possible, to place the available facts in proper perspective in relation to the overall technology of oil and associated gas production from carbonate rocks.

This book does not represent a synthesis or broad overview of the topic, which would certainly be a commendable goal. Meaningful syntheses or summaries have perhaps been achieved only for specific portions of the subject, and it is questionable whether a general overview or synthesis is feasible at this time. To synthesize is to combine singular elements into an entity and to reach a conclusion directly from given propositions or established principles. The current fund of knowledge concerning the exploitation of carbonate reservoirs would not appear to support a general synthesis. We are still in the "building stages" of the subject, and the objective of this book is rather to piece together some of the building blocks (in the judgment of the authors) in such a way as to provide a clearer understanding of current techniques for the analysis and evaluation of carbonate reservoirs and their performance. If this book serves to accomplish this end in substantial part, the authors have achieved their goal.

THE EDITORS

CHAPTER 1

I. Introduction to Carbonate Reservoir Rocks

GEORGE V. CHILINGAR, ROBERT W. MANNON, AND HERMAN H. RIEKE, III

Status of the Subject

There appears to be less fundamental knowledge and information about the physical properties and performance characteristics of carbonate reservoirs than of sandstone reservoirs. This is largely due to the fact that limestones and dolomites tend to have more complex pore systems than do sandstones, because carbonates are usually subjected to more intricate depositional environments and postdepositional processes.[1] Historically, a large portion of the original investigations to define meaningful physical principles in the production of oil and gas utilized sandstone reservoirs having intergranular porosity. A great bulk of this knowledge can be applied with discretion to carbonate reservoirs. Some of the data, however, must be greatly revised and supplemented before it can be applied to carbonates. The authors of this book, in general, have concentrated on areas generally unique to carbonate rocks. At the same time, in order to provide adequate general orientation, they have attempted to place their discussion and treatment of problems peculiar to limestones and dolomites in a broad framework that applies to all types of reservoirs.

Classification Systems

Several systems of classification covering various aspects of carbonate rocks, their contained fluids and their performance, are presented. They have been included because good classification schemes serve to simplify and clarify by placing in proper perspective each of the myriad of elements that have an effect on the particular system under examination. On the other hand, classification systems by their very nature tend to restrict, to limit, to categorize, and to oversimplify. These features are not desirable when dealing with complicated porous systems of carbonate rocks and their contained fluids. It is hoped, however, that the proposed classification systems will be helpful to the reader in his study of carbonate reservoirs, even though both the theoretical and the practical observations made are somewhat limited.

References p. 32.

Composition and Properties of Carbonate Rocks

Limestones are composed of more than 50% carbonate minerals; of these, 50% or more consist of calcite and/or aragonite. A small admixture of clay particles or organic matter imparts a gray color to limestones, which may be white, gray, dark gray, yellowish, greenish, or blue in color; some are even black. Dolomites are rocks which contain more than 50% of the minerals dolomite [$CaMg(CO_3)_2$] and calcite (plus aragonite), with dolomite being more dominant. The pure dolomite mineral is composed of 45.7% $MgCO_3$ and 54.3% $CaCO_3$, by weight; or 47.8% CO_2, 21.8% MgO, and 30.4% CaO. Dolomites are quite similar to limestones in appearance, and therefore it is difficult to distinguish between the two with the naked eye. On the basis of CaO/MgO ratios, Frolova[2] proposed the classification presented in Table 1.

TABLE 1. Frolova's Classification of Dolomite-Magnesite-Calcite Series*

Name	Content (%)			CaO/MgO ratio
	Dolomite	Calcite	Magnesite	
Limestone	5–0	95–100	...	> 50.1
Slightly dolomitic limestone	25–5	75–95	...	9.1–50.1
Dolomitic limestone	50–25	50–75	...	4.0–9.1
Calcitic dolomite	75–50	25–50	...	2.2–4.0
Slightly calcitic dolomite	95–75	5–25	...	1.5–2.2
Dolomite	100–95	0–5	...	1.4–1.5
Very slightly magnesian dolomite	100–95	...	0–5	1.25–1.4
Slightly magnesian dolomite	95–75	...	5–25	0.80–1.25
Magnesian dolomite	75–50	...	25–50	0.44–0.80
Dolomitic magnesite	50–25	...	50–75	0.18–0.44
Slightly dolomitic magnesite	25–5	...	75–95	0.03–0.18
Magnesite	5–0	...	95–100	0.00–0.03

* After Frolova, Ref. 2, p. 35.

The origin, occurrence, classification, and physical and chemical aspects of carbonate rocks are presented in detail by Chilingar *et al.*[3,4] Some geological terms commonly used in carbonate rock literature are defined in the Glossary at the end of this chapter.

It is very important to evaluate as correctly as possible various properties of carbonate rocks. A good case in point is the Fullerton Clearfork dolomitic limestone reservoir in the Permian Basin. Bulnes and Fitting[5] reported that 82% of the core samples had permeabilities of less than 1 md. As discussed in Chapter 5, the problem of what to use for minimum productive permeability becomes very acute in such instances. According to Bulnes and Fitting, if 1 md were used as the minimum productive permeability instead of the actual value of 0.1 md, the resulting estimated ultimate recovery would be 70% in error. The core analysis of some carbonate rocks is complicated by the presence of fractures and solution cavities. In order to analyze such rocks, "whole" or "large" core analysis, whereby the entire core is analyzed instead of small plugs, was developed.

Porosity and Permeability

The diagenesis of carbonate rocks, including dolomitization and consequent formation of porosity, has been discussed by Chilingar *et al.*[1] The replacement of calcite by dolomite involves a contraction (an increase in porosity) of about 12–13%[6] if the reaction proceeds as follows:

$$2CaCO_3 + Mg^{++} \rightarrow CaMg(CO_3)_2 + Ca^{++}.$$

Obviously, dolomitization will give rise to porosity, providing a solid framework (such as crinoid stems) is available to minimize the effects of subsequent compaction. Subsequent precipitation of carbonates in pores may also destroy the porosity formed as a result of dolomitization. On the basis of extensive statistical studies,[1, 6, 7] it may be said that dolomitization increases porosity. Consequently the plotting of isopleths of equal Ca/Mg ratios and determination of the directions in which Ca/Mg ratios decrease are of value in exploration for petroleum.

The relationship between porosity and degree of dolomitization for the Asmari Limestone in Iran is presented in Fig. 1. Inasmuch as at 20°C the density of calcite is 2.71 g/cc and that of dolomite is equal to 2.87 g/cc, there is also a definite relationship between the density and the porosity of Iranian carbonate rocks.[1] There is a gradual trend of density from 2.7 g/cc for the low-porosity (0–4.1%) to 2.8 g/cc for the high-porosity ($\geqslant$12.1%) group.[1]

As shown in Figs. 2 and 3, the total porosity is related to the density and the electrical resistivity of the rocks.

The original sedimentary facies variations often had a profound effect on the subsequent development of porosity and permeability; therefore, the paleogeographic studies should be stressed. Analysis of the original environment (e.g., basin, shelf, bank, reef, back-reef, barrier beach, lagoon) is of great help in the identification of porosity characteristics and the prediction of porosity trends.

Some correlation between porosity and permeability was found to exist in many carbonate reservoirs, and the shape and position of porosity versus permeability curves depend largely on lithology. For example, in Fig. 4 the relationship between the porosity and the permeability of algal carbonates and shelly calcilutites is presented.[9] The data were obtained from analysis of full-diameter cores; permeabilities were measured parallel to bedding. As

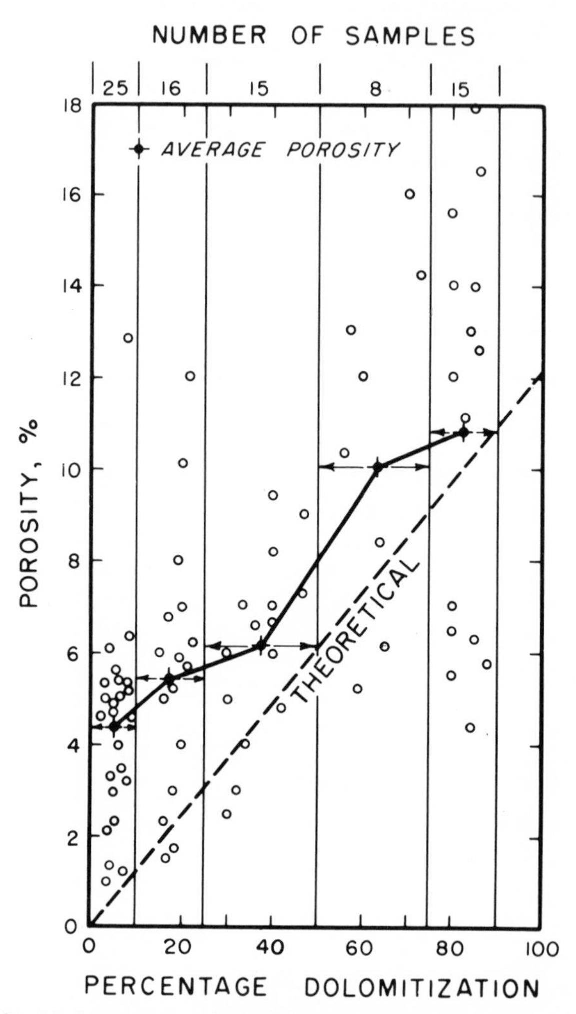

FIG. 1. Relationship between porosity and degree of dolomitization for Asmari Limestone in Iran.[6] (Courtesy of *Petroleum Engineer*.)

can be deduced from Fig. 4, the algal rocks have the larger and better-connected pores (chiefly interparticle and/or vuggy). The calcilutites contain small molds, due to leaching, which are interconnected by intercrystalline pores. According to Choquette and Traut,[9] the best porous and permeable zones in the Ismay Field, Utah and Colorado, are found in the carbonate buildups, mainly in grain-supported algal rocks which had good primary porosity and permeability. Much of the porosity now present in the buildups and in associated carbonate facies, however, is attributed to leaching.[9] Some

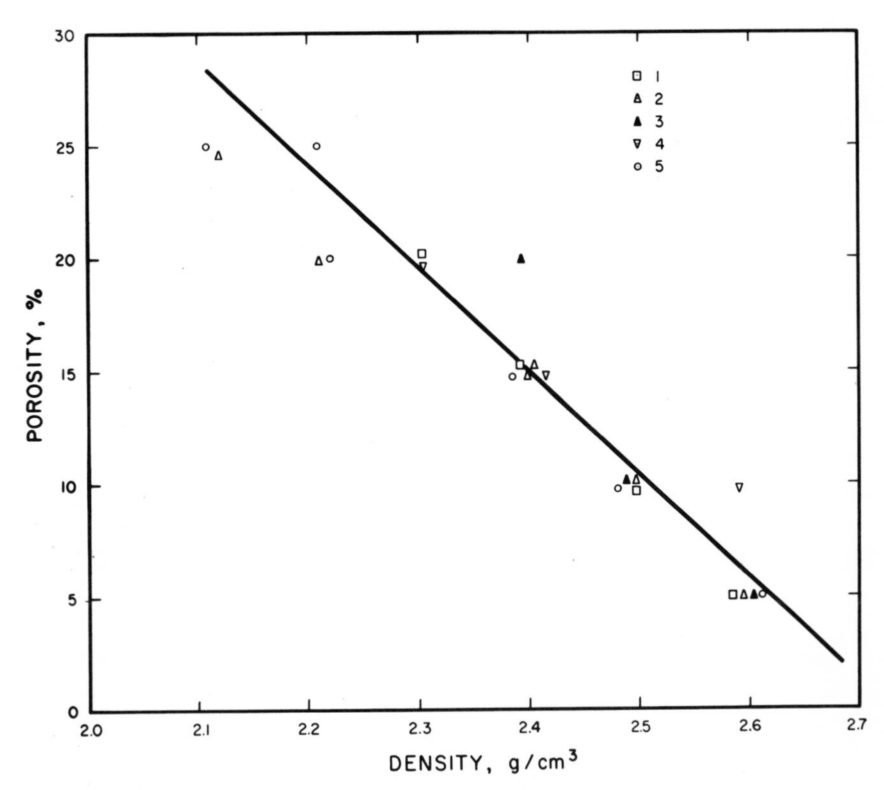

FIG. 2. Relationship between density and porosity of rocks. (After M. L. Ozerskaya,[8] p. 125.) 1—shale, 2—limestone, 3—marl, 4—dolomite, and 5—sandstone.

diagenetic processes gave rise to zones having optimum porosity and permeability, and include dolomitization and leaching. In some cases porosity was destroyed by the infilling of voids with coarsely crystalline calcite and with anhydrite.

Another example of the influence of lithology on the porosity-permeability relationship of carbonate is shown in Fig. 5 for the Cretaceous Edwards Limestone, which exhibits intercrystalline porosity. As the texture ranges from microgranular to coarse-grained, the permeability increases for a given porosity. Craze[10] also noted a similar influence of texture on porosity-permeability trends for samples from the Ellenburger Group (Ordovician) and some Permian carbonate formations in west Texas.

Types of Porosity in Carbonate Reservoirs

On the basis of pore size, the porosity in carbonate rocks may be classified as follows[11]: (1) cavernous (>2 mm), (2) very coarse (1.0–2.0 mm), (3) coarse

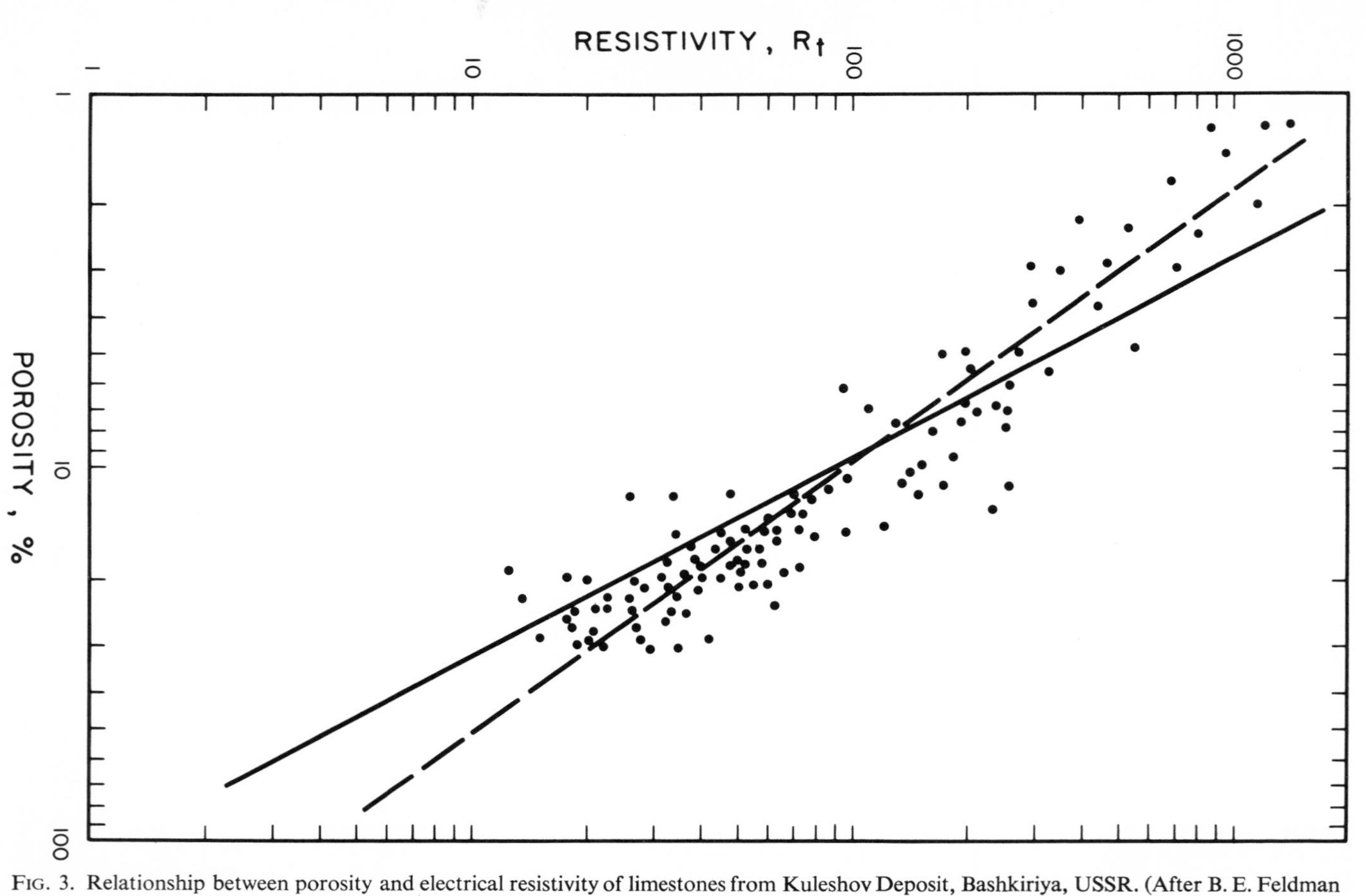

Fig. 3. Relationship between porosity and electrical resistivity of limestones from Kuleshov Deposit, Bashkiriya, USSR. (After B. E. Feldman and A. T. Boyarov,[8] p. 128.) *Solid line*—calculated from Archie's formula; *dashed line*—calculated using the Humble formula.

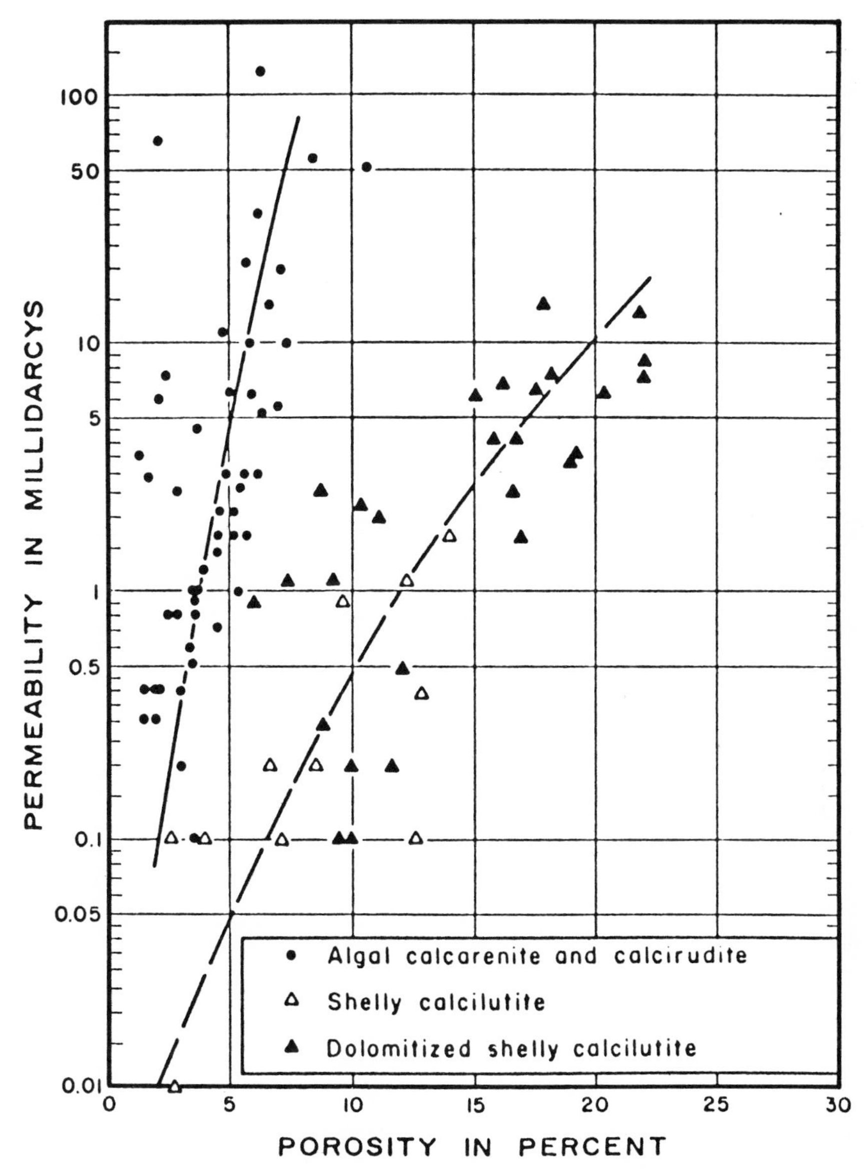

FIG. 4. Relationship between porosity and permeability of Pennsylvanian carbonate rocks, Ismay Field, Utah and Colorado. (After Choquette and Traut,[9] p. 182.)

(0.5–1.0 mm), (4) medium (0.25–0.50 mm), (5) fine (0.1–0.25 mm), (6) very fine (from 0.01–0.02 mm to 0.1 mm), and (7) extremely fine (<0.01–0.02 mm; average 0.015 mm).

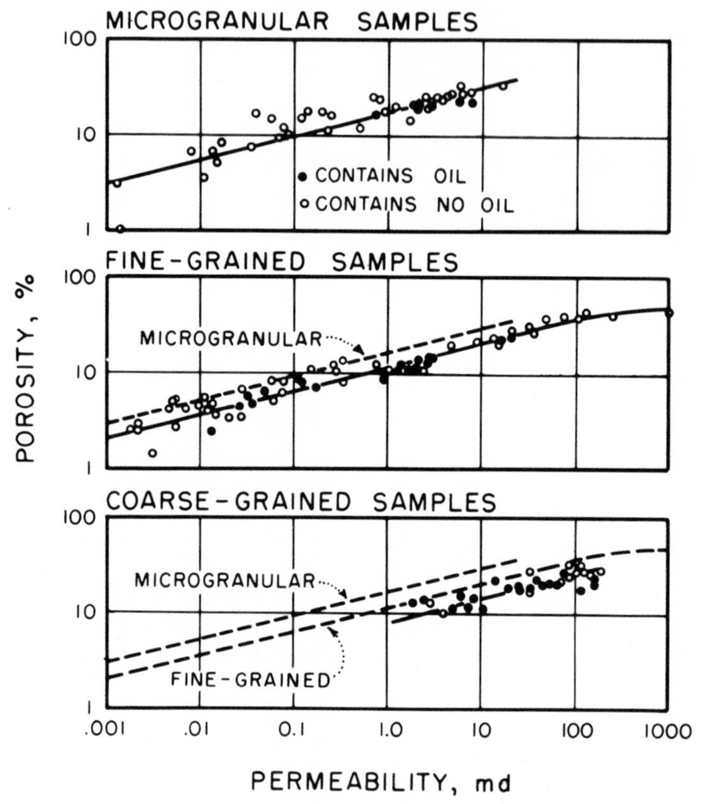

FIG. 5. Relationship between porosity and permeability for various textural types of Cretaceous Edwards Limestone. (After Craze,[10] courtesy of AIME.)

The rock voids are usually grouped in two main classes: (*a*) original openings which were formed before or during lithification, and (*b*) secondary openings which came into existence after lithification, such as solution openings and joints.

The primary interstices in limestones include[12, 13] (1) openings between the individual constituent particles of detrital carbonate rocks (oolites, coquinas, etc.); (2) openings between the individual crystals and along the cleavage planes of crystals, in crystalline limestones; (3) openings along the bedding planes resulting from differences in the deposited material and differences in crystal arrangement and size; and (4) openings within the skeletal and protective structures of invertebrates and within the tissues of algae.

The secondary interstices in carbonate rocks can be grouped as follows[12]: (1) fractures due to the contraction of sediment during consolidation or because of mineralogic changes, or resulting from crustal movements; (2) solution passages related to former and present erosion surfaces, and to leaching in general; and (3) intercrystalline pores produced by mineralogic change, such as dolomitization.

The joints caused by contraction of the sediment during consolidation are sometimes classified as primary openings. In addition, dolomitization which causes porosity could be diagenetic (formed before or during lithification) or epigenetic (formed after the complete lithification of limestones by downward-percolating meteoric solutions or rising hydrothermal solutions). The Ca/Mg ratio of the epigenetic dolomites varies widely over short distances, both vertically and horizontally.

The pore-space structures of carbonate rocks have been classified by Teodorovich[11, 13, 14] into six types:

TYPE I. The pore spaces of the first type consist of pores and of rather

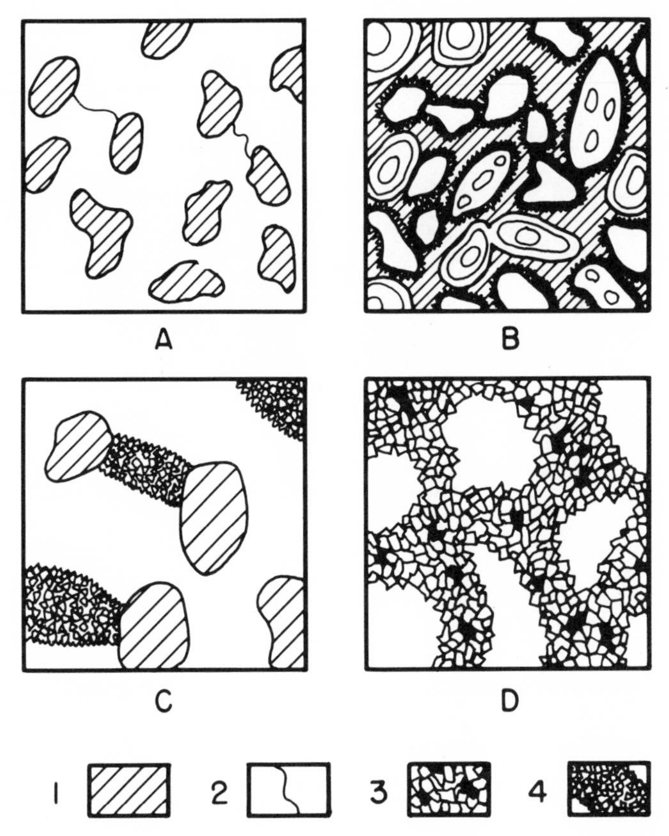

FIG. 6. Four types of pore-space structure in carbonate rocks. (After Teodorovich,[11] p. 233.) 1—pores, 2—conveying canals, 3—intergranular pores, and 4—conveying branches.

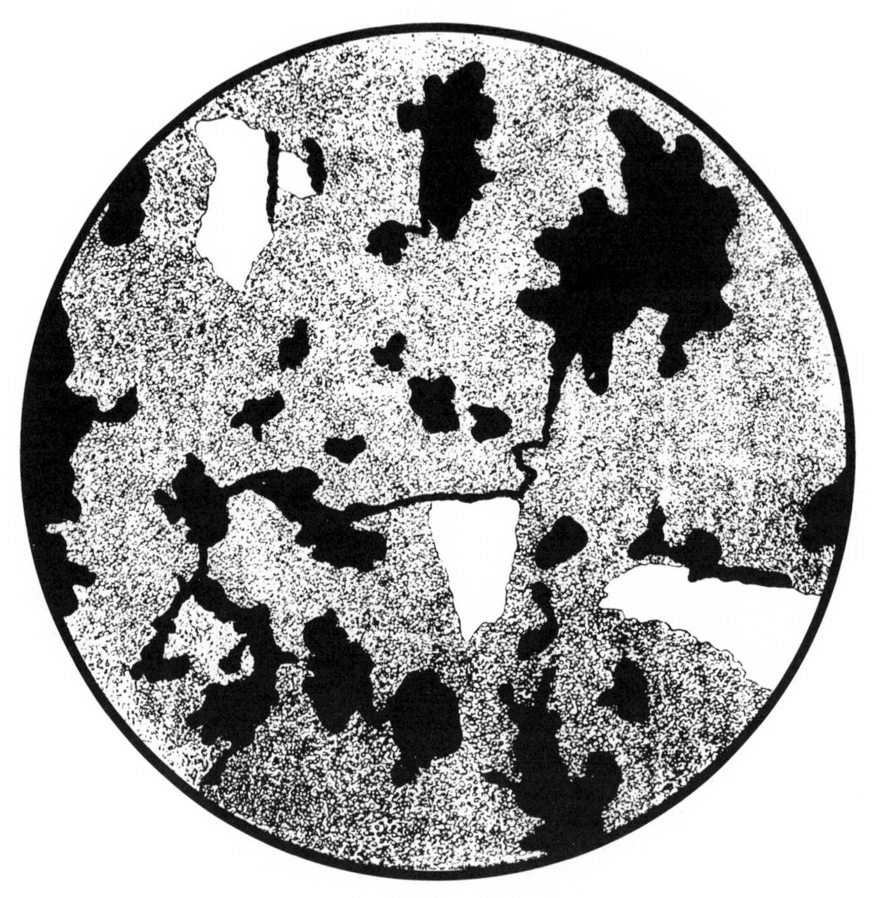

FIG. 7. Type I pore-space structure: (A) With narrow (but visible on a thin section under microscope) conveying canals. Very fine-grained dolomite from Ishimbay Region, Eastern Massive, Lower Permian, Sakmarskian (P_1^S zone), k = 127 md, ×42,[11] p. 282.

isolated, more or less narrow conveying canals (Figs. 6A, 7ABC). Commonly the narrow canals (inner diameter of 0.01–0.005 mm) which connect the pores of this type are not visible on a thin section. If the diameter is larger, however, this canal can be detected in a transparent thin section.

TYPE II. The communicating ducts of the pore spaces of the second type consist merely of constrictions in the pore space, which become wider and pass gradually into the pores proper (Figs. 6B, 8, 9).

TYPE III. The third type of structure is characterized by the presence of pores intercommunicated by means of finely porous broad canals, which are observed in the form of branches in thin sections (Fig. 6C). Occasionally, the

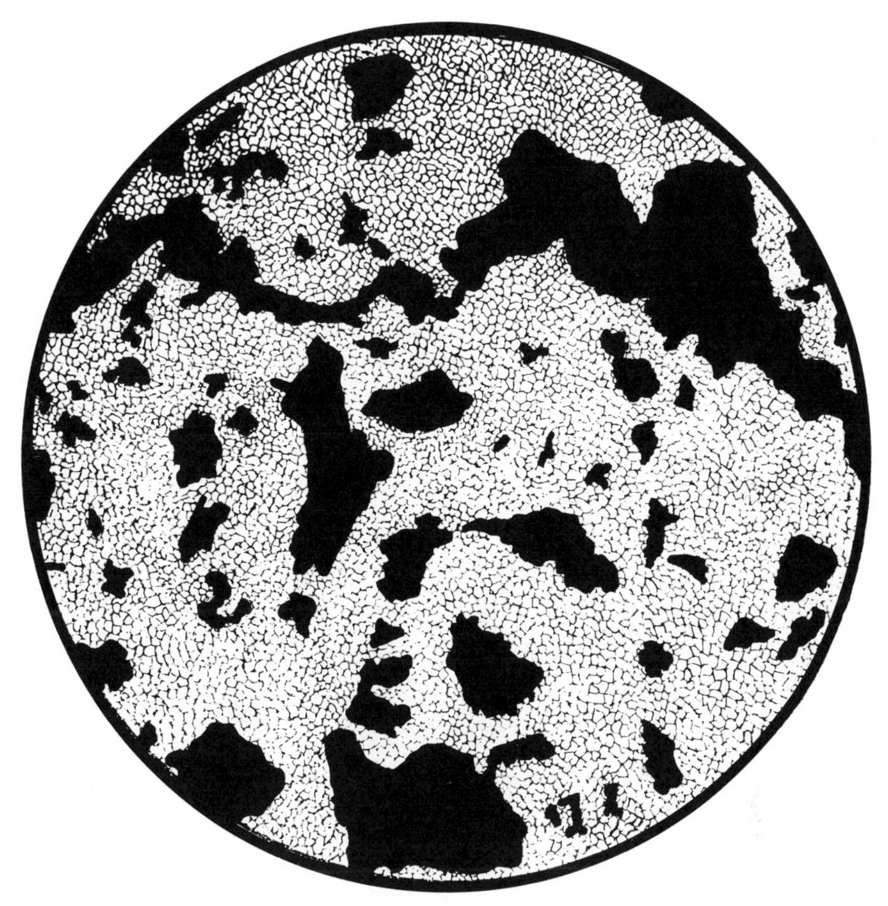

FIG. 7. (B) Same as A, but with some wide conveying canals; k = 567 md, ×42,[11] p. 283.

conveying canals may be built of coarser pores; in this case the permeability is greatly increased. The pore spaces of the third type are usually found in dolomites. Less frequently, they are observed in dolomitic limestones.

TYPE IV. The fourth type of pore-space structure is characterized by a system of pores distributed between the grains of the main mass of the dolomite or its cement. These pores reflect the outlines of the greatest part of the grains (intergranular pores). The interrhombohedral porosity in dolomites may serve as an excellent example (Fig. 6D, 10). The Type IV pore-space structure is encountered in two varieties: (1) a macroporous subtype having good permeability and (2) a finely porous subtype, characterized by medium and even high porosity but very low permeability. The number of readily communicating pores in the second subtype is small and defies

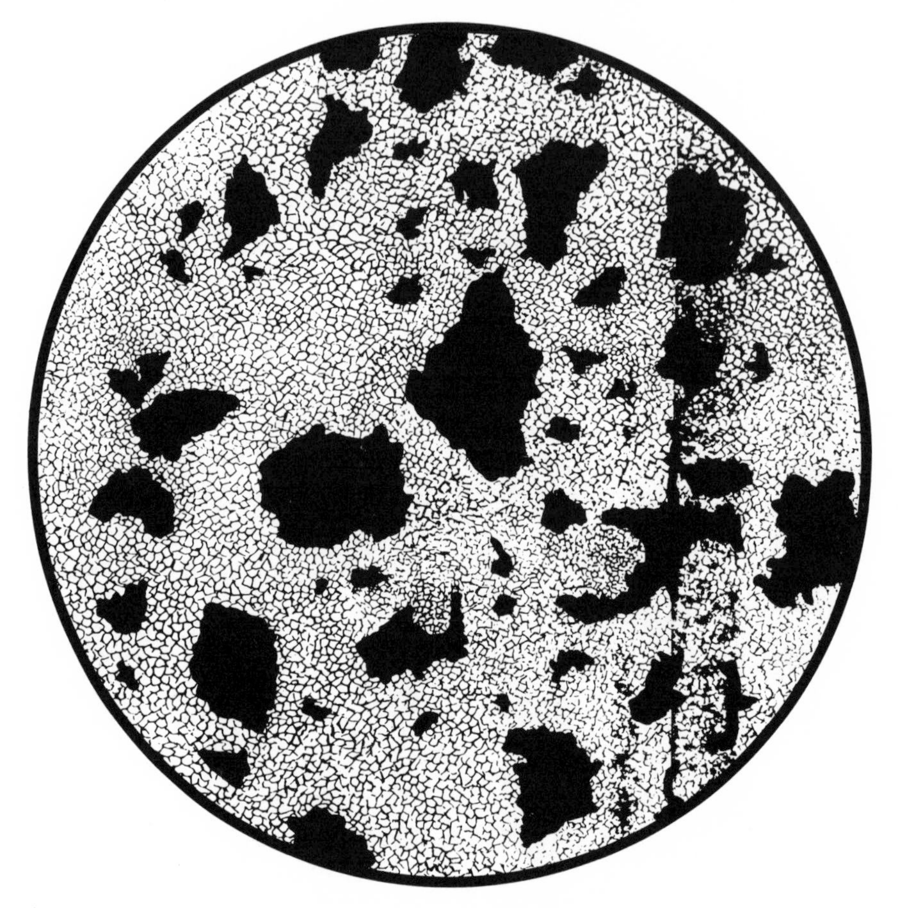

FIG. 7. (C) Conveying canals are not visible in thin section. Very fine-grained dolomite, Ishimbay Region, Southern Massive, Lower Permian, Sakmarskian (P_1^s zone), $k = 567$ md, ×42,[11] p. 281.

accurate evaluation; and, although saturated with oil, the finely porous reservoirs, in most cases, are not commercial.

TYPE V. The fifth type of pore-space structure refers to fissured reservoirs, which are made porous as a result of fracturing.

TYPE VI. The sixth type is characterized by two (or more) elementary types of pore-space structures. In a compound term such as "porous-fissured," the last term designates the prevalent type of porosity.

Teodorovich[11, 13, 15] also developed a quantitative (empirical) relationship between pore-space structure (petrographic examination) and permeability; his method is presented in Appendix A.

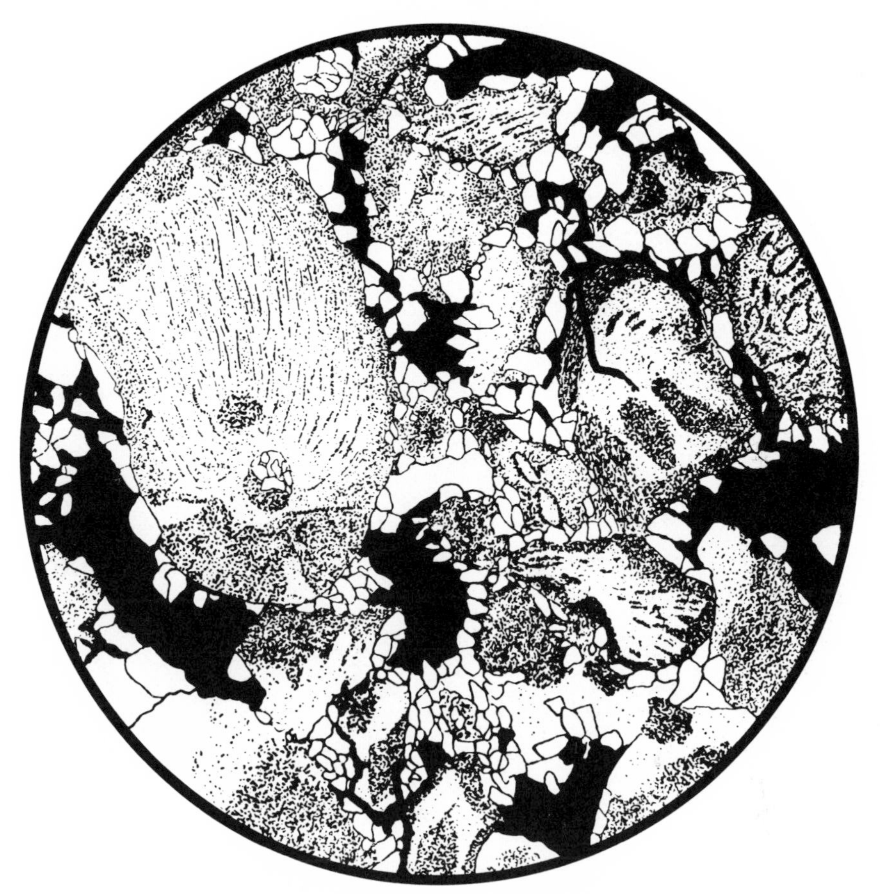

Fig. 8. Type II pore-space structure, with fine pores. Detrital, crinoidal-bryozoan limestone from Ishimbay Region, Eastern Massive, Lower Permian, Artinskian ($P_1{}^A$), $k=230$ md, $\times 42$,[11] p. 284.

Effect of Fractures on the Permeability of Carbonate Reservoirs

Although the permeability of many limestone and dolomite reservoirs is very low, their production rate is often considerably higher than one would expect from the permeability of the cores, because of the fractured nature of many of these rocks. The fractures, even those having very small width, have high effective permeabilities. Consequently, the flow capacity of these carbonate rocks is often greater than that of an average sandstone.

At depths below 3,000 ft, the fracture widths may not exceed 0.1 mm[8] (p. 123). According to many investigators,[8] the fracture porosity usually constitutes 0.5–0.6% and does not exceed 1–2%. E. C. Romm[8] (p. 123)

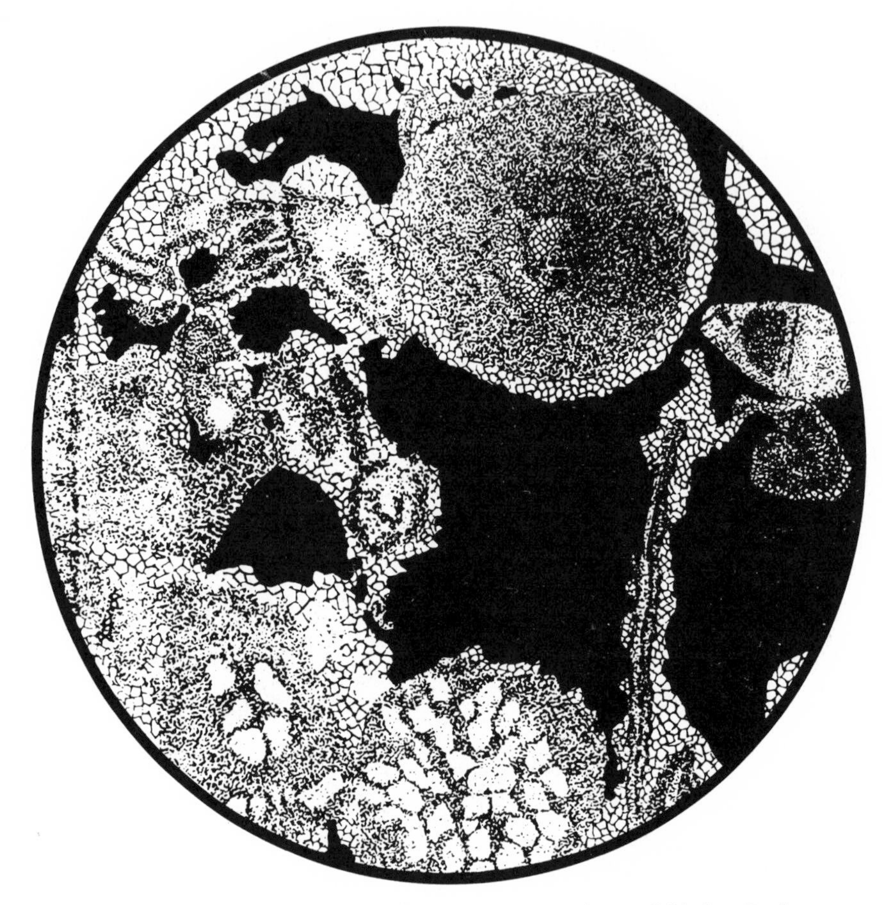

FIG. 9. Type II pore-space structure, with coarse pores and vugs. Ishimbay Region, Eastern Massive, Lower Permian, Artinskian (P_1^A zone), $k = 6{,}750$ md, $\times 20$,[11] p. 285.

calculated that the fracture porosity of a 1,000-cc cube with ten 0.1-mm wide fractures is only 1%.

A mathematical description of a general fractured limestone system, containing a number of fractures distributed at random, is indeed very difficult. It is possible to obtain some idea of the significance of the fractures as fluid carriers, however, by considering a single vertical fracture extending for some distance into the body of rock and opening into the well bore.[16] Muskat[16] (pp. 246–249) determined the fluid-carrying capacity of the fracture by using a classical hydrodynamic equation for narrow linear channels. Figure 11 shows the results of the calculations. In his example, Muskat demonstrated that for fracture widths exceeding 0.035 mm the production capacity of the fractured formation exceeds that of the simple radial-flow system without

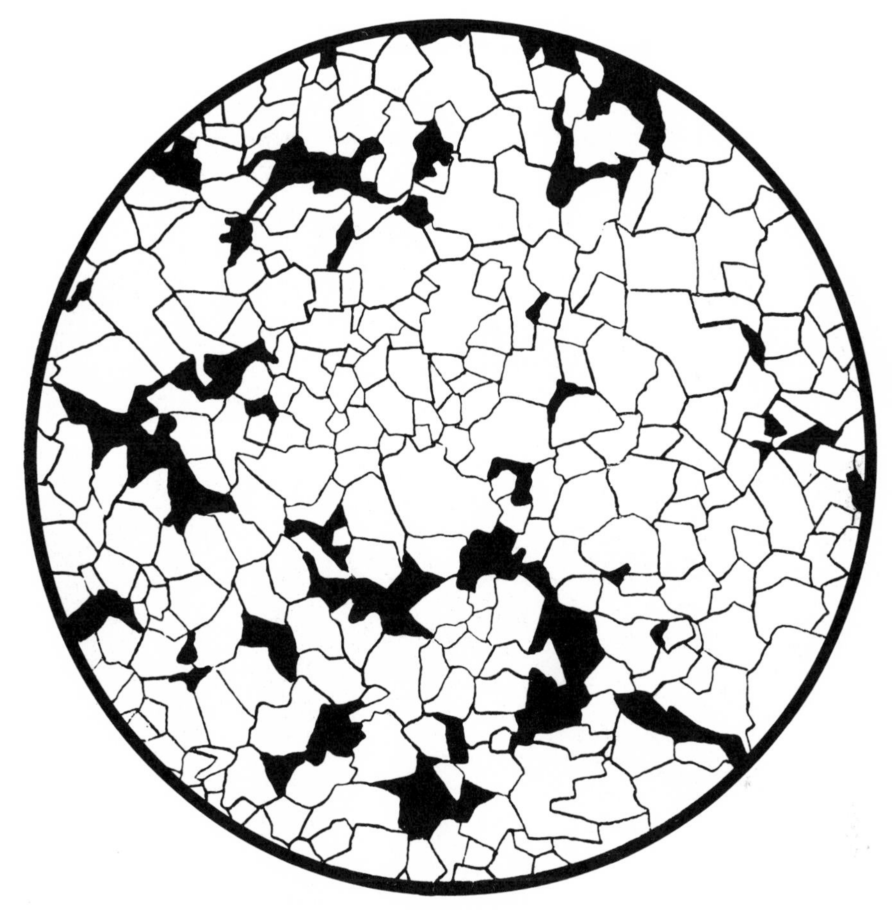

FIG. 10. Type IV pore-space structure (intergranular porosity). Medium-fine-grained, rhombohedral dolomite, Buguruslan, USSR (C_2^1 zone), ×20,[11] p. 286.

fractures. Even for fracture widths of 0.5 mm, the fracture alone could account for more than 90% of the total capacity of the composite limestone-fracture system. As shown in Fig. 11, with fracture widths of 1 mm or greater, most of the oil is carried by the fracture (q_f/q ratio approaches 1).

Reefs and their Detection

The term *reef* as employed here refers to a geologic structure erected originally by frame-building or sediment-binding organisms which resisted the destructive action of the sea. *Bioherm* refers to a moundlike mass built up by sedentary organisms (e.g., algae, corals, stromatoporoids) and embedded in rocks of different lithology. A *biostrome* is a bedded structure composed of

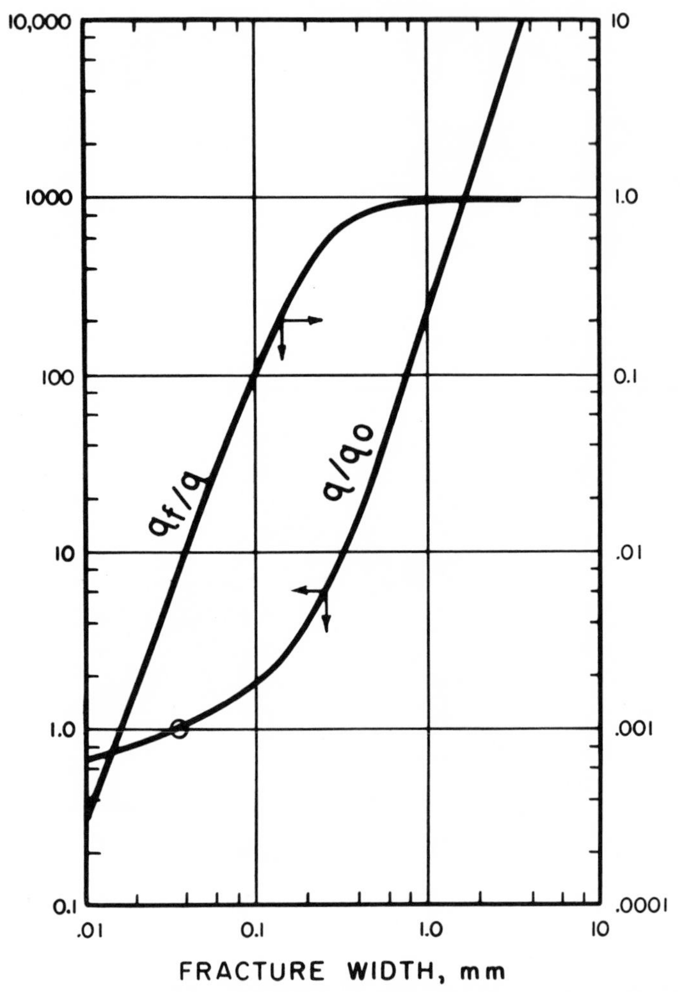

FIG. 11. The calculated steady-state homogeneous-fluid production characteristics of limestone-fracture system as functions of fracture width. (After Muskat,[16] p. 249. Courtesy of McGraw-Hill Book Company, Inc., Copyright, 1949.) q = production capacity of composite limestone-fracture system; q_0 = production of corresponding radial-flow system; q_f = production capacity of a single fracture. Permeability of limestone proper = 10 md; linear or radial extent of system = 300 ft.

skeletal and protective structures of organisms: shells, corals, crinoids, etc. It does not possess a positive structural feature like a bioherm, but is rather a stratiform deposit. The term *pinnacle reef* denotes a reef that has a very small area and grew almost vertically (high-relief).[3]

In reef reservoirs, the distribution of geological rock types is controlled by the environment of deposition, which is sometimes predictable.[17] The basis for using geologic and geophysical studies to define clearly the spatial distribution of reef reservoir rock properties is that reservoir characteristics can be associated with particular facies. The delineation of reefs is imperative for proper reservoir development. Inasmuch as most reef facies prove to be extremely complex, it has been necessary for many oil companies to employ computer simulation studies for adequate reservoir description.[18, 19, 20]

The seismograph appears to offer the most direct method for locating and defining buried reefs, because a facies change in reefs is accompanied by a sonic velocity change. In southwestern Ontario and north-central Ohio the gravity meter has been used successfully to locate shallow productive reefs.[21] In the Salina-Guelph sequence (Silurian, northern Ohio) the bioherms

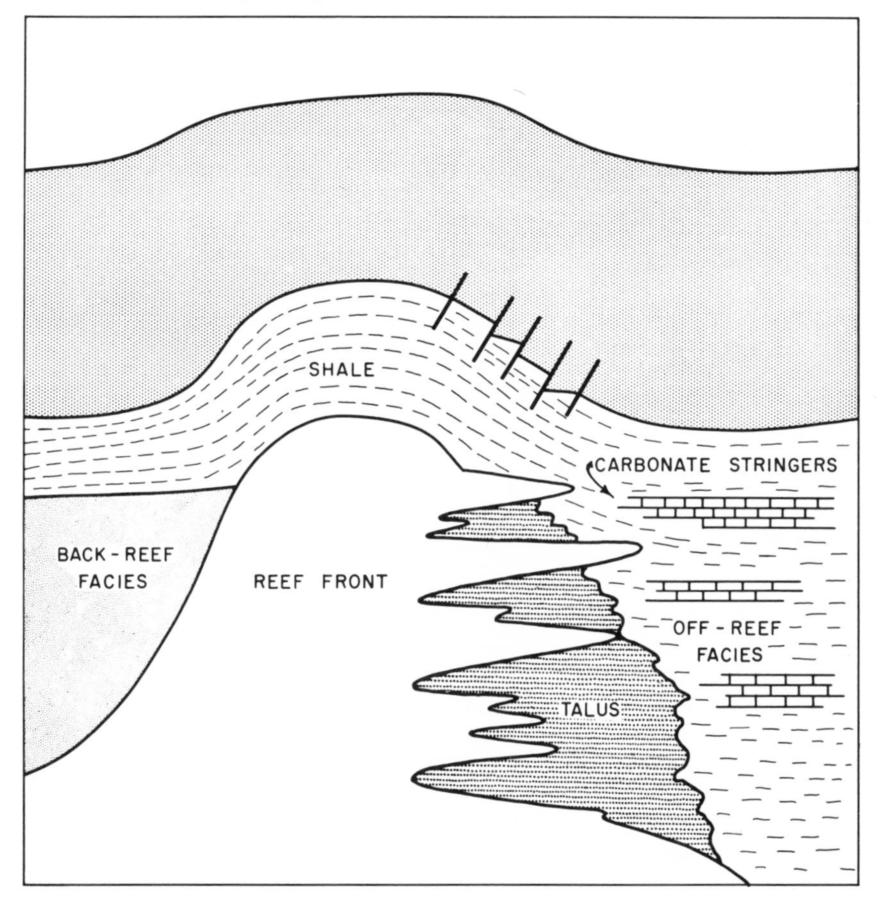

FIG. 12. Example of bioherm-shale facies transition. (Modified after Pallister[22] and other authors.)

produce a gravity contrast in the surrounding salt beds. These reefs form positive gravity anomalies, in contrast to the negative anomalies over nonstructural porous limestones and dolomites.

Figure 12 illustrates a reef-shale facies transition. Back-reef and fore-reef facies have different geophysical and structural attributes. The seaward side of the reef has high porosity, whereas the back-reef sediments are generally nonporous. The reef front is a structural protrusion, possibly resulting from compaction of the back-reef sediments. Differential compaction of carbonate sediments over a reef or "shoal" also forms productive traps (compaction domes).

Acoustic travel times in reef sediments are quite variable and probably depend on the following factors: (1) porosity, (2) depth of burial, (3) infilling of pore spaces and (4) shaliness. The velocity of reef material generally ranges

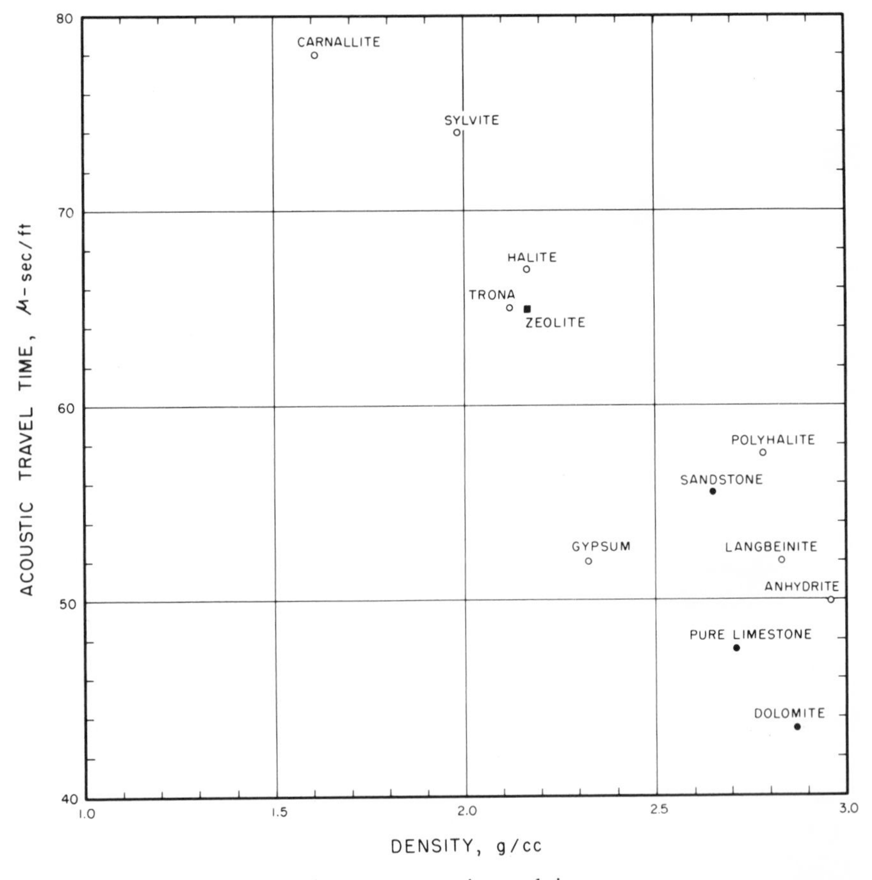

FIG. 13. Theoretical grain density versus acoustic travel time.

from 18,000 to 20,000 ft/sec, whereas the velocity of off-reef facies can vary from 8,000 to 22,000 ft/sec. Pallister[22] reported that velocities in reefs in western Canada range from 16,400 to 21,200 ft/sec (data from sonic logs of 18 wells). The relation between the acoustic travel time and the density for various minerals and rocks associated with carbonates is shown in Fig. 13.

Modern geophysical techniques, including common depth point stacking, can locate the crestal portion of pinnacle reefs with almost 100% success.[20] The areal extent of these reefs and their size-shape relationships, however, are not as easily defined by seismic methods. The risk associated with locating reefs by exploratory drilling is low, whereas the risk in developing pinnacle reefs is high. Development wells are drilled only when the results of seismic exploration show the reef to be of sufficient areal extent. In the Keg River reef pools of northern Alberta this situation resulted in many one-well pools. To solve the problem, Pritchard[20] used uncertainty analysis (Monte Carlo simulator and a history-matching technique) in evaluating both volumetric and pressure data obtained from cores to provide a most likely hydrocarbon pore volume estimate.

Hydrostatic and Hydrodynamic Pressures

Many carbonate reservoirs are surrounded by very large aquifers which will offset a pressure drop by providing an influx of water into the reservoir. The water influx into the reservoir is governed by hydrostatic and hydrodynamic pressures in the aquifer. Hydrostatic pressure is defined as the static weight of a column of water, increasing vertically downward from the ground surface to the point of production in the reservoir. Hydrodynamic pressure is the difference between a higher and a lower elevation of the piezometric surface in the aquifer. Figure 14 illustrates various hydrostatic pressure gradients observed in carbonate reservoirs. The hydrostatic gradient for the Viola Limestone is less than that of normal sea water (0.465 psi/ft), whereas both the Smackover Limestone and the Ellenburger limestones have larger gradients than 0.465 psi/ft. Deviation from the sea-water trend line is common and depends on the variation in the fluid density, the amount of solution gas, and the imposed hydrodynamic pressure gradient. The last of these occurs when aquifers rise to the surface (e.g., to the edge of the structural basins, which lies considerably above the reservoir) so as to provide artesian flow.

In highly permeable and extensive aquifers (in full communication with the reservoir), pressure gradients developed during production can extend a considerable distance outward from the pool. Moore and Truby[23] made an analog study of five fields completed in the Ellenburger Limestone of West Texas. The Andector Field suffered a total pressure loss of 1,095 psi after 9 years of production. Eighty percent of the pressure drop was due to

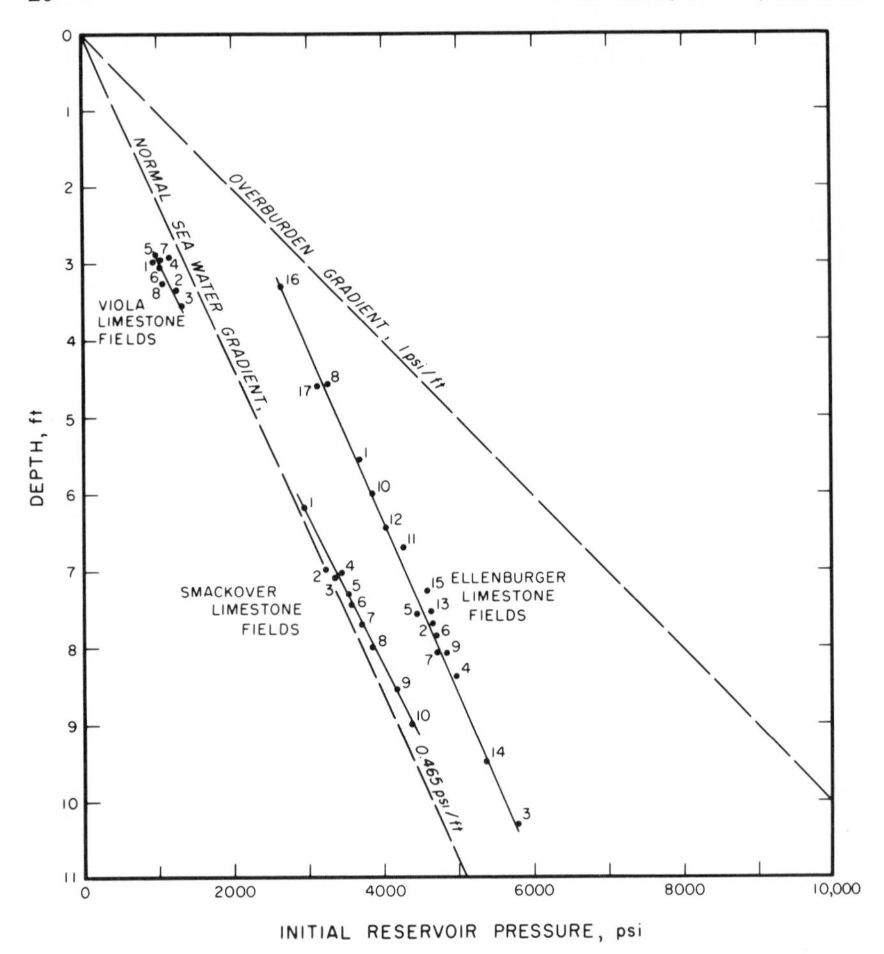

FIG. 14. Relation between initial reservoir pressure and depth for various carbonate reservoirs. Data from the following fields were used: Viola Limestone (Kansas): (1) Wilmington, (2) Wichita, (3) Strahm, (4) Newbury, (5) Mild Creek, (6) John Creek, (7) Comiskey, and (8) Ashburn. Smackover Limestone (Arkansas): (1) Midway, (2) Buckner, (3) Village, (4) Texarkana, (5) Magnolia, (6) Schuler, (7) Big Creek, (8) Atlanta, (9) Macedonia Dorcheat, and (10) McKamie (after Bruce[42]). Ellenburger Limestone (Texas): (1) Martin, (2) Yarbrough and Allen, (3) Sweetie Peck, (4) Shafter Lake, (5) South Fullerton, (6) Block 31, (7) Bedford, (8) Embar, (9) University-Waddell, (10) Jordan, (11) Keystone, (12) TXL, (13) Wheeler, (14) Wilshire, (15) Monahans, (16) Todd, and (17) Elkhorn.

production from the surrounding fields. Pressure interference has also been noticed in the reservoirs of the fractured Asmari Limestone of Iran and in the oolite Smackover Limestone in southern Arkansas. Pressures in sealed carbonate reservoirs such as bioherms and biostromes could be caused by the

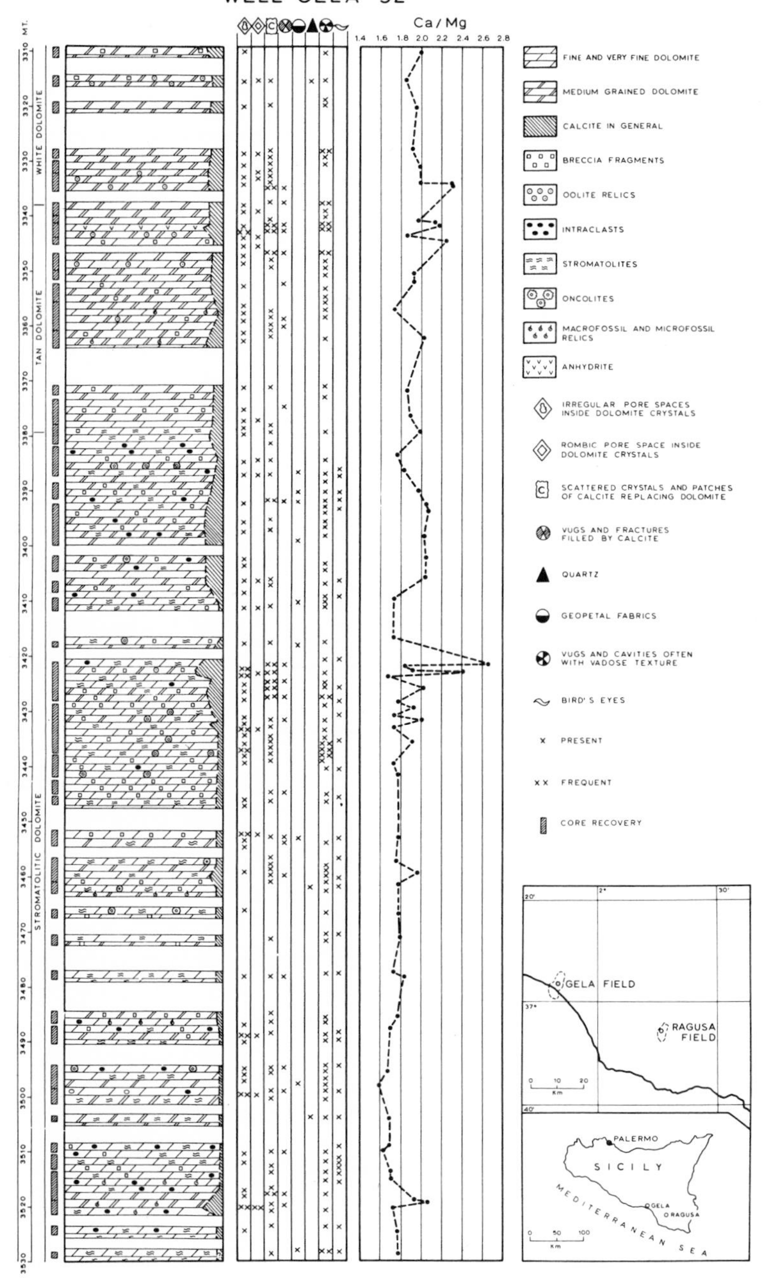

Fig. 15. Petrographic-petrologic log and Ca/Mg ratio line of the Taormina Formation, Gela Oil Field, Sicily, Italy; well Gela 32.[26] (Courtesy of Elsevier Publishing Co., Amsterdam.)

weight of the overburden. For highly permeable and continuous oolite reservoirs, it may be necessary to consider the effects of pressure interference from surrounding pools in order to make reliable material balance calculations.

Elemental Composition as an Aid in Correlation

The elemental composition of carbonate rocks (major, minor, and trace elements) can be used for correlation purposes.[24, 25, 26] Elemental ratios, such as Ca/Mg and Sr/Ca, and isotope ratios, such as $^{18}O/^{16}O$, hold much promise. For example, the Ca/Mg ratio has been used successfully for correlation by Chilingar and Bissell[25] and by Mattavelli *et al.*[26] (Fig. 15).

II. Resumé of Methods Employed in the Estimation of Oil and Gas Reserves

NICO VAN WINGEN

Reserve Calculation Methods

Two basic methods of approach, the empirical and the analytical, can be used to determine the oil and gas reserves for proven lands. The empirical method is based on an analogy to older pools for which, because of an advanced stage of depletion, the ultimate recovery can be determined with relatively great accuracy. Reserve estimates, which are based on an approach by analogy, do not lend themselves to the attainment of accuracy, however, inasmuch as the reservoir characteristics of two oil pools seldom are sufficiently alike for this method to yield reliable results. Thus, studies of this nature usually lead only to qualitative generalizations, commonly expressed in terms of anticipated recoveries per acre. The analytical methods can be divided fundamentally into two categories: (*a*) extrapolation of established performance trends, and (*b*) initial oil- or gas-in-place methods.

The fact that the data available for estimating reserves frequently are inexact and are insufficient to permit rigorous calculations is a source of particular concern. It is desirable, therefore, that reserves be determined by as many of the available methods as can be applied. When differences are indicated, a judgment must be made to arrive at the most probable value.

Extrapolation of Established Performance Trends

Decline Curves

The most accurate method for estimating oil reserves involves an extrapolation of demonstrated production decline trends to their economic limit of the production rate, or to the point where the net revenue (working interest) before income taxes equals the operating cost. This decline curve method can be used only when the wells in question have produced at their capacity over a sufficient period of time to adequately define the decline trend. Even in oil provinces where production is not controlled by regulatory bodies, in recent years capacity production has seldom been resorted to until the oil pools reached a relatively advanced stage of depletion. Accordingly, this method for determining reserves has a limited application and must be used with caution and understanding. Experience, combined with a history of the wells in question, is a prerequisite to proper interpretation. The only assumption involved is that all factors which have influenced the decline curve in the past will remain effective during the projected future life.

Oil production decline curves can be plotted as a function of time or as a function of the cumulative oil production. Experience has indicated that rate-time curves generally can be classified into three types: exponential, hyperbolic, and harmonic.[28, 29, 30] Exponential decline (constant percent) is the type wherein the change in production per unit time is a constant percentage of the production. In hyperbolic decline the drop in production per unit of time, expressed as a fraction of the production rate, is a function of the production rate raised to a power between zero and one. Harmonic decline is a specific case of hyperbolic decline in which the fractional power equals one. The corresponding classifications of the curves of rate versus cumulative oil produced are Cartesian and exponential.

For nonassociated gas reservoirs, which are not subject to water encroachment, the most reliable method for estimating gas reserves is based on an extrapolation of the demonstrated relationship between the reservoir pressure divided by the gas compressibility factor and the cumulative gas production when plotted on Cartesian coordinates. The absence of water encroachment may at times be demonstrated by a 45° slope for a plot on logarithmic paper of the cumulative reservoir pressure decline versus the cumulative gas production.[31]

Gas-Drive Oil Reservoirs

For oil reservoirs in which the expulsive energy is provided by gas, oil reserves can be determined by an extrapolation of a logarithmic plot of the cumulative gas produced versus the cumulative oil production.[32] Experience has indicated that such a relationship generally plots as a straight line. In addition to an adequate production history, this method requires a knowledge of the original gas-in-place and of the percentage of this gas which ultimately will be economically recoverable. The latter figure typically will range from 75% to 95%.

An alternative graphical method is available for estimating oil recovery. It involves the use of two separate plots of predicted gas/oil ratio performance. One curve is an extrapolation of a Cartesian plot of the actual cumulative produced gas/oil ratio versus the cumulative oil production. The other is a plot of calculated cumulative gas/oil ratio at abandonment as a function of assumed ultimate oil recovery. The latter curve is prepared by making an estimate of the ultimate gas recovery. The point at which the two curves intersect represents the ultimate oil recovery as determined by an extrapolation of the actual field gas/oil ratio performance of the pool.

Water-Drive Oil Reservoirs

For oil reservoirs in which the expulsive energy is provided by encroaching water, oil reserves can be determined by an extrapolation of a semi-logarithmic plot of the percentage of oil in the total liquid production (oil and water) versus the cumulative oil production. The actual cutoff point at which the percentage of water makes the operation uneconomic depends on the lifting and water disposal costs as well as on the actual number of barrels of oil produced. This usually occurs between 90% and 98% water production.

An alternative method involves a similar plot and extrapolation, using the water/oil ratio in lieu of the percentage of oil.

Reserve Estimation by Initial Oil- or Gas-in-Place Methods

The oil- or nonassociated gas-in-place method involves the determination of (*a*) the amount of initial oil or gas in place, and (*b*) the percentage of such oil or gas which ultimately will be economically recoverable. This method can be applied to any pool, provided that either a sufficient number of wells has been drilled to reasonably delineate the reservoir and/or enough production history is available to indicate the reservoir's behavior characteristics.

The amount of oil or nonassociated gas in the reservoir at inception can be determined by either the volumetric or the material balance method of calculation.

Volumetric Method

The volumetric method, as the name implies, involves the determination of the total volume of oil or gas originally in place on the basis of a knowledge of the size or volume of the reservoir. The only reliable method of finding the reservoir volume is to construct an isopachous map of net formation thickness. This requires the determination of the net pay thickness of oil- and/or gas-bearing strata in each well, after the locations of the gas-oil and oil-water contacts have been established. The basic tools available to locate the fluid contacts and the net pay thickness are well logs (electrical, sonic, radioactivity, etc.), core analysis, and geological descriptions. As the use of each of these tools involves interpretive judgment, it is advisable to include as many of them as possible to arrive at a reliable value for the net pay thickness.

The oil or gas is contained in the void spaces of the reservoir rock, and it is necessary, therefore, to determine the percentage of the bulk volume constituting the void space or porosity. The basic tools available to determine porosity are again electrical and radioactivity logs and core analysis. By virtue of their characteristics, the results obtained by these methods reflect principally the intergranular or intercrystalline porosity. In sandstone reservoirs this introduces no serious error. In limestone or dolomite reservoirs containing crevices, vugs, and fractures, however, the error may be large. Consequently, thin-section examination is also advisable.

A laboratory analysis of the fluid content of core samples recovered from oil wells has indicated that in almost all instances a portion of the pore space is occupied by water. This residual water is generally referred to as connate water, and the pore space available for hydrocarbons is less than the total porosity in direct ratio to the amount of such connate water present. Water saturation normally is determined by means of the same tools as porosity, and especially by capillary pressure tests. In regard to core analysis, however, it is to be noted that the indicated water content may exceed the connate water saturation as a result of flushing of the core by the mud filtrate, except in instances where oil-base drilling fluids have been used.

Oil accumulations exist in an environment of elevated pressure and temperature. It is a well-established fact that, as the oil is removed to the surface, the resultant reduction of pressure and temperature will cause the evolution of gas from the oil, so that the reservoir liquid shrinks as it travels to the surface. Inasmuch as the purpose of estimating ultimate oil recovery is to determine the amount of oil which can be sold, a correction factor must be applied to the oil in the reservoir to compensate for its shrinkage because of the evolution of gas. Generally this is accomplished by dividing the reservoir barrels of oil-in-place by the liquid formation volume factor.

For gas reservoirs the reservoir volume for hydrocarbons is corrected to the corresponding volume measured at standard conditions of pressure and temperature by application of Boyle's and Charles' laws and the gas compressibility factor (reservoir gas volume factor).

Material Balance Method

Two of the basic principles utilized in all engineering work are the law of conservation of mass and the law of conservation of energy. Most natural processes, which can be broken down into separate unit steps, can be evaluated by either material or energy balance calculations.

The production of petroleum from an underground reservoir may be considered as a basic operation to which the law of conservation of mass must apply. As in other, similar engineering processes, the balance can be made between the materials in the reservoir and those produced.[33, 34, 35] The basic equation for evaluating the reservoir material balance can be stated most simply as follows:

Total mass of hydrocarbons originally in reservoir = Mass of hydrocarbons produced + Mass of hydrocarbons remaining in reservoir.

In order to evaluate this equation, it is necessary to have (1) oil, gas, and water production data, (2) reservoir pressure decline information, and (3) a knowledge of how the physical properties of the hydrocarbons contained in the reservoir change as a function of pressure and temperature.

In very simplified terms, the amount of oil originally in place is estimated from a knowledge of how much hydrocarbon material has been produced for a measured decline in reservoir pressure.

When the basic data are adequate, the material balance method is considerably more reliable for establishing the quantity of oil than the volumetric, as it eliminates the necessity of determining the bulk volume of the reservoir, as well as its porosity and connate water content. Indeed, in many carbonate reservoirs with complex porous systems the original tank oil- or nonassociated gas-in-place can be determined reliably only by means of the material balance method of calculation.

On the other hand, material balance calculations are often inaccurate; sometimes they cannot be made at all, principally because of a lack of reliable reservoir pressure data.

For reservoirs where this is the only feasible method, therefore, periodic and adequate determinations of the average reservoir pressure are of critical importance.

Recovery Factor

Oil Reservoirs

The percentage of the stock tank oil-in-place which ultimately will be economically recoverable is influenced to a considerable degree by the basic type of reservoir mechanism which is dominant and by the method of pool operation. In general, pools subject to a water drive or those in which gravity forces play a major role will yield a greater portion of their oil than those having gas alone as a source of energy. The recovery for the latter type of pools is particularly sensitive to the degree to which the operator is willing and able to control producing gas/oil ratios.

The recovery factor can be obtained by (*a*) laboratory experimentation, (*b*) analogy to older pools which have similar reservoir characteristics, and (*c*) analytical reservoir performance prediction calculations. The first method is capable of giving only an order-of-magnitude indication of the recovery factor, as it is practically impossible to properly duplicate the prototype on a laboratory scale. The second method may lead to gross inaccuracies, as two petroleum reservoirs are seldom sufficiently alike for the approach by analogy to give reliable results. The third method is the one most commonly used, as petroleum technology has advanced to the point that reliable and representative results can be obtained by using performance prediction calculations.

It is evident that different calculation procedures must be followed, depending on which one of the various basic reservoir mechanisms is dominant for the pool in question. In this regard, four limiting types of reservoir mechanisms are recognized:

1. Solution gas drive, in which energy for production is furnished by the expansion of oil and gas.
2. Free gas-cap drive, in which the energy of the expansion of oil and solution gas is supplemented by that of an original free gas cap.
3. Water drive, in which the energy is furnished by water encroaching into the pool from beyond its limits.
4. Gravity drainage, in which the energy for production is furnished by gravity forces.

In most naturally occurring oil reservoirs, the energy for production is not obtained from one source only but may come from two or more sources. All can act simultaneously, but the contribution of each may vary from time to time, principally as a result of variation of the rate of production or the stage of depletion. It is not uncommon for gas expansion to be the principal source of energy during the early life of the pool, and either water drive or gravity drainage to be the main driving force during the later life.

Gas-Expansion-Drive Pools

As gas is the dominant source of energy for this type of pool, it is evident that a greater ultimate production will be obtained in proportion to the degree to which gas/oil ratios can be held to low values. In regard to numerical procedures for estimating ultimate oil recovery, the prediction method for gas-drive pools with or without a free gas cap proposed by Tarner[36] is probably the best known. The variables required for the prediction calculation are (1) reservoir pressure, (2) instantaneous produced gas/oil ratio, (3) cumulative oil recovery, and (4) cumulative gas recovery.

In this technique, the future reservoir pressure performance is determined as a function of the cumulative oil production. This is done stepwise for arbitrarily selected decreasing pressure increments. The amount of oil produced is assumed when the reservoir pressure has declined to each such successive chosen value. The resulting instantaneous gas/oil ratio is then calculated by (*a*) the material balance equation to satisfy the concept of the conservation of mass, and (*b*) the Darcy equation, which is assumed to govern the flow of oil and gas through porous media. By the method of trial and error (iteration), an oil-produced value is determined so that the gas/oil ratio as calculated by both methods is the same, as it should be for the answer to be valid. The recovery factor can be obtained directly from the curve of calculated pressure versus cumulative oil production, as the economic abandonment pressure can readily be determined.

Water-Drive Pools

Experience has indicated that the reservoir performance prediction methods derived for gas-drive pools are valid also for water-drive pools provided that (*a*) an estimated net water encroachment term is included in the material balance portion of the calculations, and (*b*) the water drive is only partially effective, so that some pressure decline is occurring.

For water-drive reservoirs, other calculation methods are available, which can be used to check material balance-Darcy calculation results, such as Tarner's method, or which can be substituted for such computations.[37] These methods have the further advantage that they can be used for complete water-drive pools in which pressures are not declining.

Gravity-Drainage Pools

For pools in which gravity forces are important but not dominant, the method previously described for predicting gas-drive pool performance is used.[36] In this instance, gravity merely enhances the segregation of the gas and oil phases, so that a lower average produced gas/oil ratio and hence a higher ultimate oil recovery are possible, provided selective production can be achieved economically and equitably.

For the relatively few pools which are pressure depleted but which continue to produce economically because of the drainage of oil into the wellbores owing to the effect of gravity forces alone, a separate calculation procedure is required. The method developed by Cardwell and Parsons[38] involves essentially the use of an empirical curve which shows the percentage of the remaining oil to be recovered by gravity drainage as a function of a drainage modulus.

Nonassociated Gas Reservoir

For gas reservoirs which are not subject to water encroachment, the percentage of gas originally in place that ultimately will be economically recoverable can be determined readily by establishing the volume of gas remaining in the reservoir upon economic depletion. The wells will have reached their economic limit when they no longer are capable of delivering gas into the gas-gathering system at a rate equal to that required to defray the cost of the operation. The economic limit delivery pressure is governed by (*a*) the pressure at which the gas-gathering system is being operated or (*b*) the compressor intake pressure in the event that a booster compressor is used to compress gas for delivery to the sales line. When the minimum wellhead delivery pressure has been established, the corresponding static wellhead pressure can be determined by reference to "back pressure" test data.[39] The surface static pressure then is converted to the corresponding reservoir pressure by adding the weight of the static gas column.

The recovery factor for water-drive gas reservoirs cannot be calculated because gas wells are easily "drowned out" by the irregular advance of encroaching water. Thus the recovery factor can be determined only by analogy to older, comparable reservoirs. Generally the recovery factor for this type of reservoir will range from 30% to 60%.[40]

Improved Recovery Methods

For many pools the primary ultimate economic oil recovery can be increased by fluid injection for the purpose of augmenting the native reservoir energy. When water (hot or cold) or dry gas is injected for this purpose, the methods previously cited for water-drive pools[37] and gas-expansion-drive reservoirs[36] are applicable. For miscible fluid displacement methods other reservoir performance prediction techniques must be used.[41] In recent years, low API gravity oil reservoirs have been stimulated successfully by the application of thermal methods. Calculation procedures have been derived to determine the oil recovery by steam injection[41] and by *in situ* combustion.[41] These two methods currently are being used to introduce heat into oil-bearing reservoirs; however, they have not proved to be very successful in carbonate reservoirs.

III. Glossary of Geological Terms Used in Connection with Carbonate Rocks

George V. Chilingar, Robert W. Mannon, and
Herman H. Rieke, III

1. *Allo-:* a prefix derived from "allochthonous" and indicating that material has been transported before accumulation.
2. *Allochem:* transported particles: pellets, skeletal detritus, carbonate pebbles, oolites, etc., which represent the framework of the rock.
3. *Allochthonous:* a term used here to designate sedimentary constituents which did not originate *in situ*; they underwent transportation before final accumulation.

4. *Aphanocrystalline:* texture of most micritic limestones and dolomicrites; individual crystals less than 0.01 mm in size.
5. *Authigenic:* those lithological constituents that came into existence during or after the formation of the rock of which they constitute a part; generated on the spot.
6. *Bank: in situ* skeletal limestone deposit formed by organisms which do not have the ecological potential to erect a rigid, wave-resistant structure.
7. *Biomicrite:* biogenic rock with a microcrystalline ooze matrix. Signifies either that the fossils were sedentary or that currents were calm in the depositional area. (Folk.[49])
8. *Biostrome:* a term for stratiform deposits, such as shell beds, crinoid beds, and coral beds, consisting of, and built mainly by, organisms or fragments of organisms, mostly sedentary; biostromes do not have moundlike or lenselike forms.
9. *Birdseye:* spots or tubes of sparry calcite in limestones and some dolomites.
10. *Calcarenite:* mechanically deposited limestone composed of sand-sized ($\frac{1}{16}$–2 mm in diameter) calcareous particles.
11. *Calcilutite:* limestone composed of 50% or more of clay-sized (<0.004 mm) calcareous particles.
12. *Calcirudite:* mechanically deposited limestone composed of 50% or more of gravel-sized (>2 mm) calcareous particles.
13. *Calcisiltite:* mechanically deposited limestone composed of 50% or more of silt-sized (0.004–0.0625 mm) calcareous particles.
14. *Chalk:* a porous, finely grained, noncrystalline calcareous material which may be largely composed of foraminiferal tests and/or comminuted remains. It is friable to subfriable and is largely microtextured (about 0.01 mm or smaller).
15. *Coquina:* carbonates consisting wholly, or almost entirely, of mechanically sorted fossil debris.
16. *Coralgal:* intergrowth of algae (particularly corraline types) and corals, to form a firm carbonate rock.
17. *Diagenesis:* all the processes which change a fresh sediment into a stable rock of substantial hardness, at low temperatures and pressures characteristic of surface and near-surface environments.
18. *Dismicrite:* consists predominantly of microcrystalline calcite, but contains irregular patches, tubules, or lenses of sparry calcite, almost invariably with sharp boundaries. Rock types of diverse and obscure origin; called "birdseye" by many geologists. (Folk.[49])
19. *Dolomitized:* refers to rocks or portions of rocks in which limestone textures are discernible, but which have been changed wholly or largely to dolomite.
20. *Epigenesis:* includes all processes at low temperature and pressure that affect sedimentary rocks after diagenesis and up to metamorphism in the depocenter.
21. *Intergranular porosity:* void spaces between mineral grains, either bioclastic or lithoclastic.
22. *Intraclast:* fragments of penecontemporaneous, generally weakly consolidated carbonate sediment that have been eroded from adjoining parts of the sea bottom and redeposited to form a new sediment. The fragments vary from sand size to pebble or boulder size, as in the "edgewise" limestone conglomerate, and are usually well-rounded, equant to highly discoidal.
23. *Intragranular porosity:* pore spaces within individual particles, particularly skeletal material.
24. *Lithoclastic:* carbonate detritus mechanically formed and deposited; derived from previously formed carbonates.
25. *Marl:* semifriable mixtures of clay minerals and carbonates. Marl contains 30–70% of carbonates and a complementary content of clay.
26. *Matrix:* the natural material in which sedimentary particles are embedded; the material that fills the interstices between the larger grains.
27. *Micrite:* a consolidated or unconsolidated ooze or mud of either chemical or mechanical origin. Whether crystalline or finely grained, it is material about 0.005 mm or smaller in diameter.

1405

28. *Oolite:* a spherical to ellipsoidal accretionary body up to 2 mm in diameter, which may or may not have a nucleus, and has concentric or radial structure or both; a calcareous coated grain.
29. *Pellet:* a grain composed normally of micritic material, lacking significant internal structure; it is generally ovoid in shape but may be subovoid.
30. *Pisolite:* a grain type similar to oolite and generally 2.0 mm or more in diameter.
31. *Reef:* a structure erected by frame-building or sediment-binding organisms. At the time of deposition, the structure was a wave-resistant or potentially wave-resistant topographic feature. Reefs and heterogeneous reef-derived materials form the reef complex. Types of reefs include (1) *barrier reefs*, sublinear structures that are or were separated from nearby older land by a lagoon; (2) *bank reefs*, which grow over submerged highs of tectonic or other origin; they are large and irregularly shaped; (3) *fringing reefs*, veneering types that lie or were adjacent to the pre-existing land; (4) *pinnacle reefs*, which have very small areas and grow almost vertically; (5) *shoal reefs*, which grow on the shoals of fore-reef and back-reef areas; they are smaller than the bank type and generally grow on the debris of a larger reef; (6) *patch reefs*, small, sub-equidimensional or irregularly shaped reefs that are parts of reef complexes; and (7) *atolls*, composite structures with ringlike outer reefs that surround or once surrounded a central lagoon devoid of pre-existing land.
32. *Saccharoidal:* a descriptive term meaning "sugary" texture.
33. *Sparite:* a descriptive term applied to any transparent or translucent crystalline calcite and aragonite. It is synonymous with sparry calcite. Sparite is larger than 0.02 mm in diameter.

For further details on the classification of carbonate rocks and on terminology, the reader is referred to books by Ham,[50] Chilingar *et al.*, [1, 3] and to the classical work of Folk.[49] In the latter classification the first part of a composite term, such as *biomicrite*, *pelsparite*, and *oosparite*, refers to the allochem composition ("intra-" for intraclastic rocks, "oo-" for oolitic, "bio-" for biogenic, and "pel-" for pellet rocks). In the second part of the name "-sparite" designates sparry calcite cement; "-micrite," microcrystalline ooze matrix.

References and Bibliography

1. Chilingar, G. V., Bissell, H. J. and Wolf, K. H.: "Diagenesis of Carbonate Rocks", in: G. Larsen and G. V. Chilingar (Editors), *Diagenesis in Sediments*, Elsevier Publ. Co., Amsterdam (1967) 179–322.
2. Frolova, E. K.: "On Classification of Carbonate Rocks of Limestone-Dolomite-Magnesite Series", *Novosti Neft. Tekhn., Geol.* (1959) Vol. 3, 34–35.
3. Chilingar, G. V., Bissell, H. J. and Fairbridge, R. W.: *Carbonate Rocks*, 9A, Elsevier Publ. Co., Amsterdam (1967) 471 pp.
4. Chilingar, G. V., Bissell, H. J. and Fairbridge, R. W.: *Carbonate Rocks*, 9B, Elsevier Publ. Co., Amsterdam (1967) 413 pp.
5. Bulnes, A. C. and Fitting, R. U. Jr.: "An Introductory Discussion of the Reservoir Performance of Limestone Reservoirs", *Trans.*, AIME (1945) Vol. 160, 179–201.
6. Chilingar, G. V. and Terry, R. D.: "Relationship between Porosity and Chemical Composition of Carbonate Rocks", *Pet. Eng.* (1954) Vol. 26, 341–342.
7. Jodry, R. L.: "Growth and Dolomitization of Silurian Reefs, St. Clair County, Michigan", *Bull.*, AAPG (1969) Vol. 53, No. 4, 957–981.
8. Eremenko, N. A.: *Geology of Oil and Gas*, Izd. Nedra, Moscow (1968) 389 pp.
9. Choquette, P. W. and Traut, J. D.: "Pennsylvanian Carbonate Reservoirs, Ismay Field, Utah and Colorado", in: *A Symposium, Shelf Carbonates of the Paradox Basin*, Four Corners Geol. Soc., Fourth Field Conference (June, 1963) 157–184.
10. Craze, R. C.: "Performance of Limestone Reservoirs", *Trans.*, AIME (1950) Vol. 189, 287–294.

11. Teodorovich, G. I.: *Carbonate Facies, Lower Permian–Upper Carboniferous of Ural–Volga Region*, Izd. Moskov. Obshch. Ispyt. Prirody, Moscow (1949) Issue 13 (17) 304 pp.
12. Hohlt, R. B.: "The Nature and Origin of Limestone Porosity", *Quart.*, Colorado School Mines (1948) Vol. 43, No. 4, 51 pp.
13. Teodorovich, G. I.: "Structure of Pore Spaces of Carbonate Oil Reservoir Rocks and Their Permeability, as Illustrated by Paleozoic Reservoirs of Bashkiriya", *Dokl. Akad. Nauk SSSR* (1943) Vol. 39, No. 6, 231–234.
14. Chilingar, G. V.: "A Short Note on Types of Porosity in Carbonate Rocks", *The Compass of Sigma Gamma Epsilon* (1957) Vol. 35, 69–74.
15. Aschenbrenner, B. C. and Chilingar, G. V.: "Teodorovich's Method for Determining Permeability from Pore-Space Characteristics of Carbonate Rocks", *Bull.*, AAPG (1960) Vol. 44, 1421–1424.
16. Muskat, M.: *Physical Principles of Oil Production*, 1st ed., McGraw-Hill Book Co., New York (1949) 922 pp.
17. Hriskevich, M. E.: "Middle Devonian Reefs of the Rainbow Region of Northwestern Canada, Exploration and Exploitation", *Proc.*, Seventh World Pet. Cong., Mexico City (1967) Vol. 3, 733.
18. Langton, J. R. and Chin, G. E.: "Rainbow Member Facies and Related Reservoir Properties, Rainbow Lake, Alberta", *Bull.*, Can. Pet. Geol. (Mar., 1968) Vol. 16, 104.
19. McCulloch, R. C., Langton, J. R. and Spivak, A.: "Simulation of High-Relief Reservoirs, Rainbow Field, Alberta, Canada", *J. Pet. Tech.* (Nov., 1969) 1399–1408.
20. Pritchard, K. C. G.: "Use of Uncertainty Analysis in Evaluating Hydrocarbon Pore Volume in the Rainbow-Zana Area", paper SPE 2584 presented at SPE 44th Annual Fall Meeting, Denver, Colo. (Sept. 28–Oct. 1, 1969).
21. Dyer, W. B.: "Gravity Prospecting in Southwestern Ontario, Can.", *Can. Oil and Gas Industry* (1956) Vol. 9, No. 3.
22. Pallister, A. E.: "How to Apply Seismic Data to Western Canadian Reefs", *World Oil* (Aug., 1967) 62–67.
23. Moore, W. D. and Truby, L. G., Jr.: "Pressure Performance of Five Fields Completed in a Common Aquifer", *Trans.*, AIME (1952) Vol. 195, 297.
24. Wolf, K. H., Chilingar, G. V. and Beales, F. W.: "Elemental Composition of Carbonate Skeletons, Minerals, and Sediments", in: G. V. Chilingar, H. J. Bissell and R. W. Fairbridge (Editors), *Carbonate Rocks*, 9B, Elsevier Publ. Co., Amsterdam (1967) 23–149.
25. Chilingar, G. V. and Bissell, H. J.: "Mississippian Joana Limestone of Cordilleran Miogeosyncline and Use of Ca/Mg Ratio in Correlation", *Bull.*, AAPG (1957) Vol. 41, 2257–2274.
26. Mattavelli, L., Chilingarian, G. V. and Storer, D.: "Petrography and Diagenesis of the Taormina Formation, Gela Oil Field, Sicily (Italy)", *Sed. Geol.* (1969) Vol. 3, No. 1, 59–86.
27. Stevenson, D. L.: *Oil Production from the Ste. Genevieve Limestone in the Exchange Area, Marion County, Illinois*, Ill. Geol. Survey Circ. 436 (1969) 23 pp.
28. Arps, J. J.: "Analysis of Decline Curves", *Trans.*, AIME (1945) Vol. 160, 228–247.
29. Mead, H. N.: "Modifications to Decline Curve Analysis", *Trans.*, AIME (1956) Vol. 207, 11–16.
30. Arps, J. J.: "Estimation of Primary Oil Reserves", *Trans.*, AIME (1956) Vol. 207, 182–191.
31. Gruy, H. J. and Crichton, J. A.: "A Critical Review of Methods Used in Estimation of Natural Gas Reserves", *Trans.*, AIME (1949) Vol. 179, 249–263.
32. Heuer, G. J., Jr. and Power, H. H.: "A Correlating Device for Predicting Reservoir Performance", *J. Pet. Tech.* (Sept., 1954) 10–14.
33. Coleman, S., Wilde, H. D. and Moore, T. W.: "Quantitative Effect of Gas-Oil Ratios on Decline of Average Rock Pressure", *Trans.*, AIME (1930) Vol. 86, 174–184.

34. Schilthuis, R. J.: "Active Oil and Reservoir Energy", *Trans.*, AIME (1936) Vol. 118, 33.
35. Katz, D. L.: "A Method of Estimating Oil and Gas Reserves", *Trans.*, AIME (1936) Vol. 118, 18.
36. Tarner, J.: "How Different Size Gas Caps and Pressure Maintenance Programs Affect Amount of Recoverable Oil", *Oil Weekly* (June 12, 1944) 1–7.
37. Dykstra, H. and Parsons, R. L.: "The Prediction of Oil Recovery by Water Flood", in: *Secondary Recovery of Oil in United States*, 2nd ed. (1950) 160–174.
38. Cardwell, W. T., Jr. and Parsons, R. L.: "Gravity Drainage Theory", *Trans.*, AIME (1949) Vol. 179, 199–215.
39. Railroad Commission of Texas: "Back-Pressure Test for Natural Gas Wells" (1950).
40. Agarwal, R. G., Al-Hussainy, R. and Ramey, H. J., Jr.: "The Importance of Water Influx in Gas Reservoirs", *J. Pet. Tech.* (Nov., 1965) 1336–1342.
41. Smith, C. R.: *Mechanics of Secondary Oil Recovery*, Reinhold Publ. Corp. (1966) 314, 467, 471, 386.
42. Bruce, W. A.: "A Study of the Smackover Limestone Formation and the Reservoir Behavior of Its Oil and Condensate Pools", *Trans.* AIME (1944) Vol. 155.
43. Sinnokrot, A. and Chilingar, G. V.: "Effect of Polarity of Oil and Presence of Carbonate Particles on Relative Permeability of Rocks", *Compass of Sigma Gamma Epsilon* (1961) Vol. 38, No. 2, 115–120.
44. Gewers, C. W. W. and Nichol, L. R.: "Gas Turbulence Factor in a Microvugular Carbonate", *J. Can. Pet. Tech.* (April-June, 1969) 51–56.
45. Beveridge, S. B., Spivak, A. and Bertrand, J. P.: "Recovery Sensitivities of High-Relief Reservoirs", *J. Can. Pet. Tech.* (July-Sept., 1969) 93–97.
46. White, E. J., Marchant, L. C. and Roberts, D. K.: "Rock Matrix Properties of the Ratcliffe Interval (Madison Limestone) Flat Lake Field, Montana", paper SPE 2127 presented at Rocky Mountain Regional Meeting, Billings, Mont. (June 5–7, 1968).
47. Burnett, P., Ryan, R. and Elliott, C. L.: "Estimating Gas-Water Contacts in Aquifer Gas Storage Fields Using Shut-In Wellhead Pressures", paper SPE 1303 presented at Eastern Regional Meeting, Columbus, Ohio (Nov. 11–12, 1966).
48. Osborn, R. W. and Ryan, R.: "A Case History of Aquifer Gas Storage Development in a Silurian Dolomite at Glasford, Illinois", paper SPE 1659 presented at Eastern Regional Meeting, Columbus, Ohio (Nov. 11–12, 1966).
49. Folk, R. L.: "Practical Petrographic Classification of Limestones", *Bull.*, AAPG (Jan., 1959) Vol. 43, No. 1, 1–38.
50. Ham, W. E.: *Classification of Carbonate Rocks* (*a Symposium*), AAPG Memoir 1 (1962) 279 pp.

CHAPTER 2

Pore Geometry of Carbonate Rocks (Basic Geologic Concepts)

RICHARD L. JODRY

Introduction

In broad general terms, clastic rocks such as sorted sandstones exhibit their greatest porosity upon deposition. Lithification of the rock reduces this porosity by compaction and cementation. Although there are exceptions, porosity is limited, in general, to the voids between the dense framework particles, and this space is a function of the size and shape of the framework particles and of the degree of cementation. In many cases, the sand particles can be separated and measured by sieve analysis and the pore diameter and pore throat size computed mathematically. Muskat[1] (pp. 57 and 61) recognized this principle as early as 1937. Because of this fairly simple relationship, porosity and permeability tend to have a rather direct influence on productibility. A decrease in sorting with a resultant increase in finer-sized grains, or an increase in clay content, will affect porosity and permeability in a predictable manner. Porosity, however, will still be essentially intergranular.

Carbonate rocks do not exhibit such a simple relationship. There are clastic carbonate rocks, identical in mechanisms of deposition to the sorted sandstones. They, too, lose porosity on lithification because of compaction and cementation. The framework particles, however, are not necessarily dense and may contain significant porosity within them in addition to that developed between the framework. Porosity may also develop in the carbonate cement binding the carbonate particles together. The pore throat sizes within the framework, in the cement, and in the intraparticle voids have no direct relationship. In addition, carbonate rocks are subject to leaching to a vastly greater degree than are sandstones. In such cases, pore size and distribution may conceivably have little or no direct relationship to the particle size or density of the original sediment.

Very fine clastic rocks may be classified as shales, which exhibit a rather high porosity but very low permeability. Such rocks, when the fine particles are in the clay size range, are hardly ever considered as reservoir rocks. When

References p. 82.

the particles of very fine clastic rocks are composed of carbonates (or when the very fine particles are carbonate chemical precipitates), however, the resultant muds, upon compaction, produce a shale which is subject to leaching, with concomitant alteration of porosity and permeability.

The interstices between the framework of a carbonate rock may be voids, may be filled with cement, or may contain particles of finer size. Inasmuch as framework, cement, and void filler all tend to be the same material, it is virtually impossible to separate them to make sieve analyses, as might be done in the case of sandstones.

In addition, carbonate rocks are particularly susceptible to replacement. If calcium is replaced by magnesium, for instance, in certain circumstances dolomite may be formed with a resultant sharp increase in porosity and permeability. If this process continues to completion, porosity and permeability will be vastly altered. In extreme cases they may be almost completely destroyed.

In summary, carbonate sediments are subject to a change in porosity and permeability during compaction and lithification. These parameters may be further greatly altered by leaching, cementation, and/or replacement. In the course of diagenesis, the type and the degree of porosity and permeability may be so altered that they no longer offer a satisfactory measure of the productibility of the rock. A study of the actual pore geometry is necessary to determine whether the rock is capable of producing hydrocarbons, and, if so, under what conditions.

Figures 1–4 illustrate some of the problems. Figure 1 shows photomicrographs of two core samples. Both samples have equal porosities and essentially equal permeabilities. The upper photograph (A) illustrates a biomicrite (Core 29), a rock derived from the consolidation of fine-grained carbonate mud containing scattered floating fragments of broken shell material. A few scattered solution vugs, which are the result of leaching of the bioclastic particles, can be seen.

The lower photograph (B) illustrates a rather fine-grained dolomite (Core 17). This rock was deposited as a biomicrite, and in addition shows evidence of churning and burrowing activities before lithification. Its origin was very similar to that of Core 29, except that it was 90% dolomitized during compaction and lithification. It contains a few scattered solution vugs of approximately the same size as those in the upper photograph.

Observed under the microscope these rocks appear to have similar porosities and permeabilities, as indeed they do, and it could be assumed that they might have comparable producing capabilities. Figure 2 shows that this is not the case. Only 4% of the porosity in Core 29 could hold oil in a reservoir with 450 ft of oil column. In Core 17, 50% of the pore space could contain oil with an oil column of only 20 ft. Core 17 is potentially a reservoir rock,

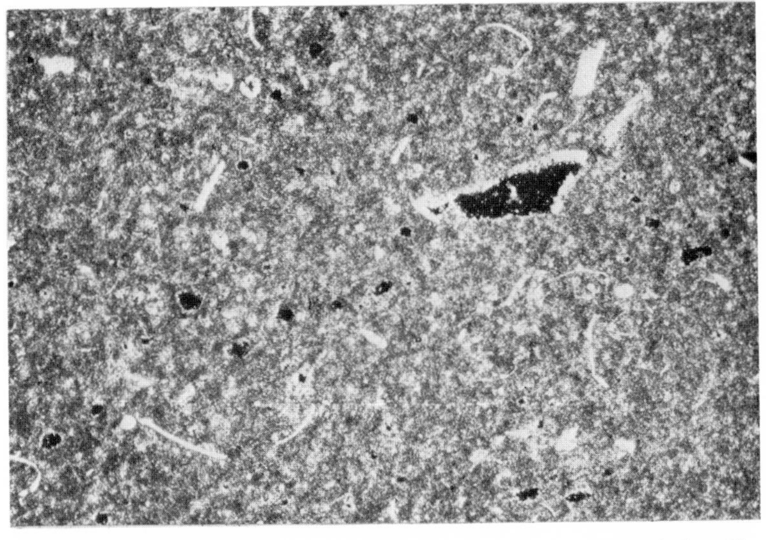

× nicols, × 35.

A

× nicols, × 35.

B

FIG. 1. Rocks with similar origin, different diagenesis, and essentially equal porosity and permeability. Core 29, A. Biomicrite, porosity = 8.5%, permeability = 0.02 md. Core 17, B. Biodismicrite, dolomitized, porosity = 8.5%, permeability = 0.08 md. × nicols, × 35.

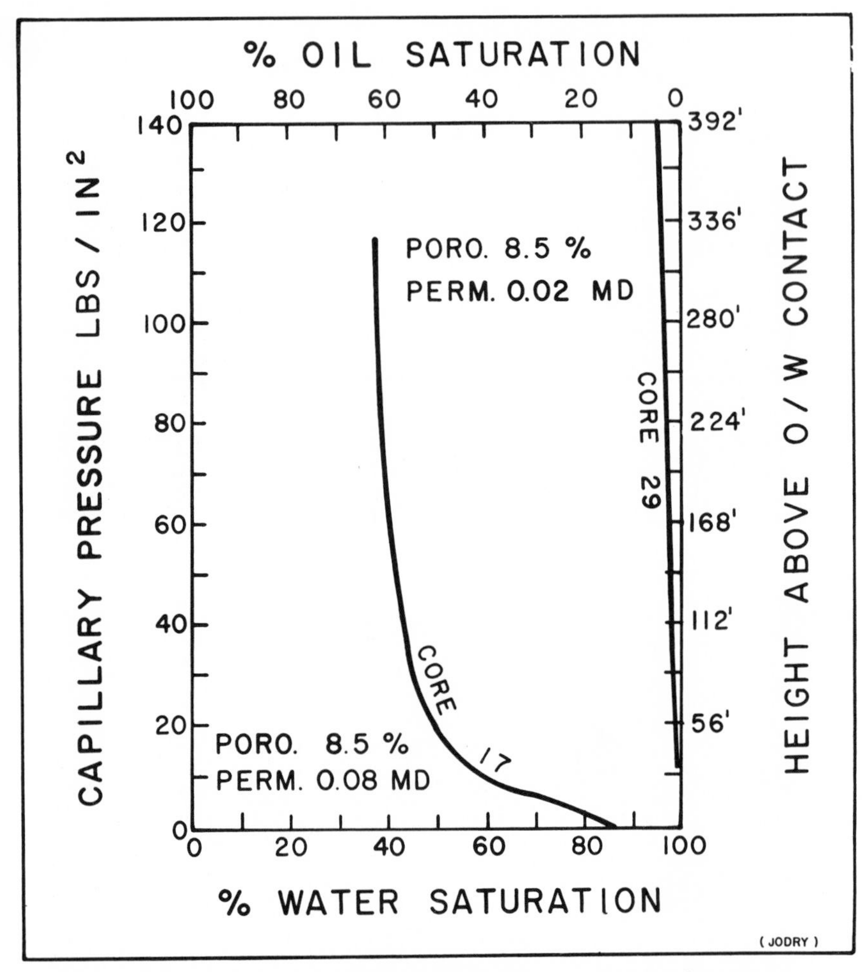

FIG. 2. Capillary pressure curves showing differences in pore geometry between two rocks having essentially equal porosity and permeability.

whereas Core 29 could not be considered a reservoir rock under any conditions likely to be found in its present habitat. Yet both rocks have identical porosity, and the permeabilities are nearly the same.

Figure 3 illustrates two sparites with very different porosities and permeabilities. Core 16 (B) has over three times the porosity, and over 60 times the permeability, of Core 130 (A). Judging from the appearance of these cores under the microscope and from porosity and permeability comparisons, one might expect Core 16 to have higher potential as a producing rock. Again, this is not the case.

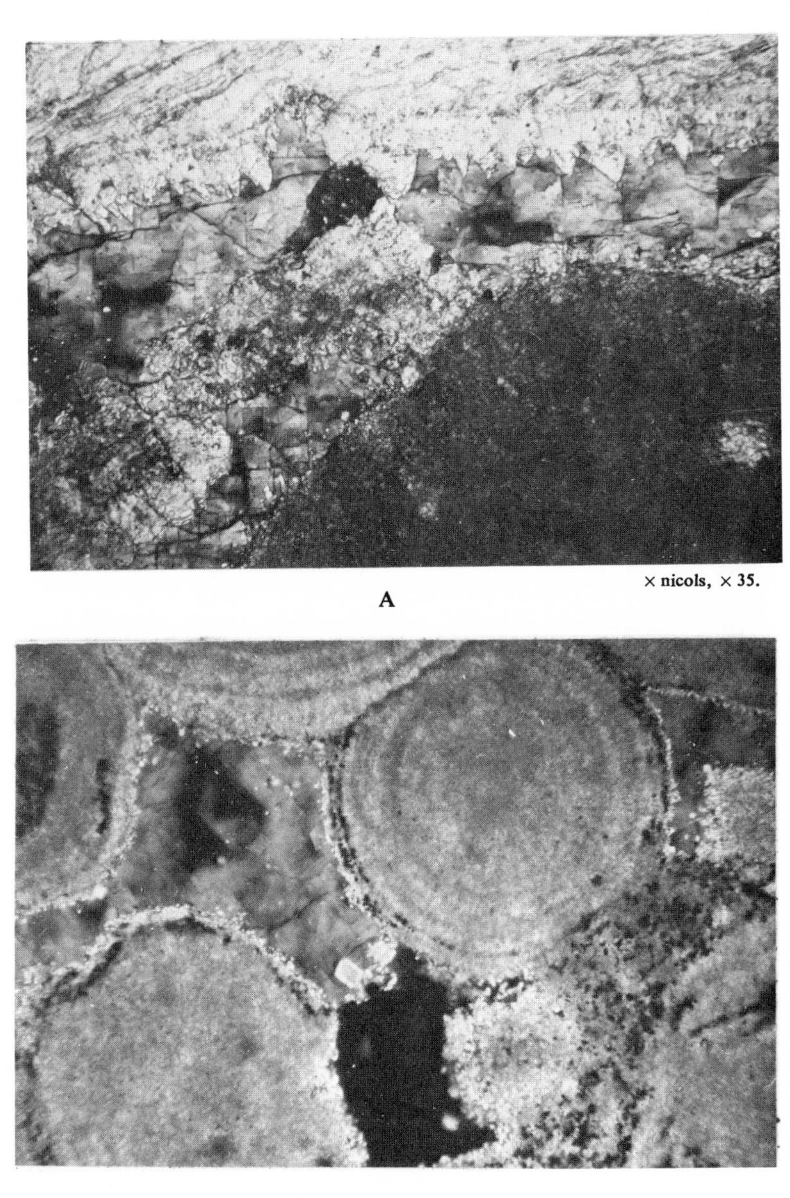

× nicols, × 35.

A

× nicols, × 35.

B

FIG. 3. Sparites exhibiting substantially different porosity and permeability. Core 130, A. Biosparite, porosity = 4.9%, permeability = 0.23 md. Core 16, B. Oosparite, porosity = 15.5%, permeability = 16.8 md. × nicols, × 35.

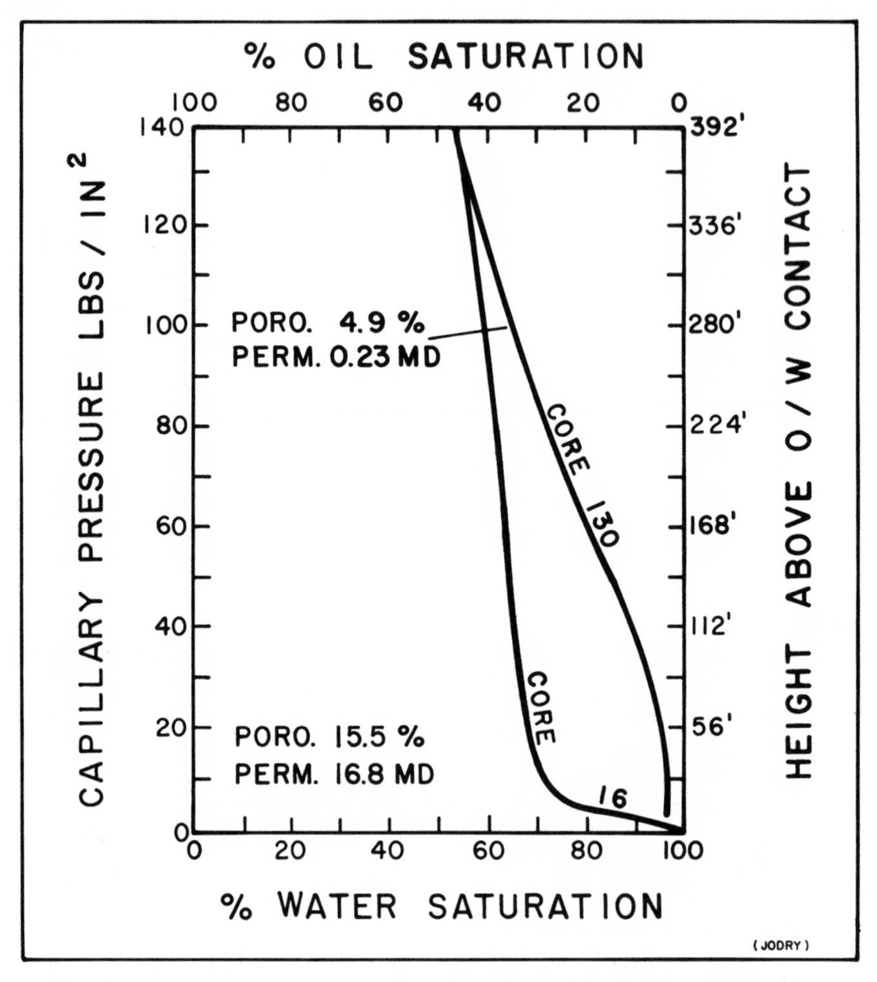

FIG. 4. Capillary pressure curves of sparites with substantially different porosity and permeability.

The mercury capillary pressure curves in Fig. 4 show that neither sample will make a good reservoir rock. Approximately 30% of the pore throats in Core 16 are relatively large and would allow oil to move at low pressures, but this is generally not sufficient to make a commercial reservoir. The visible porosity between the large oolites is deceiving, because it is the small pore throats in the sparite, separating the visible cavities, which control productibility. Core 130 has few large pore throats and would require a much greater pressure to permit the flow of oil through an equal 30% of the pore throats. In both rocks the maximum percentage of pore throats which would allow oil to pass is less than 50%, with an oil column of 450 ft.

It is obvious from these examples that something other than porosity and permeability controls the producing capabilities of the rocks illustrated here. This factor is the geometry of the pore system in the rocks.

Carbonate Rock Classification

Before investigating further the porosity development in various types of carbonate rocks, it will perhaps be helpful to explain the system of classification of these rocks which is used in this chapter. Numerous systems of carbonate rock classifications have been devised, each with some advantages. The most widely used system seems to be that of Folk,[2] and the system used here is based on his classification.

Three main groups of constituents make up carbonate rocks:

1. *Allochems.* These represent the framework particles of the rock (analogous to the quartz sand of a sandstone) and include such material as shells, oolites, carbonate pebbles, and pellets.
2. *Microcrystalline ooze, or carbonate mud.* This represents the clay-size matrix and may consist of very fine clastic fragments or chemical precipitates. As the presence of clay minerals in a sandstone represents poor sorting, so does the presence of ooze in a limestone signify the lack of vigorous currents.
3. *Sparry calcite cement.* This is simply the calcite cement that fills up pore space in the rock and cements the particles together.

As discussed in Chapter 1, rocks may be named by combining the type of allochem (framework) with the type of void filler (cement) or matrix. For example, an oolite which has been cemented together with sparite cement is an oosparite. If instead of cement the voids between oolites are filled with carbonate mud, the rock is called an oomicrite. If more than one allochem is present, this fact is indicated in the name. For instance, if both fossil fragments and oolites are present in a rock cemented with sparite cement, the rock is called a bio-oosparite.

In addition to classification by constituent types, it is sometimes desirable to classify limestones by size of particles. Folk's[2] classification system is used here also, with the terms *calcirudites*, *calcarenites*, and *calcilutites* describing transported constituents of decreasing size.

Classification of Carbonate Porosity

In addition to classification of the particles which comprise carbonate rocks, it is also helpful to classify the types of porosity found in the rock. As mentioned in preceeding paragraphs, carbonate rocks may possess porosity in or between the matrix, allochem, and sparry calcite constituents.

In order to facilitate petrophysical evaluations of limestone and dolomite reservoirs, it is helpful for well-site engineers and geologists to describe porosity in a manner that is mutually understood. Archie's[3] classification of carbonate rocks is equally suitable for describing porosity in both cores and cuttings. It incorporates a classification of the matrix texture, which indirectly gives information on the minute pore throat structure not visible under normal field magnification (10 ×). Recognition of the minute pores and of their contribution to total porosity is important because they have a decided influence on the irreducible water content of reservoir rock. The size and the percentage of the visible pore spaces in the rock also constitute an essential part of the description. These pore spaces tend to make up that part of total porosity which holds hydrocarbons. A modification of Archie's system is outlined below.

Two porosity terms are used. The first describes the texture of the matrix, and furnishes information concerning the porosity in the rock matrix or in the matrix of the sparry cement. This minute pore structure cannot readily be seen, even under the microscope at ordinary magnifications, yet it is important in the fluid distribution within reservoirs. This porosity term may also be used in describing the porosity in the matrix of the allochem or framework constituent of a rock. For example, a pelmicrite might have one class of matrix porosity in the micrite portions of the rock, and another class in the interior of the pellets that constitute the framework.

The second porosity term describes the visible pore structure. Again, a rock may contain visible pores having two or more different sizes. For instance, very fine intergranular porosity may be present in the matrix of a rock, whereas the allochem portion may be largely leached away and a system of vugs developed.

Texture of the Matrix

It is desirable to incorporate the texture of the matrix in a porosity classification because this term gives information on the minute pore throats that link the visible porosity as well as the "hidden" porosity in some rock types. In "chalky"-type reservoirs the minute pore spaces between crystals, which are not readily visible, may add 25% or more to the pore space. The matrix term also acts as a partial lithologic description of the rock.

Matrix terms are limited to three types in order to make the classification as general as possible:

TYPE I: Compact Crystalline. Matrix made up of tightly interlocking crystals with no visible pore space between them. Contains approximately 1–5% of nonvisible, often noneffective porosity. Permeabilities are almost always less than 10 md and often less than 0.1 md. Rock surface has a smooth to resinous appearance on fresh surfaces and sharp edges form on breaking.

TYPE II: Chalky. Matrix composed of small crystals or particles (less than 0.05 mm in size) which are less tightly interlocked than those in Type I. The minute intercrystalline and interparticulate pore spaces are just visible at 10× magnification under the binocular microscope. Matrix porosity often reaches 15% and sometimes is much greater. Permeabilities rarely reach the 10–30 md range, and frequently they are less than 0.1 md. Freshly broken rock surfaces have a dull, earthy or chalky appearance.

TYPE III: Granular or Sucrosic. Matrix composed of crystals or grains only partially in contact with each other, leaving interconnected pore space between the particles. The total porosity in this type under ideal conditions is in the 15–30% range. Permeabilities are usually very high, ranging up to several hundred millidarcies. Freshly broken rock surfaces have a sandy or sugary appearance.

When the rock appears to have a combination of two of the types just described, it should be designated by using both numbers in the order of importance. For example, I–III would indicate a rock composed principally of Type I with a minor amount of Type III.

For greater detail, the crystals or grains composing the matrix may be described according to size as follows:

Crystal and/or Grain Size	*Symbol*
Large (coarse) >0.5 mm	L
Medium 0.25–0.5 mm	M
Fine 0.125–0.25 mm	F
Very fine 0.0625–0.125 mm	VF
Extremely fine <0.0625 mm	XF

For example, a finely sucrosic dolomite could be classified as Type III_F.

Pore Size Classification

Visible porosity is divided into four classes:

A. Porosity is not visible to the naked eye or under 10× magnification (<0.01 mm in diameter).

B. Porosity cannot be seen without magnification, but is visible under 10× magnification (0.01–0.1 mm—approximately extremely fine to very fine size range).

C. Porosity is visible to the naked eye but is less than the size of ditch cuttings (0.1–1 mm—approximately fine to coarse size range).

D. Pore size is greater than most ditch cuttings and is usually indicated by secondary crystal growth on faces of cuttings (1 mm and above—very coarse and larger).

To use this classification system, the texture and porosity terms are combined. A compact crystalline limestone (Type I texture), with porosity so small that it is not visible with a 10× magnification (Class A), would be classified as a Type IA limestone. From this basic classification, more exact subdivisions may be made. The same rock might be further classified by crystal or grain size, as described previously. If the compact crystalline structure was extremely fine, the rock could be classified as $I_{XF}A$.

Pore Distribution

In addition to classifying pore size, the frequency of pore occurrence (i.e., the percentage of the rock occupied by pore space) may be determined and inserted in the porosity classification. This is done by estimating the percentage of pore space exposed on a broken or slabbed core surface, or the numbers of drill chips containing porosity compared to those with no visible porosity. Figure 5 shows a series of surfaces containing from 1% to 25% porosity. This illustration can be used as a basis for comparison in estimating porosity on a core surface.

The percentage value of the porosity can be written after the porosity figure in the system of classification. For example, $I_{XF}B_5$ would indicate a compact crystalline rock with extremely fine crystal size containing 5% porosity that is just visible under 10× magnification.

Examples

In this porosity classification system, nonreservoir carbonates are classified as IA, indicating a compact crystalline rock with no visible porosity. Many of the best dolomitic reservoir rocks belong to IIIC/D types with varying porosities up to 30%. Chalky-type rocks are usually classified as IIB, indicating that the porosity, although often fairly high, is of a very fine size not easily seen with the naked eye.

This summary of Archie's classification system is presented here because, regrettably, porosity classification systems do not seem to be generally used. An understanding of the genesis of different porosity types within a rock is necessary, however, to the understanding of pore geometry.

Capillary Pressure Curves

The pore structure of carbonate rocks is a complicated maze of various-sized openings, which may or may not be interconnected. Such an interconnection, when it exists, often forms a tortuous network. Several testing procedures are available to help describe this network. First, there are means to measure the quantity of void space, or the porosity, normally expressed as a

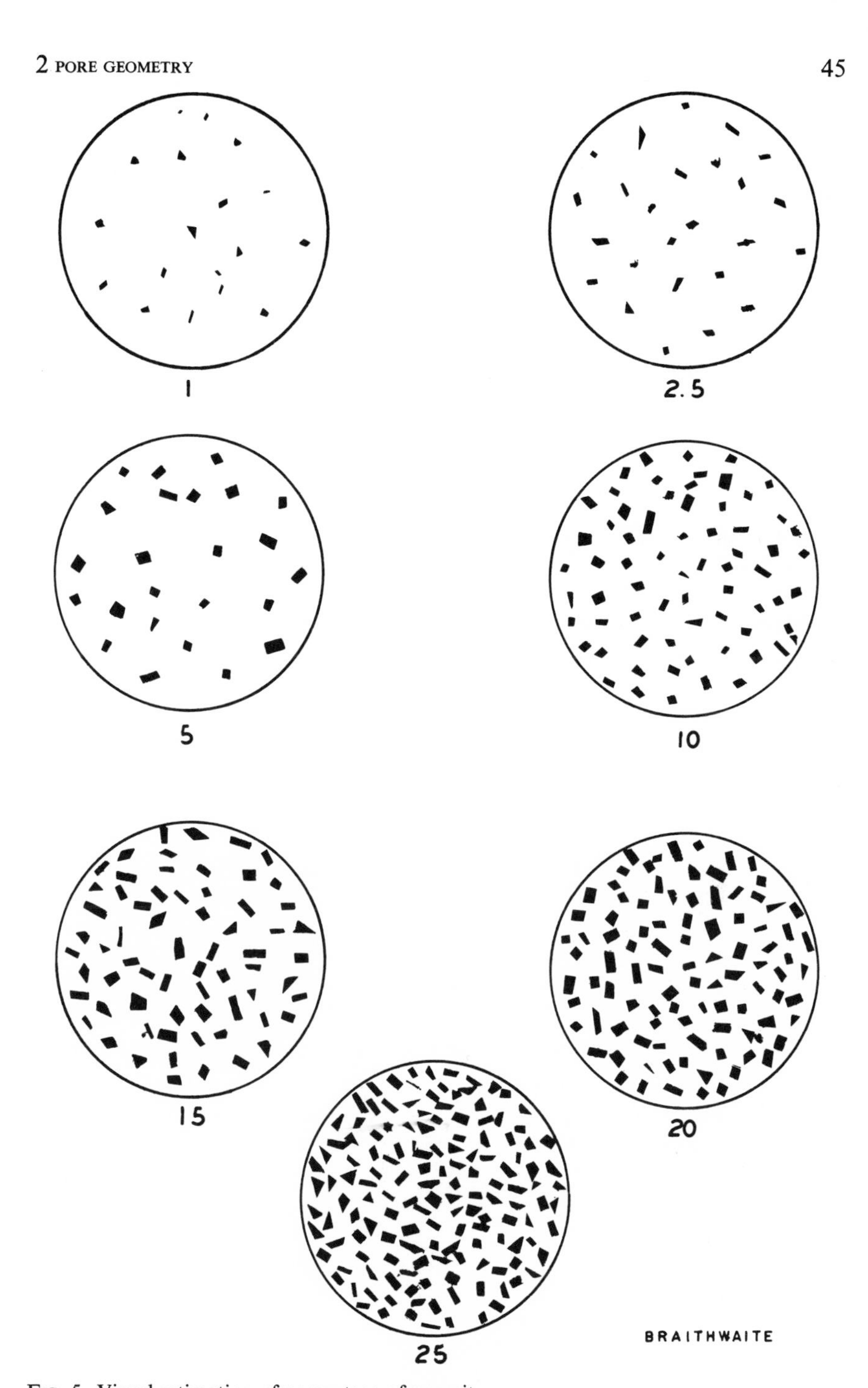

FIG. 5. Visual estimation of percentage of porosity.

percentage of total apparent volume of rock sample. A second type of test determines the extent of interconnection between an *unknown number* of pores, or the permeability. This term is used to describe the capacity of the pore system to transmit a fluid. These two terms, porosity and permeability, are commonly used by geologists, but perhaps are not always fully understood. A third test, much less frequently used and understood, provides data to determine the size of the throats connecting the pores and the percentage of total pores effective at successive levels in a reservoir. Pore throat geometry is basically the key to hydrocarbon-producing potential.

Because this chapter deals with applied carbonate petroleum geology, only the most generalized fundamentals of the capillary pressure method of studying pore geometry are discussed here. Numerous papers for petroleum engineers have been published, and Arps[4] and Stout[5] have written excellent papers on the subject of capillary pressure curves for the geologist.

In a reservoir capable of holding oil, water tends to rise because of capillary action, and to fall because of gravitational pull. If all pore throats in a reservoir were of the same size, the reservoir would have a sharp, well-defined boundary between oil and water.

Figure 6 pictures a dolomite, which is an excellent reservoir rock. Arrows indicate restrictions between large pores which, if present in two dimensions, would be pore throats. If the rocks shown in this photomicrograph were water wet, a thin film of water would coat every carbonate grain. Where

FIG. 6. Pore system of a dolomite. Arrows point to pore throats.

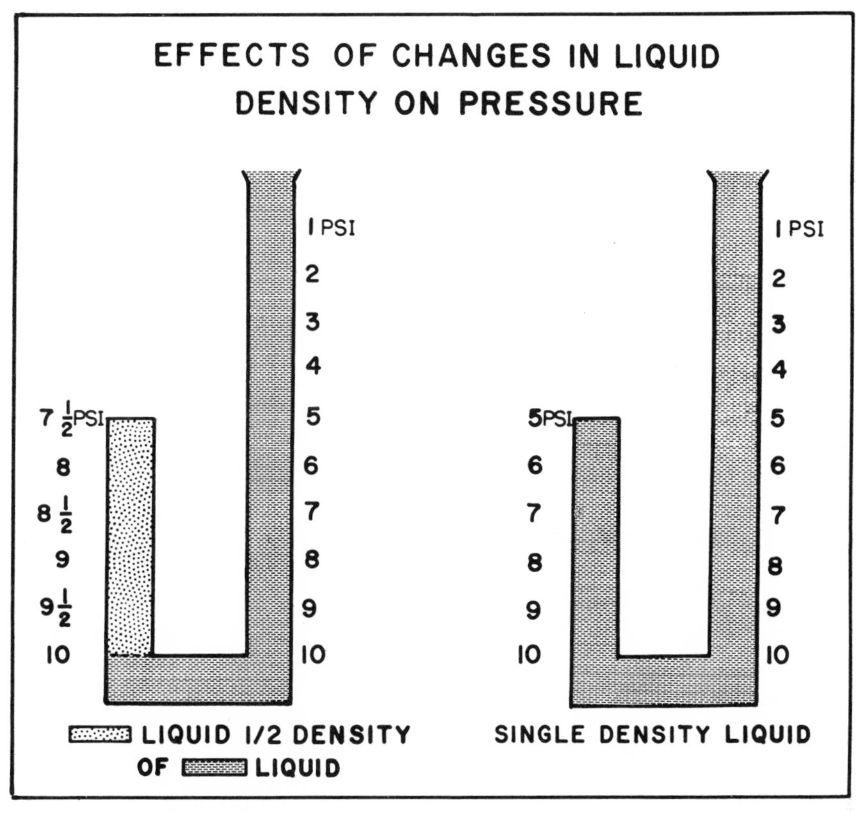

FIG. 7. Effect of a two-fluid system on pressures in a J tube. Substitution of a less dense liquid in the closed leg results in an increase in pressure proportional to the change in density.

grains are close together, capillary pressure would draw water into pore throats, blocking them. Oil could not enter the large pores until sufficient pressure was available to force the water from the capillary throats connecting the pores.

The presence of two fluids of differing densities, such as oil and water, in the reservoir has a profound effect on the distribution of the fluids in the pore space. The difference in densities creates a force balance in the zone, while the hydrocarbons are accumulating in the trap; this serves to displace water from the pore throats of the reservoir, allowing oil to enter.

A simplified explanation of how this force is initiated can be illustrated by considering a tube constructed in the shape of a J, closed at the top of the short leg, and open at the top of the longer leg (Fig. 7). The tube in the illustration is constructed in such a manner that one vertical unit contains sufficient height of water to exert 1 psi of pressure. A pressure gauge inserted five units from the top of the tube would, therefore, register 5 psi; a gauge ten units from the top would measure 10 psi; and so on. If the top of the basal cross member were ten units below the top of the water level, 10 psi would be exerted here, and the pressure at the base of the short leg would also be 10 psi. One unit up in the short leg, the pressure is 9 psi because of the hydrostatic water gradient. Five units up, at the top of the short leg, the pressure is 5 psi.

For the purposes of illustration, a hypothetical fluid with half the density of water is substituted in the closed leg in order to examine pressure distribution in such a case. Water in the longer open leg still exerts 10 psi of pressure across the basal leg of the tube. Likewise, 10 psi of pressure is exerted at the base of the closed leg, because the *same* fluid is at static equilibrium, intercommunicating and at the same level. Inasmuch as the hydrostatic gradient in the closed leg with the lighter fluid is only one-half the gradient with water, the pressure at a point one unit above the base is 9.5 psi and not 9 psi, as would be the case if the leg were filled with water. The five units of fluid column of the closed leg exert a hydrostatic head of only 2.5 psi at the base of the closed leg. The pressure at the top of the closed leg is, therefore, 7.5 psi, a gain in pressure of 2.5 psi resulting from the substitution of the less dense liquid.

If this J tube is considered equivalent to an anticlinal system (Fig. 8), it is apparent how the introduction of a second fluid (hydrocarbons) can cause an increase in pressure in the reservoir. Other relationships enter into this process, but the force generated as the result of differences in fluid density acts to force water from capillary pore throats, allowing hydrocarbons to move through the pore system during the migration and accumulation of the oil. At the base of the reservoir, only the pores connected by the largest pore throats are filled with oil, but with an increase in capillary pressure, on moving upward in the reservoir, oil enters increasingly smaller pore throats. The minimum effective pore throat size is defined as the smallest pore throat size that will permit the entry of oil at a given capillary pressure. The ultimate distribution of oil and water in the reservoir reflects this difference in pressure between the oil and the water phases, which is balanced by the capillary pressure.

A thorough understanding of absolute and relative permeability concepts is of utmost importance to petroleum geologists and engineers. Permeability, as defined previously, is a measure of the capacity of a rock to transmit a fluid. In measuring permeability, a core is cleaned of all fluids (or as nearly cleaned as possible) and a single fluid (normally air) is passed through it. The usual permeability figure simply gives a measure of the ability of the rock to transmit air (a single fluid). In nature, however, most rocks contain

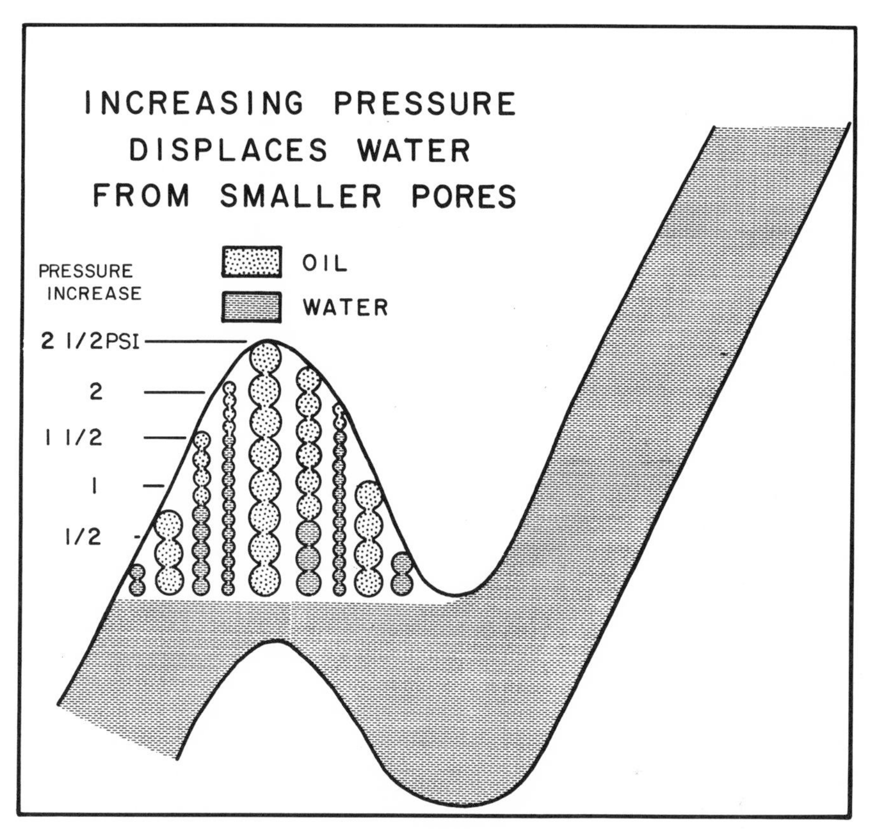

FIG. 8. Pressure buildup in an anticline caused by the presence of a two-fluid system.

at least two fluids, for example, oil and water. If the rock is water wet, most rock particles are surrounded by a thin film of water. If oil is present, it usually does not come into contact with the rock, but it is in contact with the water which surrounds the rock particles. When the rock particles are closely packed (small pore throats), the films of water will coalesce at points of contact. In such a case water may move through the rock, but oil will not move until it develops a sufficient pressure to overcome the capillary forces. Relative permeability is the ratio of the effective permeability of a rock to a fluid at a given saturation to its permeability at 100% saturation with this fluid. Thus, fluid movement in a rock is a function of relative permeability, which, in turn, for a given rock at a given time, is a function of the fluid saturation in this rock.

Fluid saturations, in turn, are affected by the pore geometry of the rock responding to reservoir pressures. Pressure is a function of height above oil-water contact, and with an increase in differential pressure, increasingly smaller pore throats are invaded by oil.

A capillary pressure curve indicates the percentage of the total pores in a reservoir rock that is connected by pore throats of sufficient size to allow oil entry at a given capillary pressure. By taking into consideration the density and fluid properties of the water and the hydrocarbons in a reservoir, capillary pressure may be directly equated with height above oil-water contact.

There are several methods of constructing a capillary pressure curve, but two are used most commonly, particularly when working with carbonates: the mercury injection method and the centrifuge method. The centrifuge method is perhaps faster and simpler, but it is more difficult to use when vuggy porosity is present, as is so often the case with carbonates.

In the mercury injection method, a core is cleaned, placed in a chamber, and evacuated. Mercury is forced into the core at a low pressure, which is maintained until no more mercury enters the core. This mercury acts as a nonwetting agent and is the equivalent of oil entering a water-wet reservoir. Mercury vapor can be considered as the wetting phase. The volume of mercury entering the core at this pressure level is recorded; then the process is repeated through a range of pressures to the highest pressure desired. The volume of mercury injected at each pressure level is used to calculate directly the percentage of total pore space which can be saturated. The saturation at each level is plotted on a chart with the vertical scale as capillary pressure in pounds per square inch, and the horizontal scale as water saturation. Because mercury represents the hydrocarbons, its amount can be considered as oil saturation. Subtracting the percentage of total pore space filled by mercury from the total pore space gives water saturation, which is plotted on the abscissa. The points computed at each pressure level are connected, and a mercury capillary pressure curve results, as seen in Fig. 9. Because pressure is directly related to height of oil column, pressures can be converted to heights above oil-water contact and used as the vertical scale, as shown.

This method illustrates the behavior of hydrocarbons entering a capillary system. The centrifuge method, on the other hand, illustrates the behavior of hydrocarbons leaving such a system. With this method cores are cleaned, evacuated, and saturated with a light refined oil. Cores are then placed in a centrifuge and spun at a low speed until no more oil is forced from them. The oil spun off is measured; then speed is increased to develop a higher pressure differential across the core, the additional oil forced out is measured, and the process is repeated. From these readings a curve can be constructed which shows the behavior of the hydrocarbons leaving the system. This can be mathematically converted to an entry capillary pressure curve, as illustrated in Fig. 9, where curves obtained by using the mercury capillary pressure and the centrifuge methods are compared. The latter method is perhaps less

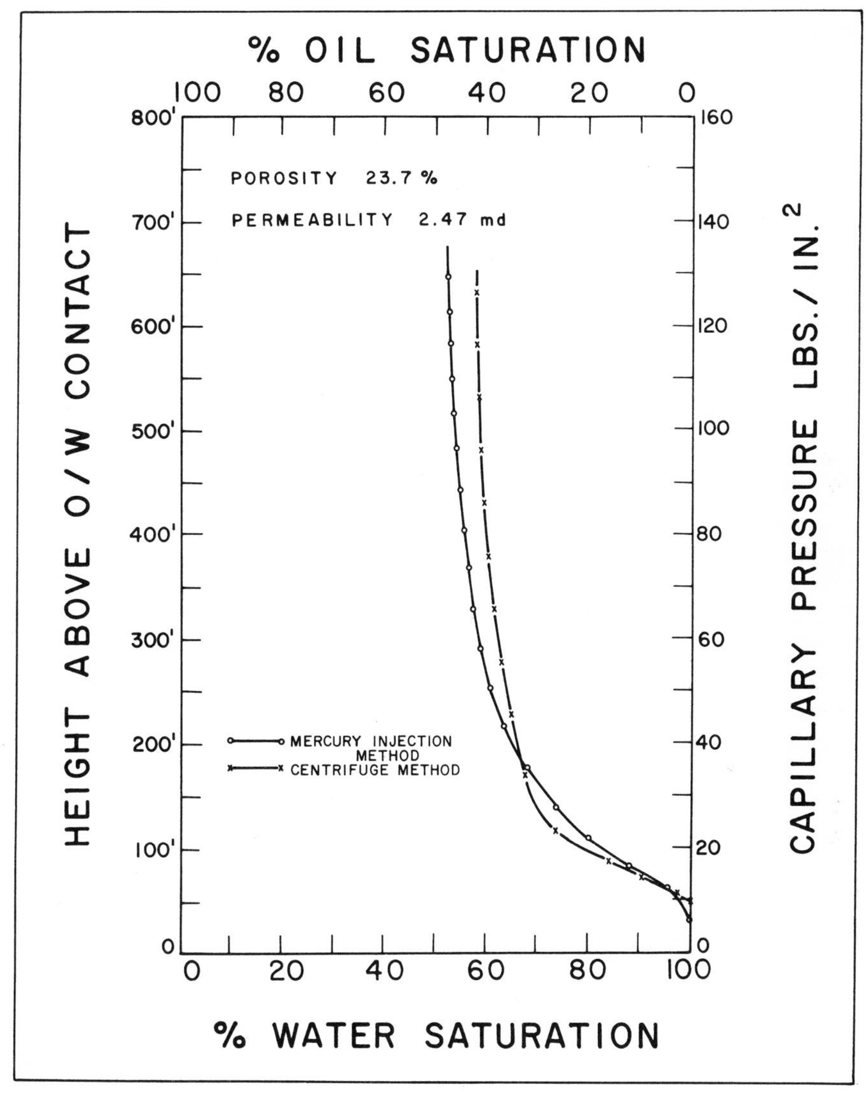

FIG. 9. Capillary pressure curves of a carbonate rock as computed by the mercury injection method and the centrifuge method.

accurate, but is acceptable in applications where a loss of accuracy can be sacrificed for an increase in speed of testing.

Capillary Pressure Curve Interpretation

The preceding discussion of the basic petroleum engineering and geologic aspects of carbonate rocks, along with the classification of porosity, indicates a close interrelation of these various aspects. By interrelating and coordinating the engineering and geologic approaches it is possible to learn much more about the role played by the size and shapes of the pore system in determining the production capability of a carbonate rock. Thus, one can use mercury capillary pressure tests to determine the general pore geometry of specific types of rocks and greatly enhance the ability to predict in advance the productive capabilities of a formation.

If relative permeability to oil and water is also determined from a core sample, in addition to the capillary pressure curve, a quite accurate picture can be constructed of fluid behavior in a reservoir composed of this rock. A typical relationship between relative permeability and capillary pressure, as well as the effects of the two parameters on productibility, is illustrated in Fig. 10. The relative permeability curve is placed in superposition to the capillary pressure curve, as suggested by Arps.[4] The rock tested here is a dolomite from the Red River (Ordovician) reservoir of the Burning Coal Field, North Dakota. In regard to production, Fig. 10 shows that 95% of the porosity in this rock is interconnected by pore throats of sufficient size to permit the flow of oil in an oil-water system, provided sufficient pressure (furnished by height of oil column) is present.

Five percent of the porosity contains irreducible water. This is represented by the location of the curve after it assumes an essentially vertical direction, or the area where oil saturation ceases to increase with rising pressure differential.

From the standpoint of geology, this curve also gives considerable information about the pore geometry of the rock. Inasmuch as the curve is a plot of the total pore volume filled by mercury as increasing pressure forces it through successively smaller pore throats, the shape of the curve is determined by the sorting and skewness of the pore throat size.

Sorting is a measure of the degree of spread from the finest to the largest pores. In general, a well-sorted pore arrangement is attributed to winnowing* of the primary sediments, or recrystallization or dolomitization of the rock, and a poorly sorted arrangement of pores is indicative of a primary unwinnowed rock condition.

Skewness of the pore distribution is a result of the grouping of pores in a particular size fraction. Fine skewness is evident in a rock with very fine-grained compact crystalline matrix porosity (IA), and is an approach to high irreducibility. Coarse skewness may be represented in a sucrosic dolomite

* The term *washed* is preferred by many earth scientists, because *winnow* is the old English word *windwian*, and has reference to exposure to the wind such that lighter particles are blown away. Ed. note.

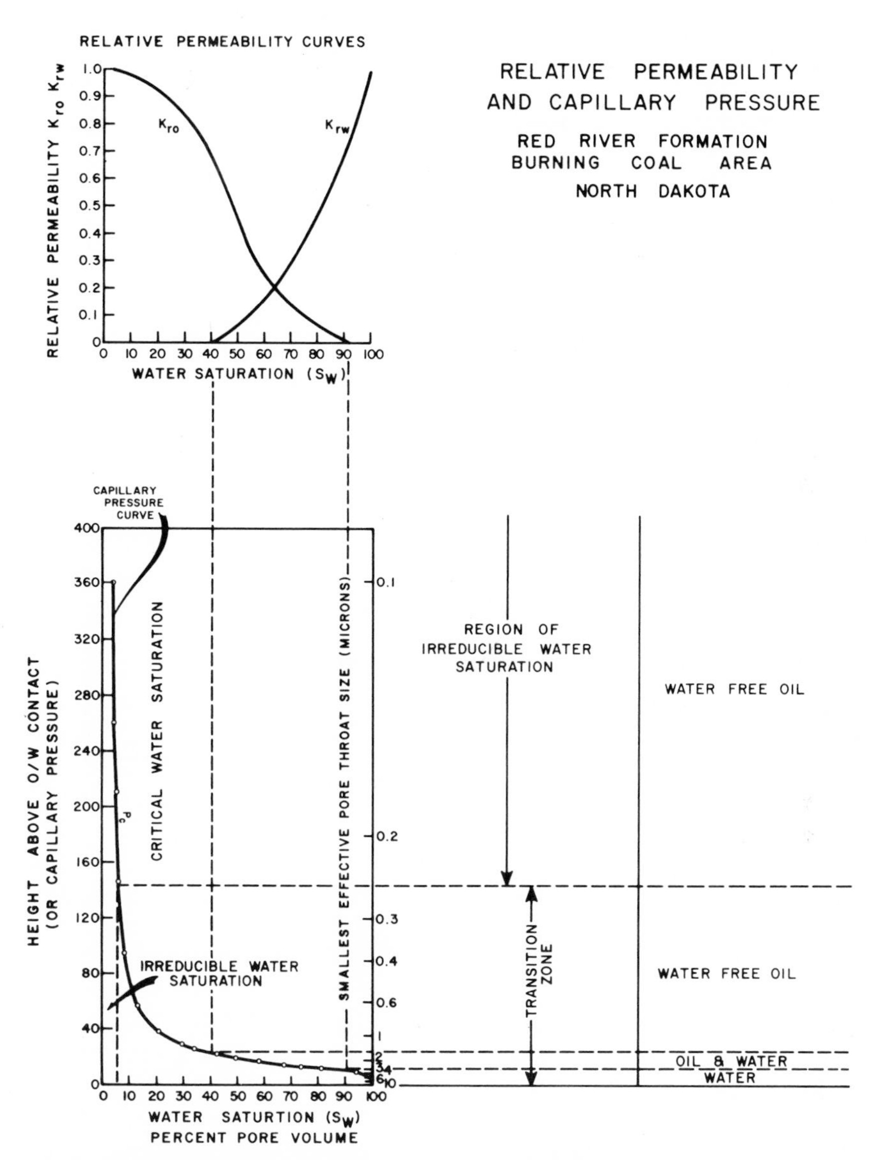

FIG. 10. Rock productibility as determined from capillary pressure and relative permeability curves. Smallest pore throats effective at any pressure level shown on scale to right of capillary pressure curve. (Modified after Arps.[4])

with vuggy porosity (IIID). The phenomenon of irreducibility is related to the reduction of effective porosity due to ineffective pore throats. As the pore-size distribution becomes increasingly fine-skewed, the pore space of

FIG. 11. Typical capillary pressure curves resulting from variations in sorting and skewness.

the rock occupied by the non-wetting phase is reduced because of the greater saturation of wetting phase, which is trapped in minute interstices. The displacement pressure, or the height of a column of oil required to offset capillary pressure and displace the interstitial water of a reservoir rock, is a function of the size of the throats between interconnected pores and is independent of total porosity.

A rock containing completely unsorted porosity should have a capillary pressure curve consisting of a straight line ranging diagonally from 100% water saturation to 100% oil saturation, provided sufficiently high pressures are reached to fill all pores with mercury.

A rock with perfectly sorted porosity should consist of a horizontal line running across the chart. If the sorting has fine skewness, the line will be at the top of the chart; if the skewness is coarse, the line will be at the bottom. Such perfection is, of course, never reached in nature. In a water-wet rock some water will always be present as a grain or particle coating, so oil saturation can only approach, but never reach, 100%. At the point of irreducible water saturation, the line should assume a vertical direction.

When these rules are applied to Fig. 10, it becomes apparent that this example exhibits a very high degree of porosity sorting, and that it has definite coarse skewness. Figure 11 illustrates by a series of simple sketches the types of capillary pressure curves that would result from various degrees of sorting and skewness.

Mathematically, it is possible to compute the size of pore throats through which oil will move in an oil-water system of given densities, at pressures equivalent to any specific height of oil column. This has been done with the core illustrated in Fig. 10. (Also see Appendices A and B.)

The minimum size of pore throat effective at each pressure level (height above oil-water contact) is shown to the right of the capillary pressure curve in Fig. 10. Using this scale, one can see that approximately 28% of the pore throats are larger than 3 μ. Almost 38% of the pore throats are between 1.5 and 3 μ, and 34% are smaller than 1.5 μ. These figures confirm that this pore system is coarsely skewed, and is well sorted with maximum sorting in the 1.5–3-μ range. Figure 12 shows pore entry radius (pore throat size) plotted against cumulative percentage of pore volume for the same core sample.

The degree of sorting and skewness which the pore throat system exhibits is of more than academic interest. It should be quite apparent that, if a rock exhibits markedly fine skewness, excessive heights above oil-water contact are required before it is capable of holding oil. Thus, even though a rock may be porous, it may be incapable of holding oil if the closure of the potential trap, although substantial, is less than the required height above oil-water contact.

Examination of Fig. 10 shows that the thickness of the oil-water transition

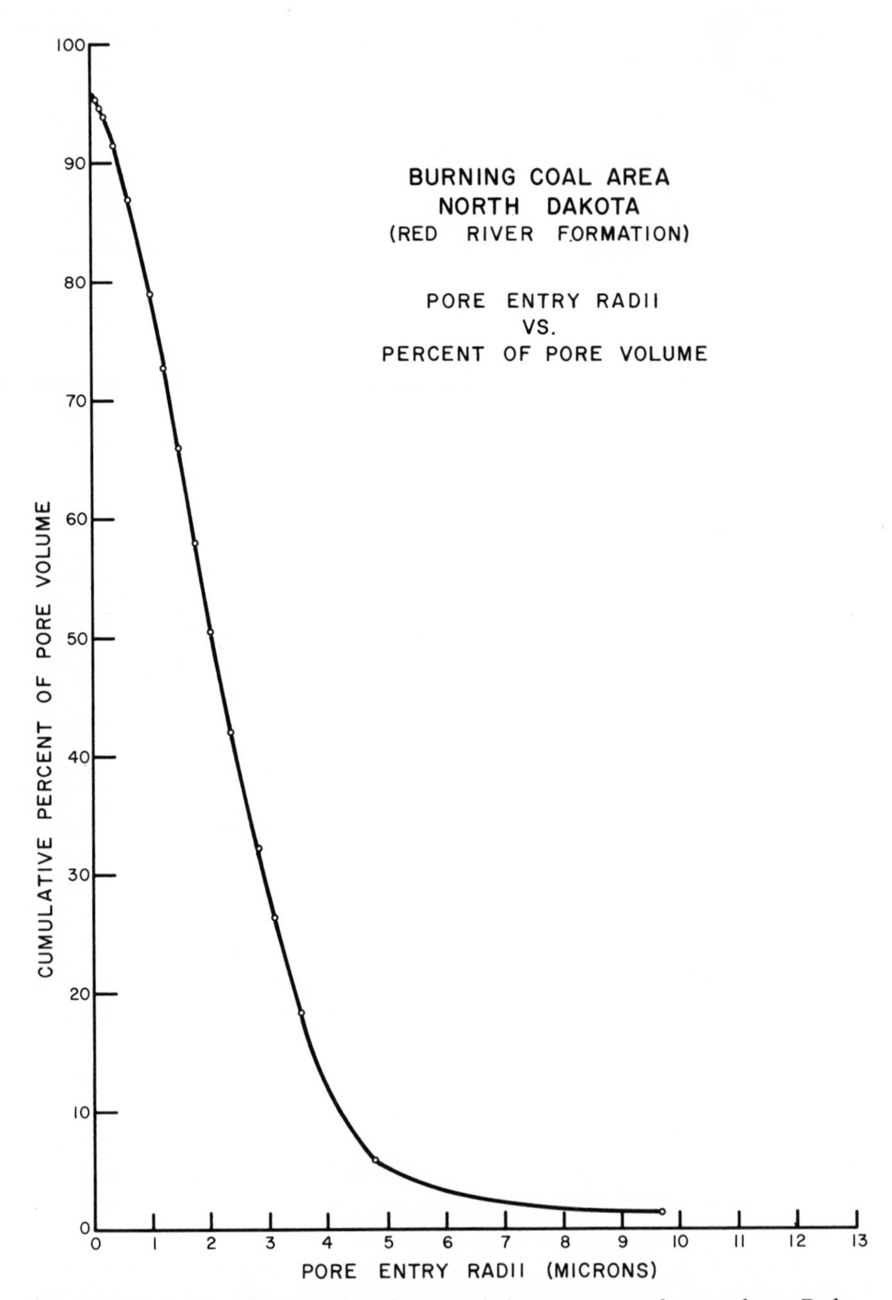

FIG. 12. Pore entry radius plotted against cumulative percentage of pore volume, Red River Formation, Burning Coal Area, North Dakota.

zone depends on the slope of the capillary pressure curve. This slope is a function of porosity sorting. A well-sorted, coarsely skewed rock may produce water-free oil with only a minimum oil column. A coarsely skewed but poorly sorted rock, on the other hand, may have a thick zone where both oil and water are produced before a sufficient height is reached to yield clean oil.

Interrelation between Pore Geometry and Rock Types

The principles discussed above have a practical application. On examining thin sections and capillary pressure curves of cores from hundreds of wells, it became evident that rocks grouped together by lithologic, petrographic, and diagenetic characteristics also have surprisingly similar mercury capillary pressure curves. Permeability and porosity may vary widely, but productibility remains strikingly the same. Mud-supported, grain-supported, and leached rocks have distinct families of curves. Departures from these curves depend on pore throat sizes within each class of rock.

The Montana-North Dakota carbonate-bank limestones (Mississippian) present an excellent area to study this interrelation of lithology, diagenesis, and productibility. More than 200 cores from wells in this area were examined, and grouping and comparison of the capillary pressure and lithologic data showed a high degree of correlation.

Limits of the Study

Before discussing this study, some finite limits must be established, and it should be constantly kept in mind that the results and conclusions mentioned here apply only within these limits. They may be extrapolated to other areas only as conditions can be shown to be comparable. Principles, however, should be universally applicable.

This discussion is limited to an area extending for 20 miles on both sides of the North Dakota-Montana border, and for a distance of 120 miles south of the Canadian border. It is further limited to the Charles Formation of Mississippian age. This is an area of limestone bank and lagoon deposition, and it appears that uniform conditions existed throughout the area during deposition and diagenesis of the rocks. Hence it is considered that the physical properties of formation water and hydrocarbons found in the area hereafter will not vary widely from those discovered there to date.

Classification of Rocks by Productibility

By using porosity, permeability, and capillary pressure data, thin sections of

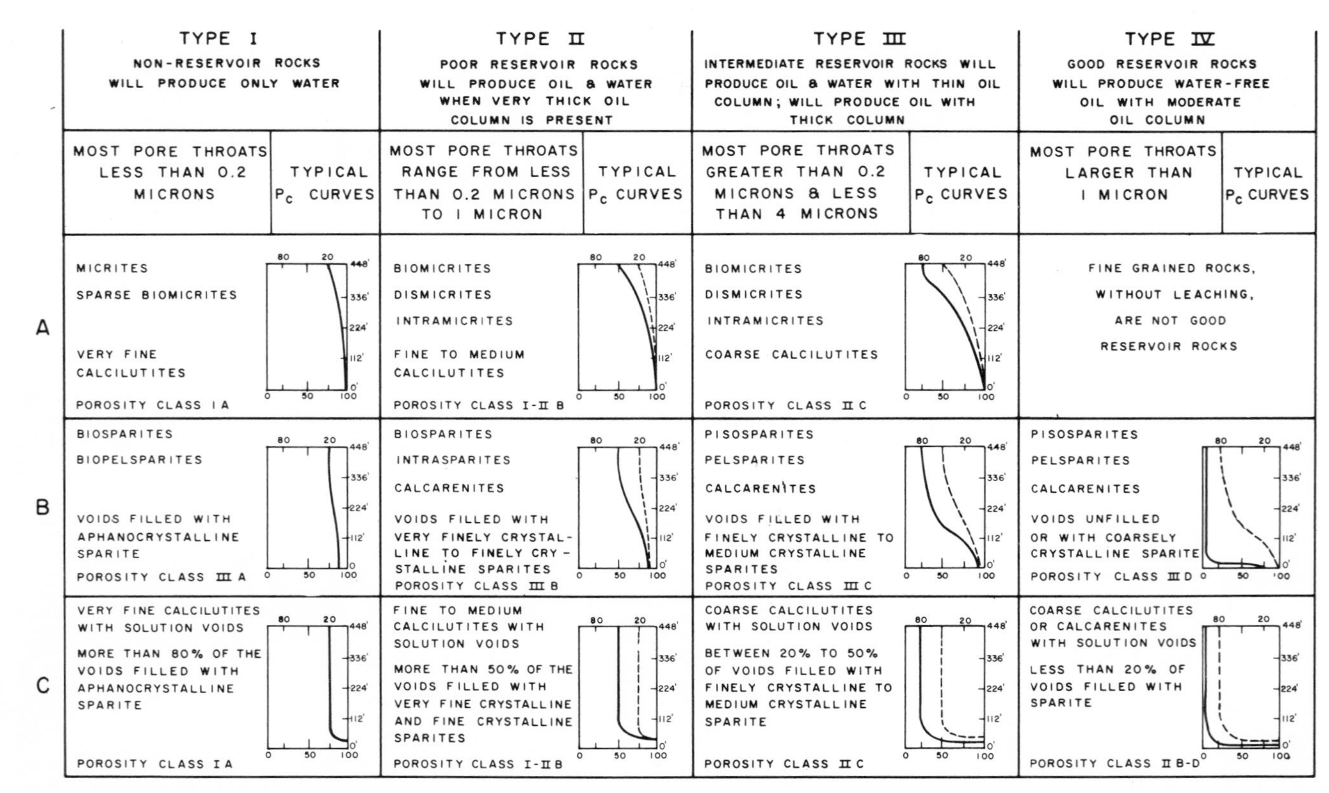

FIG. 13. Relationship of families of mercury capillary pressure curves and rock types in the Mississippian of the western Williston Basin.

the cores, and a mathematically computed pore throat size curve, families of capillary curves can be very closely matched with families of carbonate rocks. Figure 13 shows this interrelationship. It should be emphasized that this chart is not presented as a general classification system for the productibility of carbonate rocks. It merely attempts to show how carbonate rocks in a specific habitat may be classified. Once devised, it is extremely useful in deciding where to explore, where to test, and where to attempt to produce oil in this habitat. Such a classification is very useful in understanding reservoir behavior; however, great care must be exercised in extrapolating this chart to other areas.

Eleven families of curves are shown. They are divided into four types, grading from those indicative of nonproductive rocks to those representing good reservoir-type rocks. Each type is subdivided into families on the basis of the shape and departure of its capillary pressure curve. The genesis of the rock and the size and shape of pores and pore throats within the rock determine the shape and departure of the curves.

The term *family of curves* is used because each curve within the family is related in shape and has finite limits of departure from the *X*- and the *Y*-axis. Each family of curves is related to a distinct family of rocks. In Family IA, for example, the typical curve illustrated is characteristic of a rock near the coarse extreme in the family. As the rocks become finer grained, the curve will maintain its shape, concave to the left, and its intersecting point with the *Y*-axis will move upward, indicating that the upper limit of pore size is decreasing. At the same time, the entire curve will move to the right, indicating increasingly higher irreducible water content. In this family the curve never reaches zero on the *Y*-axis, nor does it reach lower than 80% irreducible water on the *X*-axis. Similar limiting behavior patterns exist for all the curve families.

From this study certain distinct constant curve characteristics are apparent:

1. *Fine-grained rocks* have no large pore throats. The point at which the capillary pressure curve reaches the *Y*-axis is always above zero. The family of curves is always concave to the left (fine skewness). With an increase in pore throat size the curve moves to the left, maintaining its fine skewness.

2. *Coarse-grained rocks* have capillary pressure curves which typically intersect the *X*-axis, indicating that at least some large pore throats are present. The curves are concave to the right (coarse skewness).

3. In this area no coarse-grained rocks which had been leached and then infilled (*altered rocks*) were recognized, perhaps because of the difficulty of distinguishing them from the coarse-grained rocks. All rocks showing significant solution and then recrystallization belonged in the fine-grained class. No large pore throats were found in fine-grained rocks, although the

class contained the best reservoir rocks. The capillary pressure curves invariably show a sharply coarse skewness, with the curve always meeting the *Y*-axis well above zero. This must indicate that all pore throats in such rocks are remarkably uniform in size. The question is unresolved as to whether during leaching the smaller pores are enlarged first, until all pore throats reach uniform size; or whether during infilling sparite crystals develop first in the largest pores, until all pore throats have been reduced to a uniform size. In either event, a remarkable uniformity of pore throat size exists in rocks of this class.

A definition of each type of rock, with a discussion of every family found within it, is given in the following paragraphs. The separation between types is somewhat arbitrary, but is in basic agreement with subdivisions of porosity, grain size, and general limits of productibility. The examples shown are among those with the largest pore throats in each type. Frequently, even in a leached, fine-grained rock (Class C), the capillary pressure curve will meet the *Y*-axis at a point considerably above those shown, so that relatively great intervals above oil-water contact are required to produce hydrocarbons.

Type I. Nonreservoir Rocks

Type I curves are indicative of nonproductive rocks. They will produce only water under conditions that might be expected to be commonly found in nature in their Williston Basin habitat. These rocks show 80% or more irreducible water saturation up to pressures equivalent to an oil column 448 ft in height. (This figure of 448 ft represents the maximum equivalent pressure used in testing rocks for this study and will appear throughout this chapter. Presumably it will far exceed any possible height of oil column in the study area.) Most (80% or more) pore throats in all rocks of this type are less than 0.2 μ in size.

Three broad classes of rocks produce the following three distinct families of Type I curves.

IA. Nonreservoir, fine-grained. This family of curves is found to be concave to the left. The irreducible water saturation is never less than 80%, and this is at the maximum height of oil column. The point at which the capillary pressure curve intersects the *Y*-axis is always above zero. Micrites and sparse biomicrites with grains in the size range of very fine calcilutites produce this family of curves.

Figure 13 illustrates a typical curve, generally showing an example of the best pore geometry, from each family. Curves in any family will fall in the area between the curve shown and the example from the preceding family. For example, curves from Family IIA will fall in the area between the solid line and a broken line equivalent to the curve shown in IA.

Curves from Family IA will fall between the line shown and the 100% water saturation line. Because of the reduced nature of the capillary pressure curves in Fig. 13, an enlargement of a specific curve from Family IA is shown

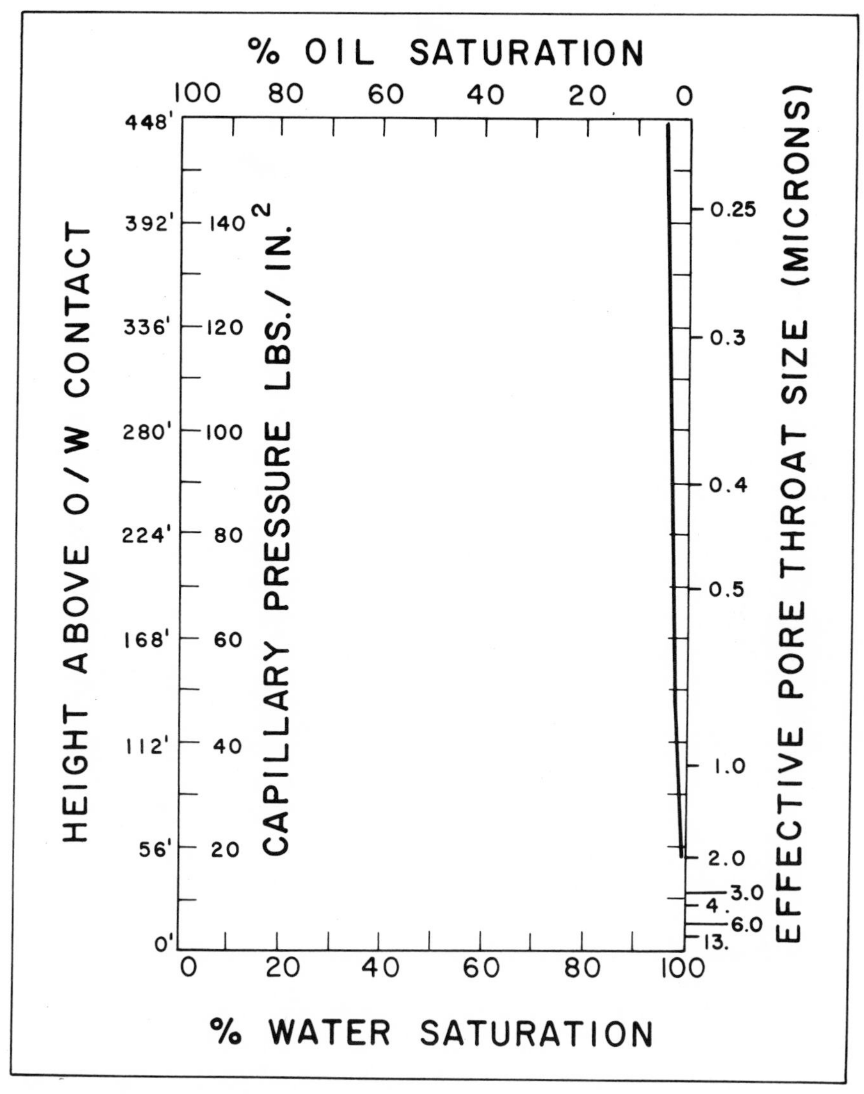

FIG. 13A. Enlarged capillary pressure curve from Fig. 13, showing scales used for capillary pressure, height above oil-water contact, and the smallest pore throats effective at any pressure.

in Fig. 13A. This represents the rock in this group with perhaps the least production potential. The curve is almost a straight line and nearly parallel to the 100% water saturation line. If any fine skewness is present, pressures developed in testing were not enough to detect it. Capillary pressures in pounds per square inch and also as heights above oil-water contact are shown. The minimum sizes of pore throats effective at any given height above oil-water contact, under the conditions found in the study area, are also shown. The relative scales used on this figure and on all the reduced curves in Fig. 13 are exactly the same, so that extrapolations of scale may be freely made.

Figure 14 shows a photomicrograph of a typical rock in Family IA. Even though scattered solution vugs are visible, they are interconnected only by the matrix pore system. The capillary pressure curve indicates that less than 5% of the pore throats in this rock are larger than 0.2 μ. Of this 5%, apparently no significant number exceed 1 μ in size.

Many rocks within this class produce a capillary pressure curve which is an almost vertical straight line, very close to the right side of the chart. As grain size increases the curve becomes increasingly concave to the left, until it reaches the 20% irreducible water saturation point at a height of 448 ft. In the rock samples found to produce curves in this family, no permeability above 0.02 md was found. Porosities ranged from 2% to 10.2%.

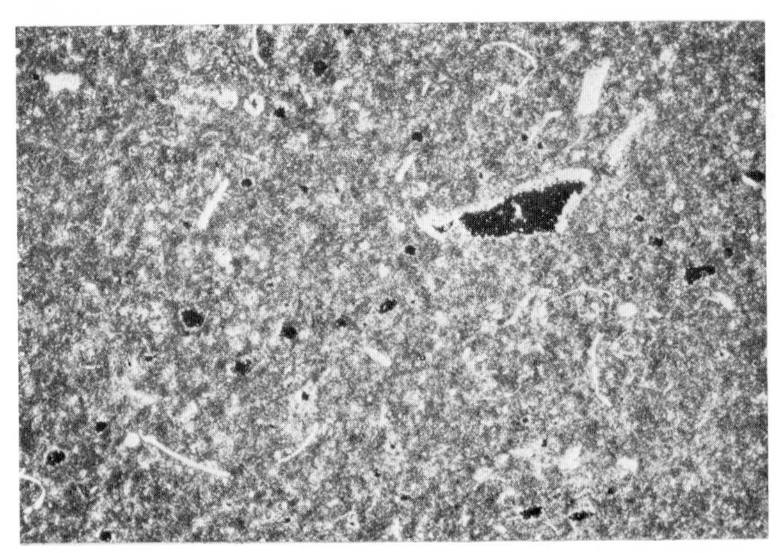

× nicols, × 35.

FIG. 14. Sparse biomicrite, very fine calcilutite, nonreservoir rock, porosity = 8.5%, permeability = 0.02 md, Family IA. × nicols, × 35.

IB. Nonreservoir, coarse-grained. Rocks producing curves in this family are composed of coarse fragments (calcarenite or calcirudite grain sizes) cemented together and having intergranular voids filled by aphanocrystalline sparite. All examples found in the study were biosparites or biopelsparites. Inasmuch as this type of rock is produced by an infilling of void space, some larger pores may be left at least partially unfilled, and also porosity may be preserved in the allochem portion. Even though 80% or more of the pores are less than 0.2 μ in size, a small percentage of large pores and large pore throats may remain unfilled by sparite.

Such conditions produce a family of curves which typically reach down to meet the *X*-axis, and which are concave to the right. Figure 15 is a photomicrograph of a typical rock of this class. Rocks producing this family of curves ranged from 2.0% to 2.6% in porosity, and had permeabilities of less than 0.02 md.

IC. Nonreservoir, altered rocks. All Class C rocks shown in Fig. 13 are of orthochemical origin, or have a micrite matrix, were leached, and then were infilled with sparites. In theory, at least, allochemical sparite matrix rocks which have been leached and infilled should also be present. None were recognized, however, in the study area, perhaps because they were not

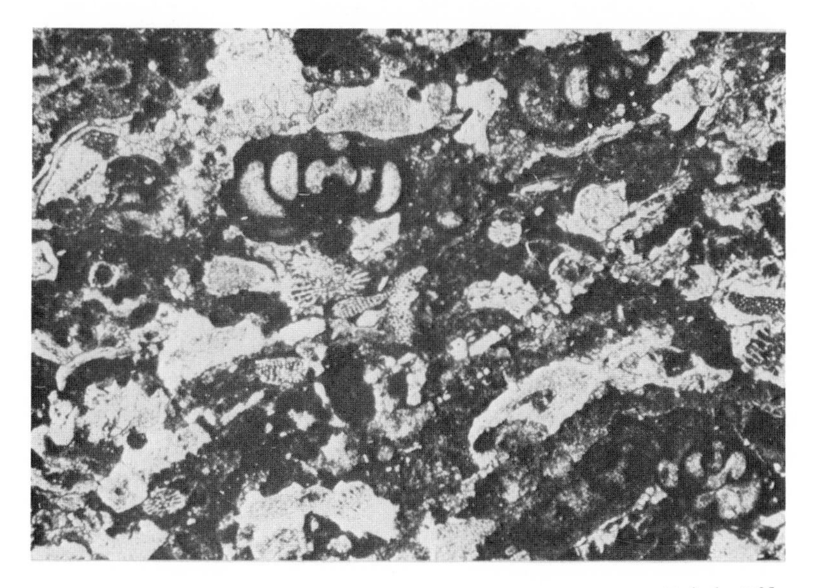

× nicols, × 35.

FIG. 15. Biosparite, coarse-grained, voids filled with aphanocrystalline sparite, non-reservoir rock, porosity = 2.6%, permeability = 0.02 md, Family IB. × nicols, × 35.

formed, or because they are too difficult to distinguish from Class B rocks. It should be kept in mind that these rocks may be present in other areas and may produce additional families of curves.

Only one good example was found of Family IC rock. Shown in Fig. 16, it is a very fine calcilutite (peldismicrite) which has been leached; more than 80% of the voids thus created are filled with aphanocrystalline sparite and anhydrite. The remaining voids are interconnected only by matrix porosity. This example is at the coarser limit of the family. As the voids are filled with finer crystals, the entire curve should move to the right, the rocks still exhibiting highly unsorted pores in the less than 20% fraction larger than 0.2 μ. The single example found of Family IC had porosity of 5.9% and permeability of less than 0.02 md.

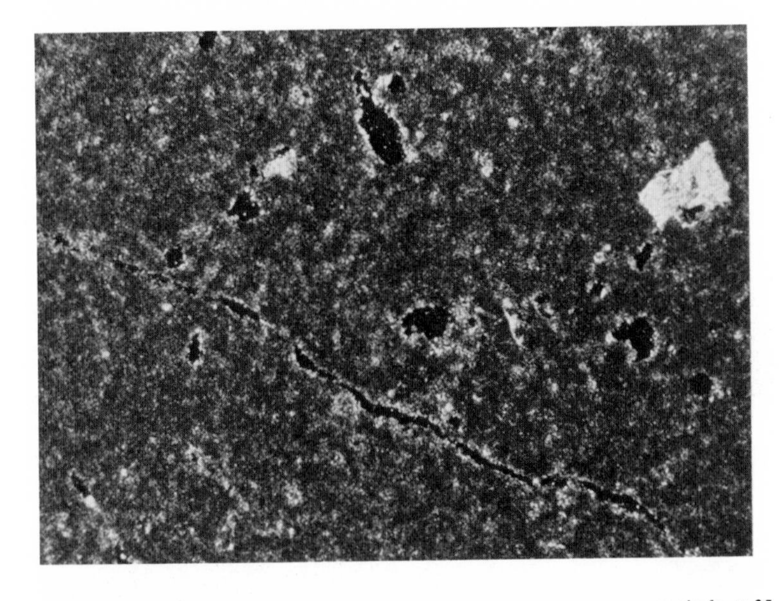

× nicols, × 35.

FIG. 16. Peldismicrite, leached, voids filled with aphanocrystalline sparite, nonreservoir rock, porosity = 5.9%, permeability = 0.02 md, Family IC. × nicols, × 35.

On examining curves produced by all the rocks in Class C, a strong tendency for the pore systems to be highly sorted, with coarse skewness, was noted. This seems to suggest two possible origins: (1) As this type of rock is leached, solution occurs first in the smaller pore throats and progresses to pore throats of a constantly increasing size, until a uniform system is created with all pore throats being of a more or less even size. In these fine-grained rocks few pores with throats larger than about 4 μ seem to have been created. (2) As infilling occurred, the largest pores and pore throats may have been filled first, filling pores of a constantly decreasing maximum size, and producing marked skewness. In either case, the resultant rocks exhibit a typical vertical curve, breaking sharply between the 25- and the 50-ft height of oil column which typifies all families in this class.

In all of the rocks classified as Type I, porosities ranged from 2% to 10%, and permeabilities were never more than 0.02 md.

Type II. Poor Reservoir Rocks

Curves grouped in Type II are indicative of poor reservoir rocks. They may produce oil and water when a sufficiently thick oil column is present, and only under very exceptional conditions will they yield water-free oil.

Most pore throats (50% or more) are in the range between 0.2 and 1 μ. Only a small percentage of pore throats are between 1 and 4 μ. As in Type I, three classes of rocks produce three families of curves.

IIA. Poor reservoir, fine-grained. This family of curves is produced by biomicrites, dismicrites, and intramicrites with the matrix grains in the fine to

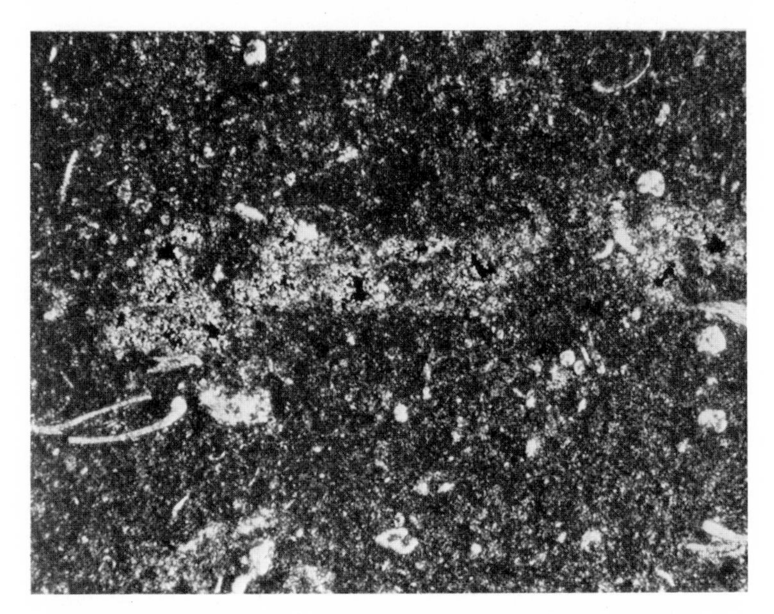

× nicols, × 35.

FIG. 17. Biomicrite, medium calcilutite, poor reservoir rock, porosity = 12.5%, permeability = 0.02 md, Family IIA. × nicols, × 35.

medium calcilutite range. Small vugs, connected only by matrix porosity, may be developed. At the 448-ft limit the curve reaches the 50%–80% irreducible water range. A concave-to-the-left shape is maintained. The curve intersects the Y-axis at or above the zero point, indicating no pore throat development in the larger range. Pore throats larger than 1 μ are not common.

Figure 17 illustrates a rock which produces a curve in this family. It is a biomicrite, and comparison with Fig. 14 shows noticeably larger matrix grain size. The capillary pressure curve is typical.

Porosities found in the rocks associated with this curve family range from 0.9% to 15.6%, and there is a range from 0.02 to 0.23 md of permeability.

IIB. Poor reservoir, coarse-grained. This family of curves is produced by biosparites, intrasparites and calcarenites, with voids filled with very finely crystalline to finely crystalline sparites. Curves are concave to the right, but skewness becomes more important. At the 448-ft extreme the curves indicate 80%–50% irreducible water content. The curves do not cross the Y-axis, but they intersect the X-axis between the 95% and 100% water saturation lines. Apparently up to 5% of the porosity is represented by original pores which were above 80 μ in size, and are still preserved in these rocks.

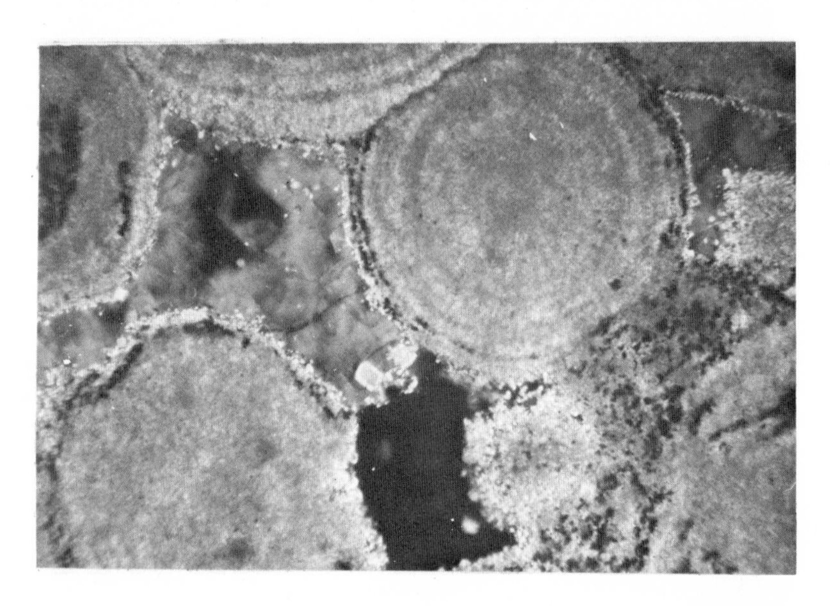

× nicols, × 35.

FIG. 18. Oosparite, coarse-grained, voids partially filled with finely crystalline sparite, poor reservoir rock, porosity = 15.5%, permeability = 16.8 md, Family IIB, enlarged. Note how fine sparites control fluid movement between interoolitic pores. × nicols, × 35.

Figure 18 illustrates the pore characteristics of a rock belonging to this family. It is a rather well-sorted pisolitic and oolitic calcarenite. It is obvious that the sorting of the allochems and the size and the number of pores have little effect on the productibility, which is controlled by the size of pore throats in the sparites between the large pores.

Porosities observed from this family ranged from 4.9% to 15.5%, and permeabilities from 0.02 to 16.8 md.

IIC. Poor reservoir, altered rocks. This family of curves is produced by fine- to medium-grained calcilutites with solution voids more than 50% filled with very finely crystalline to finely crystalline sparites. All curves from rocks of Class C exhibit a strong coarse skewness. It is difficult to distinguish this curve from that of a coarsely skewed Family IIB rock, except that a IIC curve typically crosses the *Y*-axis above the zero point, whereas the IIB curve typically crosses the *X*-axis. These rocks do not have pores with throats larger than 80 μ developed to the extent that the IIB rocks do. The pore geometry of this rock family (Fig. 19) indicates that although large voids

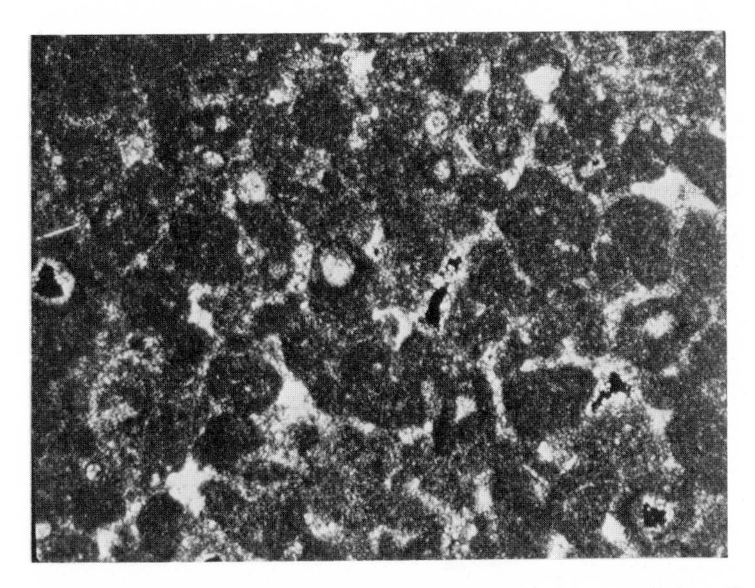

× nicols, × 35.

FIG. 19. Pelmicrite, leached, voids filled with finely crystalline sparite, poor reservoir rock, porosity = 15.3%, permeability = 2.66 md, Family IIC. × nicols, × 35.

are present (50–900 μ in size) the throats are generally filled with sparite cement, and most interconnection appears to be through matrix porosity.

Rocks of Type II exhibit a range of porosities from 0.9% to 15.6%, and permeabilities from 0.02 to 16.8 md. Some rocks of this type seem to have porosities and permeabilities which might make them good reservoirs, but they do not perform better than other rocks having much lower values for these parameters. It is very difficult to distinguish the poor reservoir rocks (Type II) by porosity or permeability measurements alone.

Type III. Intermediate Reservoir Rocks

Intermediate reservoir rocks produce families of curves classified as Type III. These rocks generally produce oil and water, but may yield water-free oil if a sufficiently thick oil column is present. The thickness of the oil column required varies with changes in rock class. Because of the concave-to-the-left nature of the typical curve in Family IIIA, it does not appear likely that these rocks will produce water-free oil in the study area with closures less than 275 ft. Rocks in Family IIIB, as indicated by the concave-to-the-right nature of their curves, may produce water-free oil with oil column heights as low as 150 ft. Rocks in Family IIIC, with their characteristic coarse skewness, may produce clean oil with oil columns as low as 40 ft. This same skewness, however, limits the decrease of irreducible water with greater oil column height, so that a large portion of total porosity is ineffective.

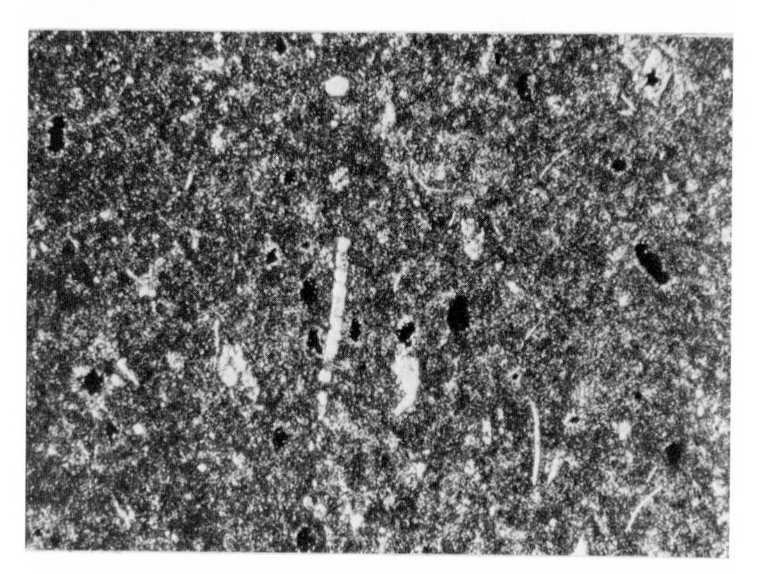

× nicols, × 35.

FIG. 20. Biomicrite, coarse calcilutite, intermediate reservoir rock, porosity = 19%, permeability = 0.38 md, Family IIIA. × nicols, × 35.

Most pore throats (50%–80%) are greater than 0.2 μ and less than 4 μ in width.

IIIA. Intermediate reservoir, fine-grained. Figure 20 shows a thin section of a rock having a curve typical of this family. Curves, which characteristically are concave to the left, meet the *Y*-axis at or near the zero point, and at the fine limit exhibit between 50% and 80% oil saturation. The rocks are biomicrite, dismicrite, or intramicrite, composed of coarse calcilutite grains. Porosities observed in this family ranged from 12.7% to more than 30%, and permeabilities from 0.23 to 4.11 md. It is obvious, however, that this family of rocks is a potential producing zone only when an extremely thick oil column is present.

IIIB. Intermediate reservoir, coarse-grained. Curves in this family represent biosparites, pelsparites, and calcarenites. Voids are filled with finely crystalline to medium crystalline sparites. Figure 21 is an atypical, but interesting,

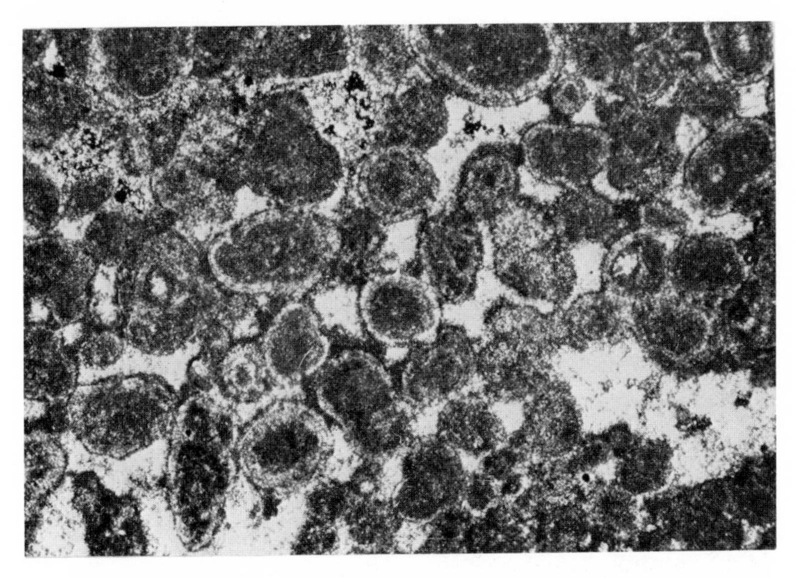

× nicols, × 35.

FIG. 21. Biooosparite, coarse-grained, voids filled with medium crystalline sparite, porosity = 11.4%, permeability = 0.83 md, Family IIIB. × nicols, × 35.

example of a curve from this family. All other curves representing these rocks have a shape closely similar to that shown in Fig. 13. In the example shown in Fig. 21 selective filling of larger pores has taken place, so that fewer than normal pore throats in the 0.5–1.5-μ range remain. The resultant curve is much more finely skewed than the typical IIIB curve shown in Fig. 13; except for the fact that it crosses the X-axis, it quite closely resembles a curve from Family IIIA. The rock example from Family IIIA (Fig. 20) had porosity slightly increased by leaching, and the rock pictured in Fig. 21 had porosity infilled, so that in both rocks pore throats of almost identical size remained. Both would require an oil column over 225 ft in height to produce clean oil.

Another interesting observation can be made by comparing the rock in Fig. 21 with a poor reservoir rock in the same coarse-grained class (Fig. 18). The rock from Family IIB has 40% more porosity, and 2,000% more permeability, yet it is not as good a reservoir rock as that shown in Fig. 21. The difference in productibility is clearly determined by the wider pore throats in the coarser sparites, which are not measured by porosity or permeability determinations.

Porosities in Family IIIB ranged from 6% to 22.1%, and permeabilities from 0.02 to 5.22 md. Productibility showed little correlation with either porosity or permeability.

IIIC. Intermediate reservoir, altered rock. Rocks in this family are coarse calcilutites with solution voids that are between 20% and 50% filled with finely crystalline to medium crystalline sparite. The curves in this family are

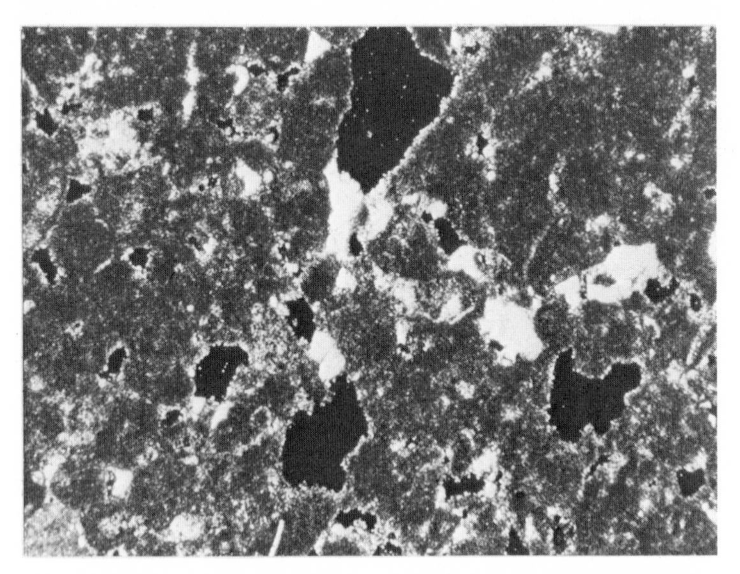

× nicols, × 35.

FIG. 22. Pelmicrite, leached, voids partly filled with medium crystalline sparite, intermediate reservoir rock, porosity = 13.6%, permeability = 9.88 md, Family IIIC. × nicols, × 35.

identical in shape to others in Class C, having merely moved further to the left on the X-axis. An example of a rock from this family is illustrated in Fig. 22.

The capillary pressure curve again confirms a characteristic of this class of rocks. When a sufficient height of oil column is reached to produce oil, an increase in oil column does not significantly improve reservoir properties.

Permeabilities in the family ranged from 0.26 to 9.88 md, and porosities from 7.4% to 15.8%.

In Type III reservoir rocks, permeabilities ranged from 0.02 to 9.88 md, and porosities from 6% to over 30%. There seems to be little correlation of permeabilities or porosities between the families, or between these parameters and productibility.

Type IV. Good Reservoir Rocks

Only two families of rocks and related curves are included in this type. It is characterized by good reservoir rocks, which will produce water-free oil with only a moderate oil column (40 ft or less). More than 50% of the pore throats exceed 1 μ in width.

IVA. Good reservoir, fine-grained. No rocks in this family have been found. It is doubtful whether rocks in Class A can be good reservoir rocks, because their grain size is not sufficient to produce large pore throats without leaching. When this class of rocks becomes leached, it belongs in Family IVC. The general conclusion might be reached that fine-grained rocks, without leaching, cannot become good reservoir rocks, regardless of porosity or permeability.

IVB. Good reservoir, coarse-grained. Rocks in this family are pelsparites, pisosparites, and calcarenites. The voids either are filled with coarsely crystalline sparite or are unfilled. Typically, these rocks produce capillary pressure curves that have coarse skewness and that cross the X-axis. As many as 20% of the pore throats may be above 80 μ in width. Figure 23 is a photomicrograph of a rock typical of this family. It is an intrapelsparite composed of fine pellets and coarse interclasts with some fossil fragments. Cementation by sparite is minimal. Almost 80% of the pore throats are above 2 μ in size, and fewer than 20% are less than 0.02 μ. The well-developed sorting and coarse skewness are very apparent on the capillary pressure curve.

Porosities in this family range from 13.7% to 19%, and permeabilities from 0.65 to 17.4 md.

IVC. Good reservoir, altered rocks. As leaching affects rocks in Family IIIA, the ability of these rocks to hold oil becomes greater, and the irreducible

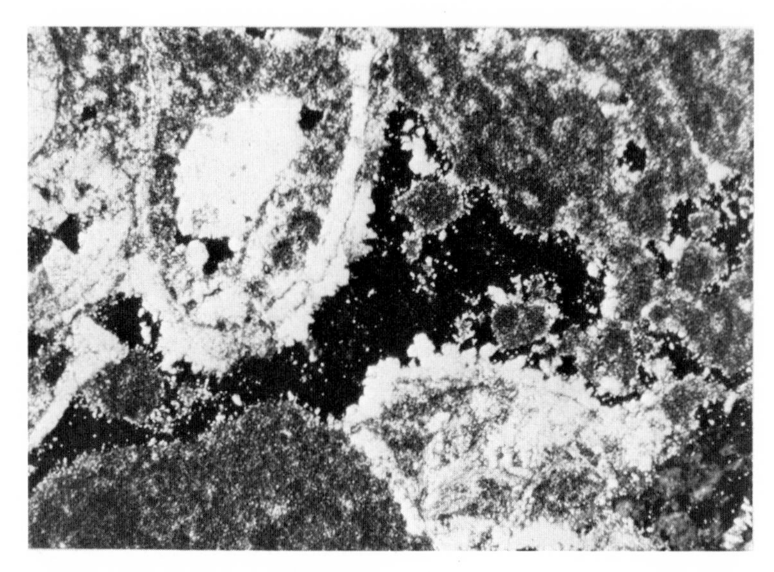

× nicols, × 35.

Fig. 23. Intrabiopelsparite, some coarse sparite infill, good reservoir rock, porosity = 13.7%, permeability = 17.4 md, Family IVB. × nicols, × 35.

water saturation lower. The point of transition between a rock in Family IIIA and one in IVC is sometimes difficult to determine. These rock families must be identified by the shape of their capillary pressure curves, rather than by skewness or irreducible water saturation. In the area considered in this study, all good reservoir rocks had less than 20% irreducible water content, and in all cases more than 50% of the pores were able to contain oil with a height above oil-water contact of less than 40 ft. Any rock which exceeded these requirements produced oil in commercial quantity.

Figure 24 illustrates a core where leaching has progressed sufficiently to produce an excellent reservoir rock. Leaching and recrystallization have proceeded to a point where little if any original porosity remains. Void interconnection is through the coarsely crystalline secondary matrix; and almost all porosity is secondary. This is an excellent reservoir rock.

Only a small percentage of rocks from this family was found in the study area. These rocks ranged in porosity from 17.7% to 37.6%, and in permeability from 2.6 to 40.1 md.

Curves in this family typically show very well-sorted pore distribution. The degree of coarse pore development determines the degree of coarse skewness. Curves in this family always meet the *Y*-axis at zero or above.

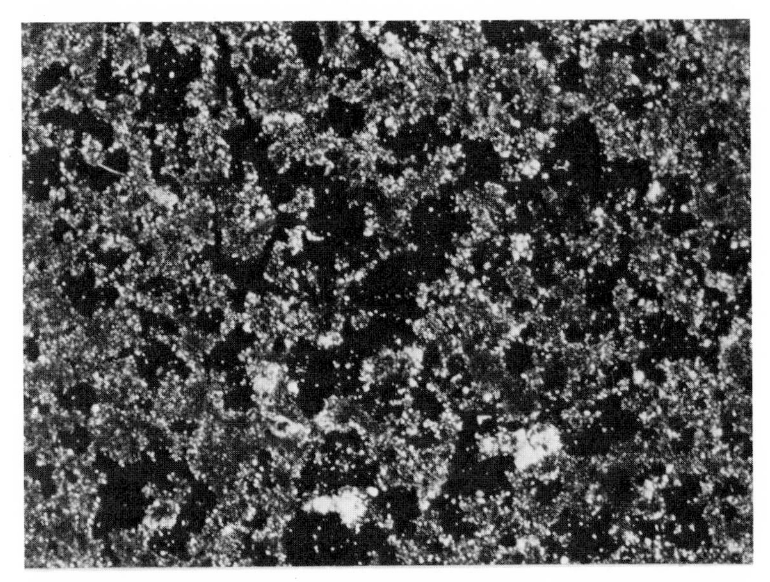

FIG. 24. Fine calcarenite, leached, good reservoir rock, porosity = 37.6%, permeability = 17.4 md, Family IVC. × nicols, × 35.

Unsorted Rocks

Some rocks are composed of such a diverse size range of grains that they show almost no sorting of pore throat sizes. They are not common, but a few were found in the study area. Figure 25 illustrates a biopeldismicrite, which has a great range of pore throat sizes. The capillary pressure curve for this rock is essentially a diagonal line from 100% water saturation to 100% oil saturation. Despite a porosity of 18% and a permeability of 4.39 md, it could contribute to oil production only in an excessively thick reservoir.

Summary of Limestone Pore Geometry Characteristics

On studying the relationship between rock types and families of capillary pressure curves it becomes apparent that the pore geometry of a rock is as important as its porosity and permeability—or perhaps even more important —in determining whether the rock has reservoir potentialities.

From the study of the Montana-North Dakota area some clear conclusions that should be universally applicable have been reached. Perhaps the most important caution is this: Beware of micrites. Unless extensively leached, these rocks may be extremely porous and permeable, but may have such small pore throat sizes that they cannot, by their nature, become good

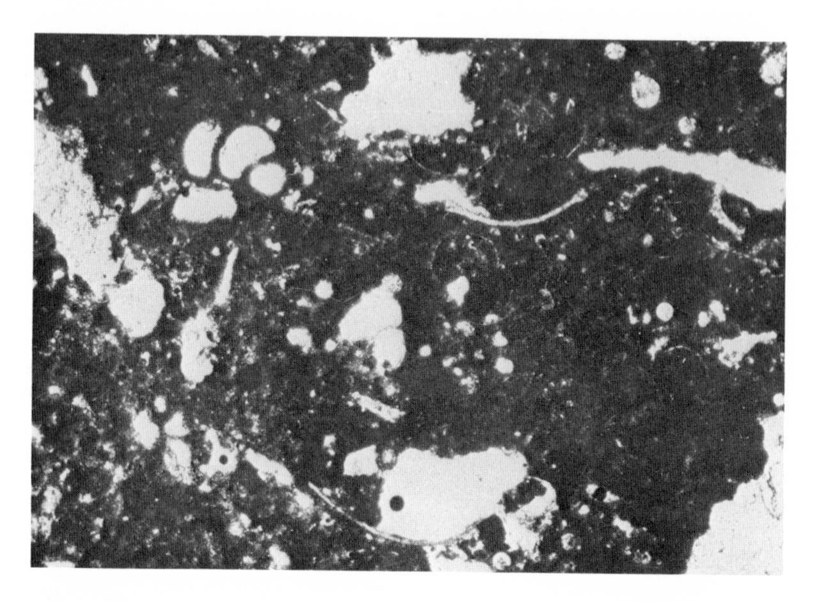

plain light, × 35.

FIG. 25. Biopeldismicrite, unsorted, poor reservoir rock, porosity = 18.1%, permeability = 4.39 md. Plain light, × 35.

reservoir rocks. Inasmuch as the prime areas for the deposition of carbonate muds, which upon diagenesis produce micrites, are central reef and central bank areas and lagoons, these should be drilled with caution. Such areas contain thick sections of very porous rocks, but require excessively high oil columns to produce water-free oil. Unless leaching, fracturing, or very great closure is known to be present, these areas should be avoided.

Another conclusion that seems widely applicable is that only a small percentage of porous or permeable limestone rocks actually makes good reservoir rocks. In order to find them, it is necessary to identify the depositional or diagenetic processes that create large pore throats and to locate areas where such processes were active. This cannot be done by simply attempting to follow porosity.

Application of Pore Geometry Characteristics to Exploration

The relationship between families of rocks and productibility can be developed into a practical exploration tool. Early in the exploration of Charles (Mississippian) carbonates in the Montana-North Dakota border area, a small field was discovered. Interest in this discovery encouraged the drilling of additional wells, and porosity logging devices soon showed areas of thick,

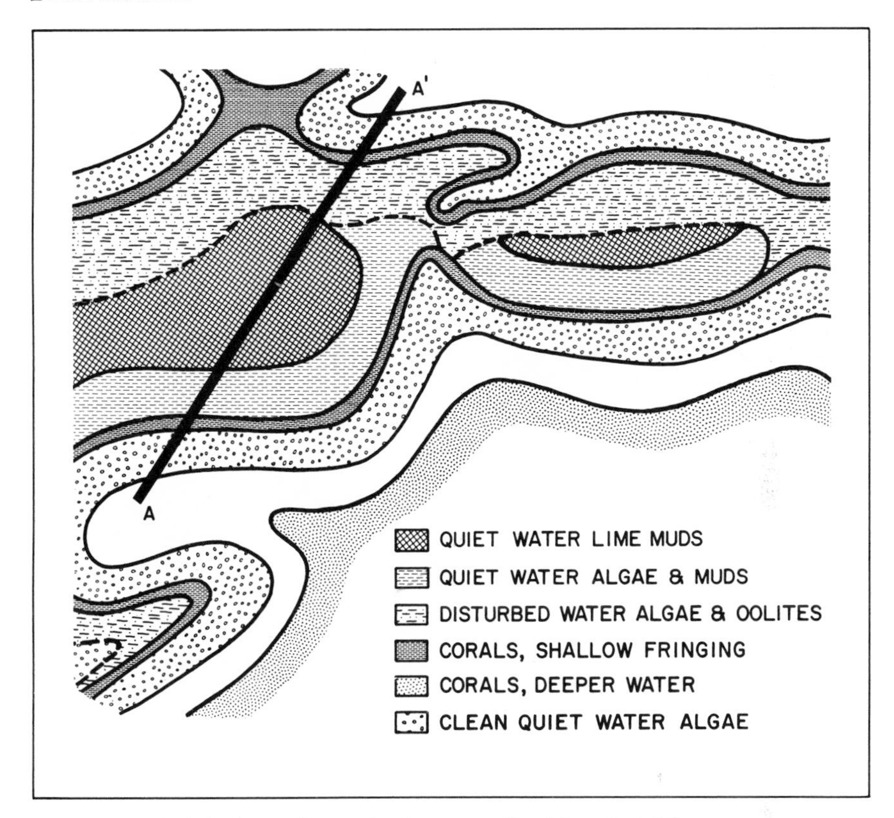

FIG. 26. A Mississippian carbonate bank area, outlined by selected faunas and lithologies.

high-porosity carbonate buildup. No significant new discoveries, however, resulted. As drilling continued, it became possible to reconstruct the paleoenvironment of the area. (Fig. 26). The field first discovered occupied a position near the top of an ancient carbonate bank, at a point where a constriction nearly divided the bank into two segments. The bank area was fringed by an area of concentrated coral growth. Inside the coral border on the windward side was a band of well-winnowed coralgal debris and oolites, with little fine material present. On the lee side there was a less winnowed band of coralgal material, containing a larger percentage of quiet-water algae, mixed with lime mud. The center of the bank was occupied by a thick accumulation of highly porous sediment, composed of the fine carbonates winnowed from the edges of the bank and also a large percentage of chemically precipitated lime muds. It was in this central area that most drilling had been done, sometimes resulting in wells with shows of oil but no production.

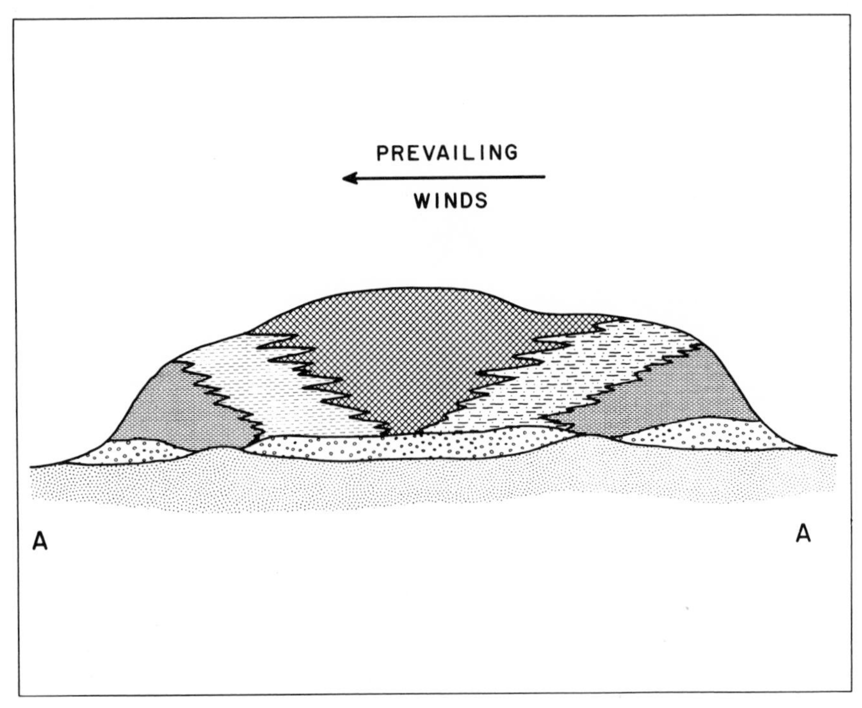

FIG. 27. Schematic cross section through Mississippian carbonate bank. See Fig. 26 for location of section and legend.

Figure 27 is a schematic cross section through this bank. The attractiveness of the thick central bank area, with porosities ranging above 30%, can be understood. Porosity was measured for all available cores in the area, and it was found that three broad rock facies had three distinct ranges of porosity (Fig. 28). The areas outside the bank had very low porosity, less than 2% almost everywhere, and ranging up to 6% only in the coral zone ringing the bank. The windward, winnowed zone and the lee, partially winnowed zone had moderate porosity. The porosity throughout the winnowed zone ranged generally from 8% to 12%, and up to 15% as it graded into the central bank area. The porosity in the lee zone averaged slightly higher, from 10% to 15%. In the central bank area porosities increased rapidly, rising from 16% near the lee edge to over 30% throughout much of the area. At the windward side, where the central bank grades into the area of wave action, leaching had raised the porosity to as high as 37%. After the central high-porosity area had been extensively drilled, with seldom more than a show of oil, it became apparent that something other than porosity controlled production.

Capillary pressure curves were determined for all cores; curves from each

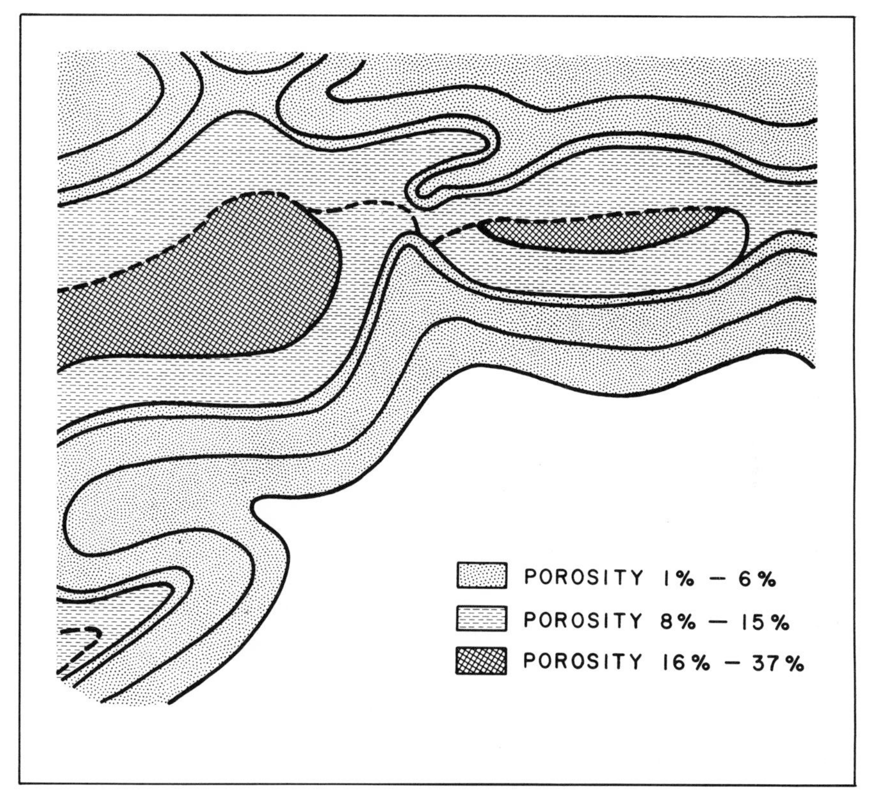

FIG. 28. Correlation between porosity and rock facies of Mississippian carbonates.

facies were compared and were found to fall in well-delineated natural groupings. A map was constructed relating facies to productibility, as expressed by the oil column required to saturate 50% of the pores with oil, using the average oil and water densities found in the area. This map (Fig. 29) shows why the central bank area is nonproductive, why the original field area was productive, and where additional production would most probably be found.

The area outside the bank accumulation, unattractive from a porosity standpoint, is still less encouraging from the standpoint of productibility. An absolute minimum of 250 ft of oil column would be required for these sediments to contain oil, with most of the area requiring over 450 ft. The central bank, high-porosity sediments are almost as unattractive; oil columns exceeding 450 ft would be required in much of the area. As the windward transition zone is approached, the required height drops to 100 ft in a narrow band. On the lee side, an oil column height of more than 150 ft would be required at the transition zone. Seismic and subsurface geology indicates

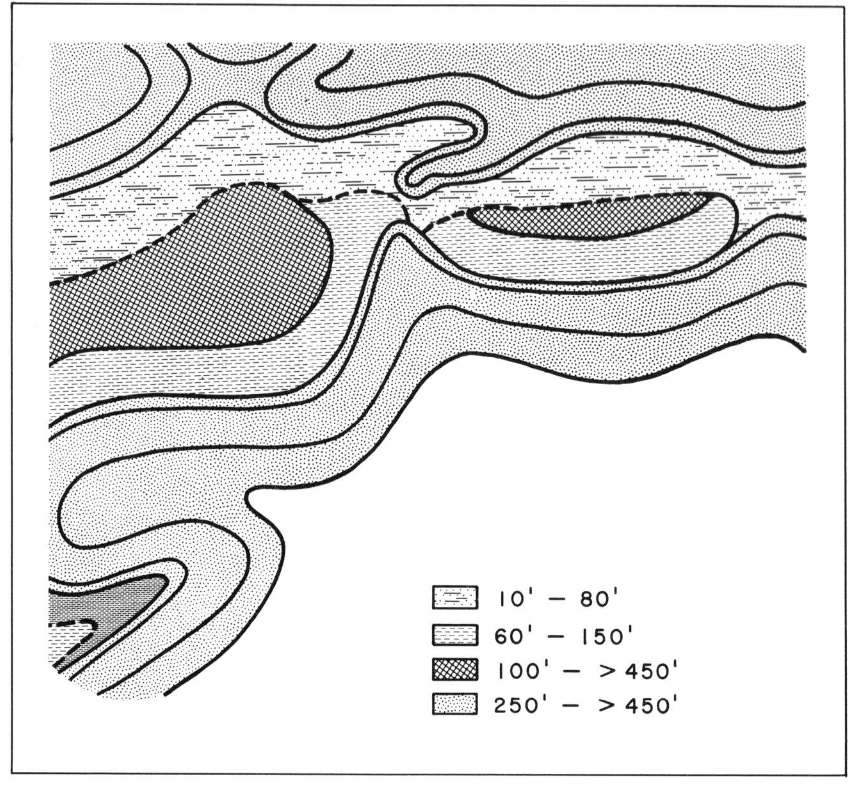

FIG. 29. Relationship between carbonate facies and productibility, expressed in feet of oil column required to saturate 50% of pore space with oil.

that in the bank area traps with 100 ft or more of closure would not be likely to occur. This fact would essentially rule out the possibility of oil being produced outside the bank area or in the highly porous central bank area. Wells drilled in these areas would be spudded with essentially no chance of success, although they might confirm the presence of structure and encounter porosities in excess of 30%.

Capillary pressure curves from the partially unwinnowed, lee sediments indicated that 60–150 ft of oil column would be required for production. This area would have to be considered marginal, with production possible only under the most favorable conditions.

Only the winnowed, moderately porous, windward bank sediments produced favorable capillary pressure curves. In this facies oil columns of 10–80 ft would saturate 50% or more of the pore space with oil. Here traps with 50–60 ft of closure, the size expected in the area, could produce oil in commercial quantities.

After this map was constructed, several single-well discoveries were made in the lee, partially winnowed facies, confirming its marginally productive nature. A series of fields, however, has been developed in the windward, sorted, moderate-porosity facies, and now a large part of the area shown is productive, as capillary pressure studies indicated it should be.

In addition to the horizontal facies zonation discussed above, there is a vertical facies zonation which is equally important.

The Charles Formation in the Williston Basin is the result of a series of sedimentary cycles, each characterized by a prolonged regressive phase followed by a short readvance of the sea. The successively younger readvances do not equal the regressions in magnitude, so as a whole the Charles Formation has a successively shallower water lithology with the passage of time.

In such a regressive environment the zone of wave action advances seaward as time goes on, and shallower water sediments also advance seaward across those of deeper water origin. The cross section in Fig. 27 illustrates how this happened in the bank area under discussion. When such a sequence is drilled on the windward side of the bank, the upper part of the section may be composed of a thick, highly porous micritic limestone gradually grading downward into a winnowed, coarse-grained limestone with moderate porosity, and finally through a coralline zone into an open marine unsorted facies with low porosity.

Such a section may be drill-stem tested with sufficient oil recovery to indicate commercial production. Logs may be run and casing set. On a porosity log a thick carbonate sequence will be indicated, with very high porosity at the top grading downward into low porosity. Unfortunately, porosity logs do not indicate pore throat size. If, in such a situation, the upper part of the section is perforated in the zone of high porosity as indicated on the logs, water production almost always results. Because of the unfavorable pore geometry, the upper, fine-grained, highly porous section may contain water, grading *downward* into an oil-water transition zone, and *further downward* into a coarse-grained, winnowed, moderately porous limestone which is an excellent reservoir rock producing water-free oil. Because of failure to understand these principles, wells in the area that should have been productive have been abandoned. It is always important to remember that in cyclic carbonates water may easily overlie oil or lie updip of oil in a continuously porous reservoir, without impairing the oil-producing capability of the oil-saturated portion of the reservoir.

Capillary pressure curves are also very useful in locating oil-water contacts without the necessity of drilling to water. Oil saturation in a rock may be determined from core or log analysis; by locating this percentage of saturation on a capillary pressure curve of the rock, the distance to the oil-water contact can be found. When oil columns are found to be thicker than known closures, a strong possibility of stratigraphic trapping may be indicated. Because large stratigraphic traps are sometimes found by drilling structures which actually are not significant to the oil accumulation, early identification of stratigraphic trapping can be very important and may put an entirely different complexion on an oil play.

Pore Geometry of Dolomites

Dolomitization in general has an extremely beneficial effect on the productibility of carbonates. Pore throats are enlarged, coarse skewness is almost always enhanced, and an excellent sorting of pore throat sizes results. In 1961, Chilingar and Bissell[6] noted that the only time dolomitization does not give rise to porosity is possibly during very early diagenesis, when no solid framework is present.

The term *dolomitization* is used here to imply the action of dolomitizing sediments which are in a late stage of compaction and lithification, or which are already lithified. Rocks classifiable as primary dolomites or dolomites created in the very early stages of compaction and lithification may have extremely poor productibility. The writer (Jodry[7]) has published a paper on the effects of dolomitization on pore geometry. In summary, this work indicates that, when limestones, which have grain or framework support or which have already been lithified, are dolomitized to a point where more than 75% but less than 100% of the limestone in the rock is dolomite, a highly favorable pore throat system is developed. Capillary pressure curves show the pore system to be (1) very coarsely skewed, indicating that production is possible with minimal oil columns, and (2) very well sorted, indicating very thin oil-water transition zones. Irreducible water saturation may be very low, frequently substantially less than 5%.

Perhaps to an even greater extent than examination of the limestones, study of a dolomite under a microscope will enable the geologist to judge reservoir capacity much more realistically than by simply accepting permeability and porosity figures.

Conclusions

The pore geometry of a limestone or dolomite has as great an influence on the ability of the rock to contain and produce oil and gas as do porosity and permeability. This has been clearly demonstrated in the limestone bank area of northeastern Montana and in other areas throughout North America. By studying the lithology and genesis of the rocks and comparing families of rocks with families of capillary pressure curves, it was possible to establish very clear mutual relationships. With experience it is possible to study a rock under the microscope and, with a knowledge of pore geometry in the area and with even rough values for the porosity and the permeability of the rock, to predict quite closely what the productibility of the rock may be. This is frequently not possible when porosity and permeability data are used without considering the pore geometry.

Using this procedure it has been possible to determine that whole classes of rocks do not make good reservoirs. This is true even if the porosities sometimes exceed 30% and permeabilities are of magnitudes entirely adequate for other classes of rocks. An example of such poor reservoir rocks are the central bank micrities from the North Dakota-Montana area. Porosity logging devices show a buildup of very porous rocks in this area, and many wells have been drilled there in the search for oil. A study of the pore geometry, however, has consistently revealed that such rocks, despite their very high porosities, would need at least 450 ft of oil column to produce oil. It thus appears rather futile to drill with rock of this type as the objective.

Pore geometry studies also enable one to better understand some of the producing characteristics of cyclic layered rocks. Frequently, in limestone formations, alternating horizons of fine, highly porous and coarse, less porous rocks are found. It is not uncommon for the more porous, fine-grained rocks to yield water, whereas the coarse but more dense rocks may produce water-free oil. The danger of perforating such a well on the basis of porosity determinations alone should be obvious. In such circumstances it is possible to develop a commercial oil reservoir, despite high water saturation in parts of it.

The relationships between rock family and curve family developed here may not always be applicable to other areas in which limiting factors are different. The general principles should apply, however; and a careful study of rock lithologies, along with capillary pressure curves, in any area should produce a much better understanding of the rock factors which limit production.

Dolomites are much less complex rocks than limestones. A uniform degree of dolomitization in a rock produces a uniform crystal development. This creates in dolomites a very well-sorted pore system, with typical coarse skewness. Thus, if dolomitization has proceeded far enough, and if a supporting framework is present, dolomites make uniformly excellent reservoir rocks. Again, a study of rock type and a knowledge of pore geometry can lead to a better understanding of the ability of dolomites to produce oil in any area.

It is increasingly apparent that a visual examination of a core by the geologist and simple porosity and permeability measurements by the engineer do not constitute an adequate basis on which to profitably explore for oil and produce it. Detailed lithologic studies by the *geologist* and detailed reservoir rock studies by the *engineer*, including capillary pressure measurements, with close cooperation between the two specialists, are becoming increasingly essential.

Acknowledgments

The capillary pressure curves used in this paper were prepared under the direction of Wilbur M. Hensel, Laboratory Supervisor, Sun Oil Company, assisted by James E. Davis. Without their help this chapter would not have been possible. Philip Braithwaite, Research Geologist, Sun Oil Company, aided in preparing material related to carbonate petrology and in photography. The writer is also grateful to Sun Oil Company for permission to publish this work.

References

1. Muskat, M.: *The Flow of Homogeneous Fluids through Porous Media*, J. W. Edwards, Ann Arbor, Mich. (1937).
2. Folk, R. L.: "Spectral Subdivision of Limestone Types", in: *Classification of Carbonate Rocks (a Symposium)*, AAPG Memoir 1 (1962) 62–84.
3. Archie, G. E.: "Classification of Carbonate Reservoir Rocks and Petrophysical Considerations", *Bull.*, AAPG (Feb., 1952) 278–298.
4. Arps, J. J.: "Engineering Concepts Useful in Oil Finding", *Bull.*, AAPG (Feb., 1964) 157–165.
5. Stout, J. L.: "Pore Geometry as Related to Carbonate Stratigraphic Traps", *Bull.*, AAPG (Mar., 1964) Part 1, 329–337.
6. Chilingar, G. V. and Bissell, H. J.: "Discussion of Dolomitization by Seepage Refluxion, by J. E. Adams and M. L. Rhodes", *Bull.*, AAPG (May, 1961) 679–681.
7. Jodry, R. L.: "Growth and Dolomitization of Silurian Reefs, St. Clair County, Michigan", *Bull.*, AAPG (Apr., 1969) 957–981.

CHAPTER 3

Fluid Flow in Carbonate Reservoirs

MARTIN FELSENTHAL AND HOWARD H. FERRELL

Introduction

Concepts of reservoir fluid flow during both primary and secondary recovery were first developed for sandstone reservoirs. The object of this chapter is to examine the applicability of these classical fluid flow concepts to the much more complex and heterogeneous carbonate reservoirs.

Fluid Relationships

Distribution of Reservoir Fluids

The questions "Where is the oil located in the reservoir?" and "How are injected fluids likely to move through the reservoir?" are germane to a discussion of fluid flow. Elkins[1] pointed out the importance of thorough geologic description of cores in a perceptive review of reservoir performance and analysis of a tight, fractured limestone reservoir in the West Edmond Field, Oklahoma. It was during the development stage of the field in 1948 that Littlefield *et al.*[2] had examined the available cores and forecast the oil-in-place volume with reasonable accuracy, largely on the basis of oil-stained porosity. According to the analysis by Littlefield *et al.*, oil was confined to fractures, solution channels, and the more porous and more permeable rock matrix of this reservoir. Core descriptions and early well performance were interpreted to mean that most of the flow in one of the major zones of the reservoir (the Bois d'Arc sequence of the Hunton Limestone) occurred in an extensive interconnected system of fractures and solution channels constituting about 10% of the total reservoir void space in rock of otherwise very low permeability. Littlefield *et al.* predicted that this system of fractures would result in severe channeling of naturally encroaching water or injected fluids with little or no benefit to the ultimate recovery of oil. This analysis was proved to be correct later on; however, in the early life of the field, other engineers disagreed with the interpretations of Littlefield *et al.* regarding fluid distribution, continuity of fractures, and effect of fractures on performance.

References p. 137.

Elkins'[1] paper describes the difficulties encountered in the application of conventional reservoir engineering concepts to the fractured, tight reservoir in the West Edmond Field, Oklahoma. For instance, it was erroneously assumed at first that the extensive production of oil at solution GOR meant that the *entire* reservoir was necessarily oil saturated. This assumption was contradicted by observations made during later operations, such as deepening and coring a down-structure well, indicating that a substantial part of the tight reservoir matrix contained free gas. Also, it was found later that initial pressure buildup tests were of insufficient duration, and that erroneous conclusions had been drawn from the resulting data.

The experience gained in the West Edmond reservoir led to the conclusion that there is definitely a need in future reservoir studies "for more quantitative description of reservoir rock oriented toward better understanding of distribution and movement of fluids within the rock. . . . The various features of this reservoir demonstrate," Elkins states, "the need for a *total* approach to reservoir engineering. Overemphasis of any one method fails to account for the many complexities introduced by the internal anatomy of the reservoir rock." This last statement is considered to express the main point of this chapter. Carbonate rocks are so complex that many approaches and analysis methods are needed for proper evaluation of oil content and for optimum application of recovery methods.

No sweeping general conclusions in regard to the distribution of oil can be made that would be applicable to all reservoirs. Each reservoir needs to be considered as a separate case. A few examples help to illustrate this point. Bulnes and Fitting[3] inferred from visual examination of cores and drill cuttings from many fields of west Texas and New Mexico that large amounts of oil were contained in limestone having a permeability of less than a few tenths of a millidarcy. They cited the performance of well D (Fig. 1) as evidence that tight sections (<0.1-md air permeability) can be oil bearing. All of the pay in this well was cored and analyzed, showing that only 8.4% had permeabilities greater than 0.1 md. During an observed period of production, oil appeared to have migrated from the tight zones into the zones of higher permeability, thereby continuously resaturating these lanes of communication with the wellbore. As a result, when the well had produced a total of 69,000 bbl of oil, its productive capacity had declined only about 20%.

Craze,[4] on the other hand, cited reservoirs in which the tight carbonate matrix was water filled and only the fractures and vugs contained movable oil. For example, a core from the Martin Ellenburger Field, Texas, had an average total porosity of 3.3%, whereas the average porosity of small samples taken from the intercrystalline matrix averaged 1.51%, leaving 1.79% porosity for the fractures and vugs. Craze noted that the ratio of matrix to total porosity (1.51/3.3 or 45.8%) was almost the same as the water saturation

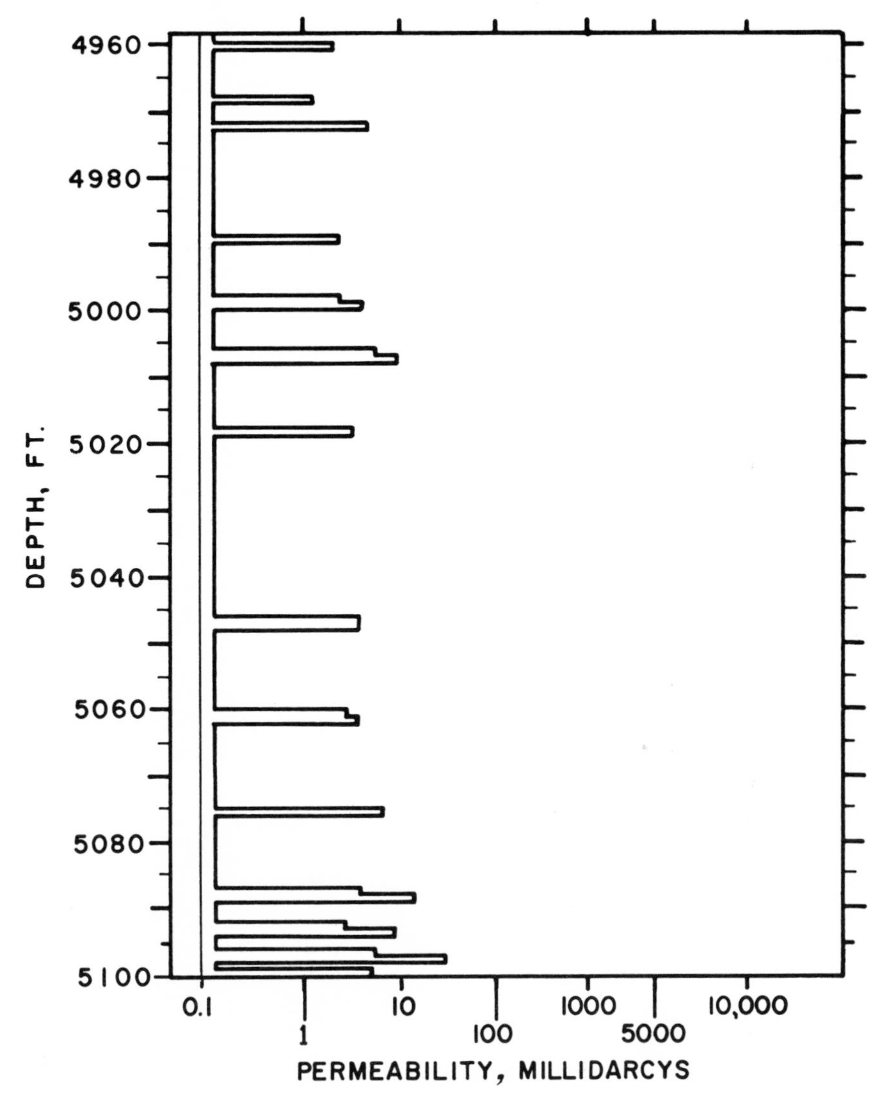

FIG. 1. Permeability profile, limestone well D, Wasson Field, Yoakum County, Texas. (After Bulnes and Fitting,[3] courtesy of AIME.)

of 47% observed in cores drilled with oil-base mud. He concluded from these data and a visual field examination of the cores that in this part of the reservoir oil occupied all of the fractures and vugs, whereas water occupied all of the matrix porosity. The matrix was composed of dolomite with intercrystalline porosity having a permeability of about 0.1 md.

Importance of Fractures

The importance of fractures in oil production from rocks having low matrix permeability has been well described in the literature (e.g., Daniel[5]). As a general rule, fluid flow in carbonate rocks is greatly affected by the presence of conductive or partially conductive stylolites, solution channels, fissures, and fractures which frequently dissect the low-permeability matrix as shown in Figs. 2 and 3.

The following equation for the permeability in a horizontal direction, k_H, through an idealized fracture-matrix system (Fig. 4) is based on the work of Huitt[6] and Parsons[7]:

$$k_H = k_m + 5.446 \times 10^{10} W^3 \cos^2 \alpha / L \tag{1}$$

where k_m = matrix permeability (md); W = fracture width (in.); L = distance between fractures (in.); and α = angle of deviation from horizontal plane (degrees). If W and L are expressed in millimeters, Eq. 1 becomes:

$$k_H = k_m + 8.44 \times 10^7 W^3 \cos^2 \alpha / L. \tag{1a}$$

For the rock shown in Fig. 2, a total permeability, k_H, of 3.4 md was measured by large core analysis. By conventional core analysis, the matrix permeability, k_m, was found to be 0.3 md. The distance, L, between conductive fractures and fracture-like features was about 1 in. (as determined by bubbling air through the underside of the water-submerged core sample). For the core, α was taken to be zero. Using Eq. 1 gave a fracture width, W, of 0.0004 in. or 0.01 mm. This example illustrates the large effect on total permeability that can be exerted by a system of fractures of extremely small width.

To further illustrate the effect of fractures, if W = 0.005 in., L = 1 in., $\alpha = 0$, and k_m = 1 md, k_H will be equal to 6,800 md. This second example demonstrates clearly the overwhelming contribution which relatively small fractures can exert on total permeability. Similar examples are given by Muskat[8] (pp. 246–255, 267–269). (See Fig. 11, p. 16.)

In practice, the petroleum engineer is more interested in the total permeability of the fracture-matrix system than in the permeability contributions of its various parts. A general insight into fracture permeability can be gained with the aid of large core analysis, as described in detail by Kelton.[9] In this analysis the entire core, cut into sections of about 1 ft in length, is tested, rather than the conventional small core "plugs" used in the analysis of most sandstone reservoirs. Large core analysis may result in better evaluation of the effects of fractures, vugs, and solution channels than does

FIG. 2. Stylolites and hairline fractures in low-porosity matrix rock (porosity = 1.1%, matrix permeability = 0.3 md; whole core permeability = 3.4 md), Paradox Formation, San Juan County, Utah. (Courtesy of Continental Oil Co.)

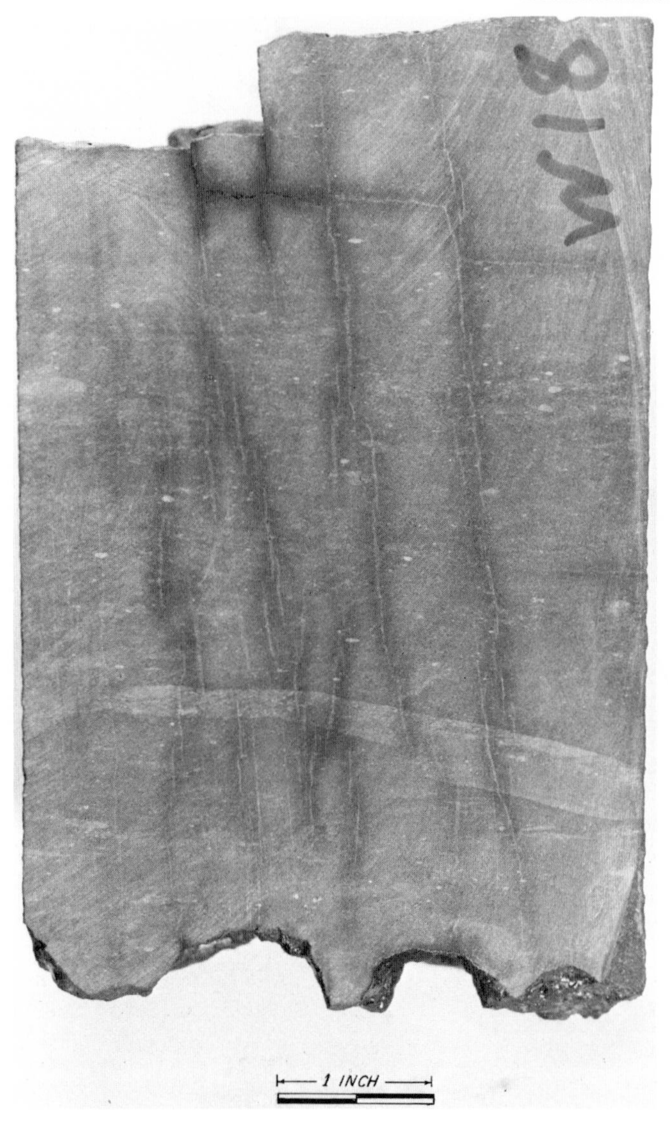

FIG. 3. Hairline fractures in fine-grained matrix (porosity = 11.5%, matrix permeability = 0.2 md; whole core permeability = 1.7 md), Paradox Formation, San Juan County, Utah. (Courtesy of Continental Oil Co.)

core plug analysis. Although large core analysis is an improvement, it does not always yield a formation capacity, k_oh, of the oil zone (where k_o is determined by multiplying k_{air} by the proper relative permeability factor, k_{ro}) that is in close agreement with k_oh data derived from well productivity

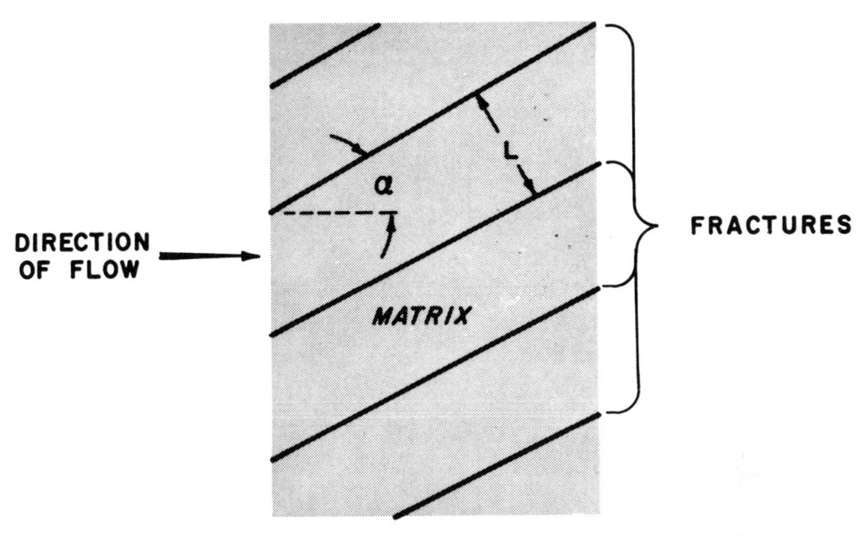

FIG. 4. Idealized fracture-matrix system. (After Parsons,[7] courtesy of AIME.)

or pressure buildup analyses. It is the total capacity, $k_o h$ [k_o = effective permeability to oil (md), and h = net effective pay thickness (ft)], of the entire oil zone, of course, that is of greatest concern in the final analysis. Nevertheless, the detailed physical and geologic analysis and description of large-diameter cores serve an important purpose in the analysis of reservoirs. Information obtained from cores is not "merely corroborative detail intended to give artistic verisimilitude to an otherwise bold and unconvincing narrative," to use the words of Gilbert and Sullivan. On the contrary, it is an important tool for unraveling the mystery of just where the oil is in a reservoir and how it may best be recovered. The best time to obtain this information is when the field is being developed initially.

Figure 1 illustrates a typical permeability profile. The erratic nature of the profile indicates a tight matrix interspersed by moderately high-permeability zones, which in carbonate reservoirs are rarely correlatable from well to well.

Comparison of Core Analysis with Pressure Buildup Analysis

Warren *et al.*[10] compared average permeabilities obtained from core analysis with equivalent values derived from pressure buildup tests in fourteen wells producing from the low-permeability McKnight Dolomite in Texas and the tight, limy Bromide Sandstone of Oklahoma. The core recovery in each well was at least 90%. Large core analysis was used for the dolomite; conventional plug core analysis, for the sand. The results are summarized in

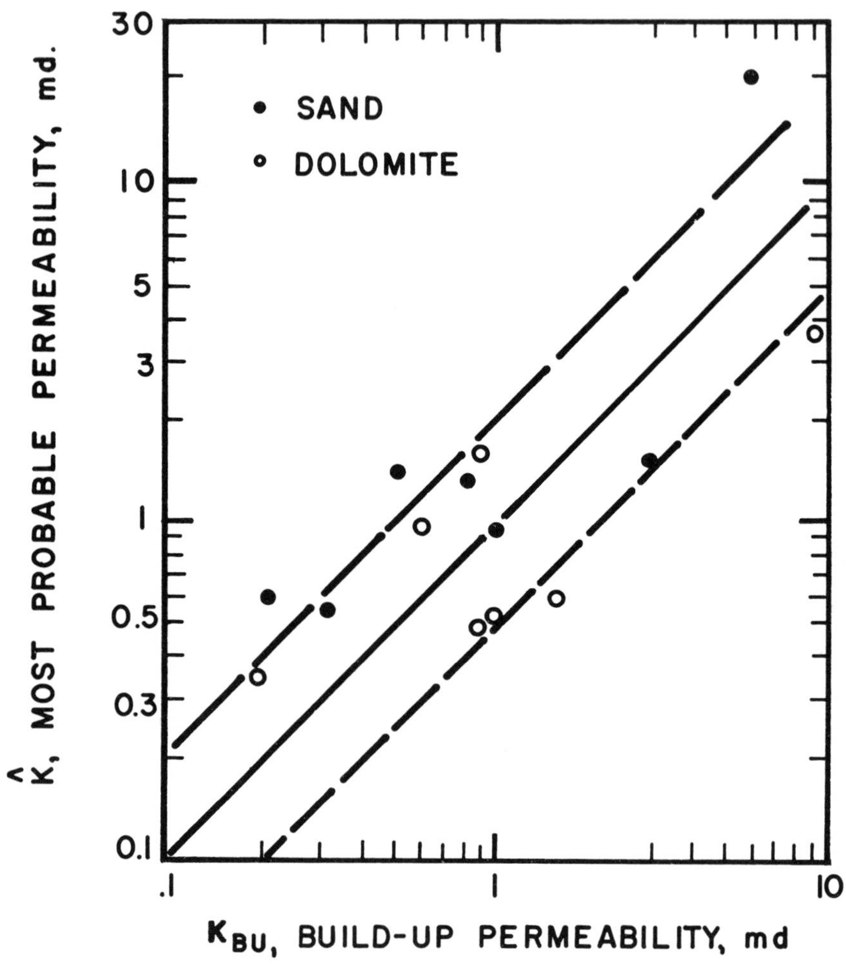

FIG. 5. Comparison of core analysis and pressure buildup permeability. (After Warren *et al.*,[10] courtesy of AIME.)

Fig. 5, which shows that the pressure buildup data agree reasonably well with $\hat{k}$ (the most probable permeability).

$$\hat{k} = \frac{1}{M} \sum_{1}^{M} k_j \tag{2}$$

where M = number of random distributions considered and k_j = effective permeability for a particular random distribution (md). The k_j term is evaluated by randomly arranging n equally weighted samples in a fixed geometry and calculating the effective permeability. As the number of random arrangements, M, increases, $\hat{k}$ approaches a limiting value. To state the concept more simply, $\hat{k}$ is derived from core analysis results by a statistical procedure involving random rearrangement of the data. Such a rearrangement implies that the reservoir is not a "layer-cake" type of system but instead is a completely disordered system, that is, one which is "homogeneously heterogeneous".

The statistical procedure to obtain "the most probable permeability", $\hat{k}$, is rather complex; fortunately, however, it appears that one can substitute the more easily determinable "median average" for $\hat{k}$ as demonstrated by the data of Fig. 6. The median is defined as the midpoint of a set of data which has been arranged in either ascending or descending order. As indicated in Fig. 6, the geometric average could also be substituted; however, the median

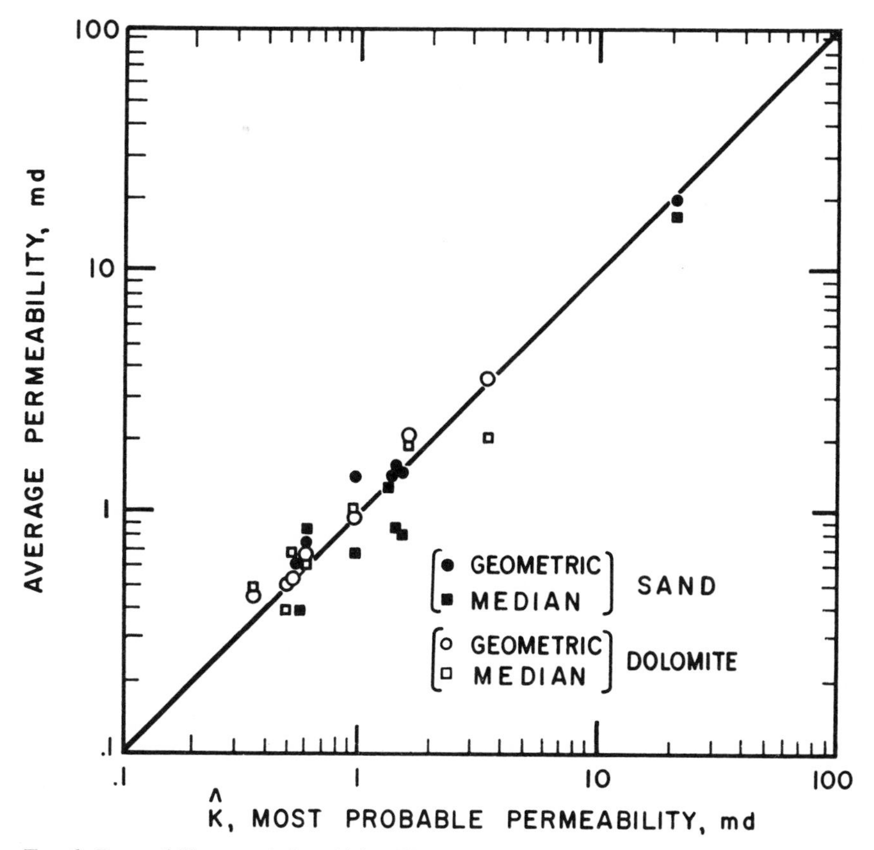

FIG. 6. Permeability correlation. (After Warren *et al.*,[10] courtesy of AIME.)

is preferred because it is more readily determined. The geometric average is defined as:

$$k_g = \left(\prod_{1}^{n} k_i\right)^{1/n} \tag{3}$$

where n = number of core samples, and k_i = core analysis permeability (md).

Core samples possessing median permeability often also possess other properties, such as porosity and irreducible water saturation, that are characteristic of the average properties of the reservoir. The water saturation evaluated in this manner applies only to the part of the reservoir that is located at a reasonably great elevation above a "water table" and, therefore, is unaffected by the "water table".

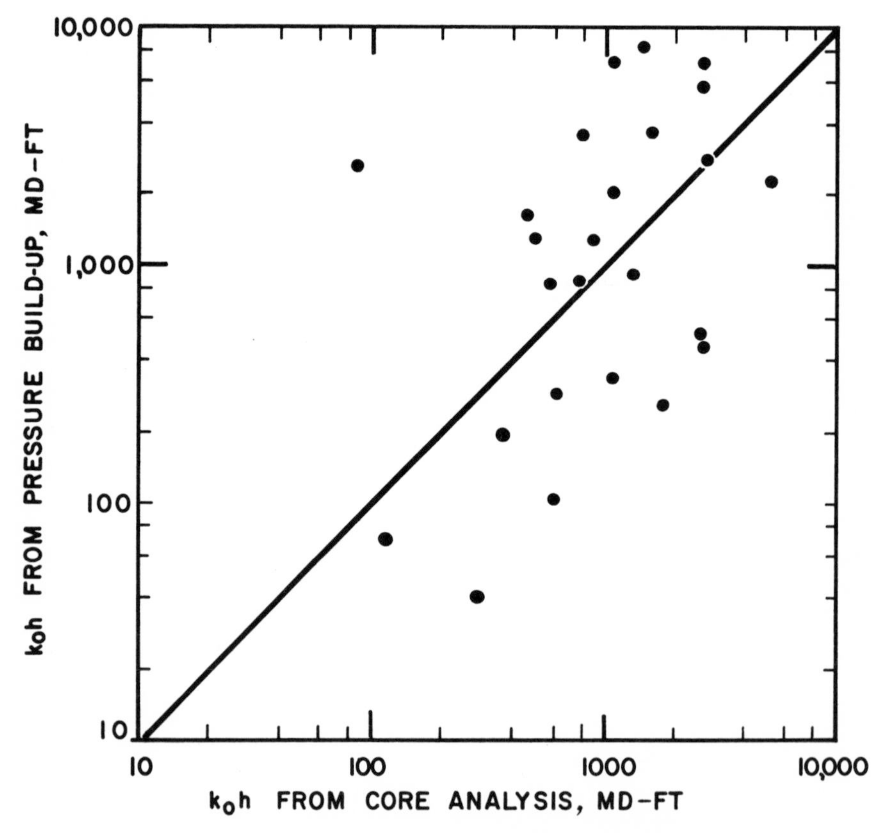

FIG. 7. Comparison of oil zone formation capacity (k_oh) from pressure buildup and core analysis in area No. 1 (24 wells), Paradox Formation, San Juan County, Utah. (After Tarr and Heuer,[11] courtesy of AIME.) (Σk_oh from pressure buildup)/(Σk_oh from core analysis) = 1.64.

Although the "median" core permeabilities agree closely with well performance data for the cases illustrated in Figs. 5 and 6, there is no assurance that this occurs in all situations. Warren *et al.*[10] reported that in the case of the Burgan reservoir in Kuwait the buildup permeabilities agreed reasonably well with the arithmetic average rather than the median average. This comparison was based, however, on only 20–40% core recovery.

It is not unusual in carbonates that the $k_o h$ value derived from pressure buildup permeabilities is larger than the $k_o h$ calculated from core analysis permeabilities.[11] This discrepancy was observed in a carbonate reservoir producing from the Paradox Formation in Utah (Figs. 7 and 8). The difference between core and buildup data is plausible for two reasons. (1) The producing wells were acidized under high pressure sometime before transient well

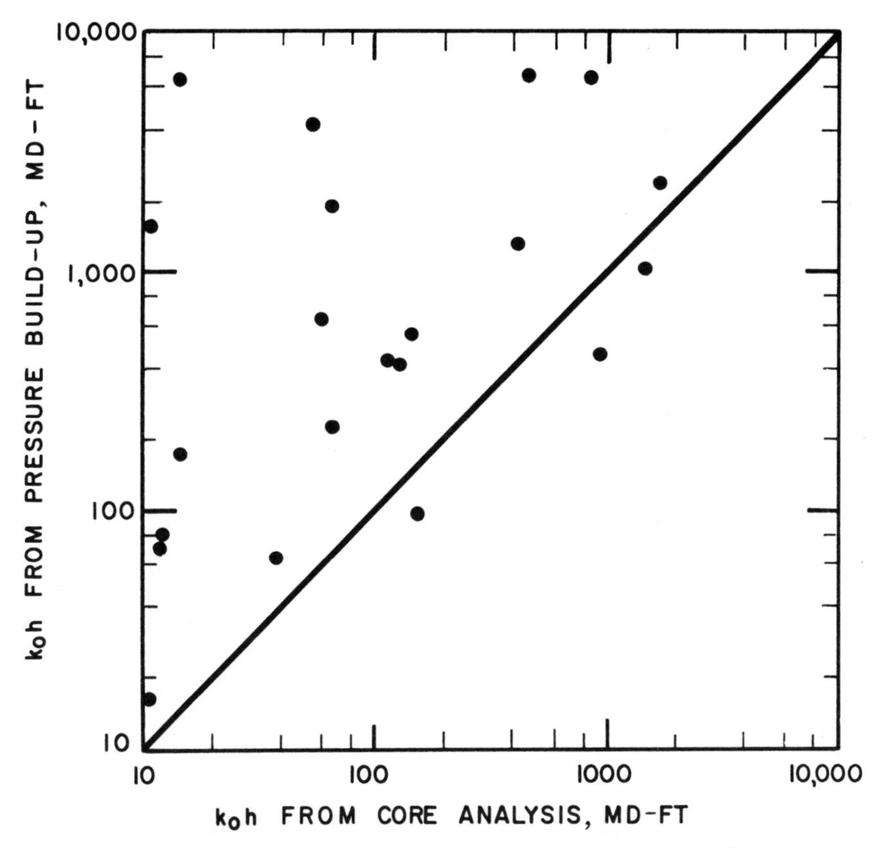

FIG. 8. Comparison of oil zone formation capacity ($k_o h$) from pressure buildup and core analysis in area No. 2 (21 wells), Paradox Formation, San Juan County, Utah. (After Tarr and Heuer,[11] courtesy of AIME.) ($\Sigma k_o h$ from pressure buildup)/($\Sigma k_o h$ from core analysis) = 5.29.

testing. The acid, which probably channeled into some high-permeability zones, numerous open or partially open fractures, and/or stylolites, effectively increased the flow capacities of the wells. (2) Some of the larger fractures that may have been originally present in the reservoir probably did not survive the rigors of coring.

In general, in reservoirs rocks which contain open or partially open fractures,[12] core analysis can be expected to give only minimum permeability values. In other words, a large difference between core analysis and pressure buildup analysis is generally diagnostic of fractures.

The state of the art of well test analysis has been comprehensively summarized by Matthews and Russell.[13] Chapter 10 of their monograph includes useful discussions and reproductions of key graphs from the work of Pollard,[14] Warren and Root,[15] and Russell and Truitt[16] concerning pressure analysis in fractured reservoirs. Figure 9 is a schematic model of an idealized fracture-matrix system in carbonate rocks. The theoretical curves derived from this model show typical parallel slopes during transient pressure tests, as illustrated in Fig. 10. In the authors' experience, however, the occurrence of parallel slopes in known fracture-matrix systems is quite rare. The reason for this rarity was given by Odeh,[17] who showed that fractured reservoirs frequently behave as homogeneous reservoirs.

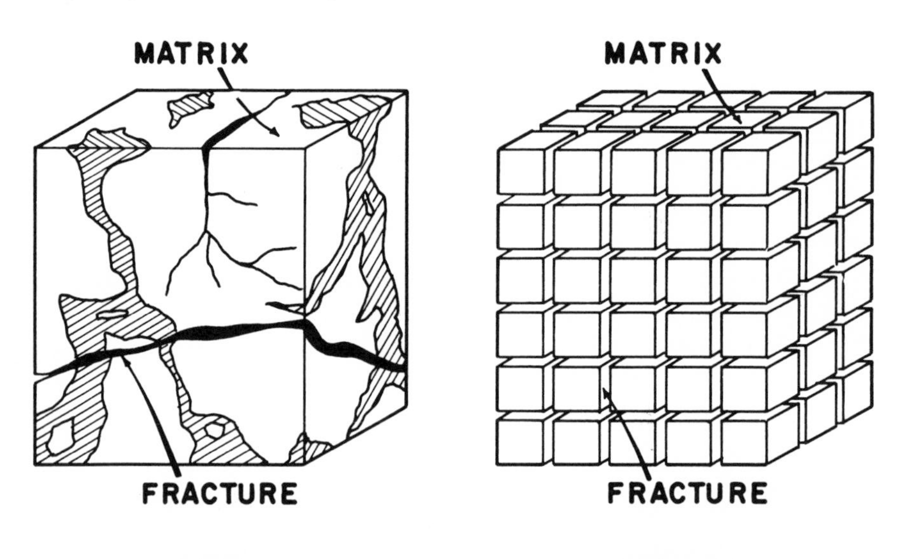

FIG. 9. Idealization of a naturally fractured heterogeneous porous medium. (After Warren and Root,[15] courtesy of AIME.)

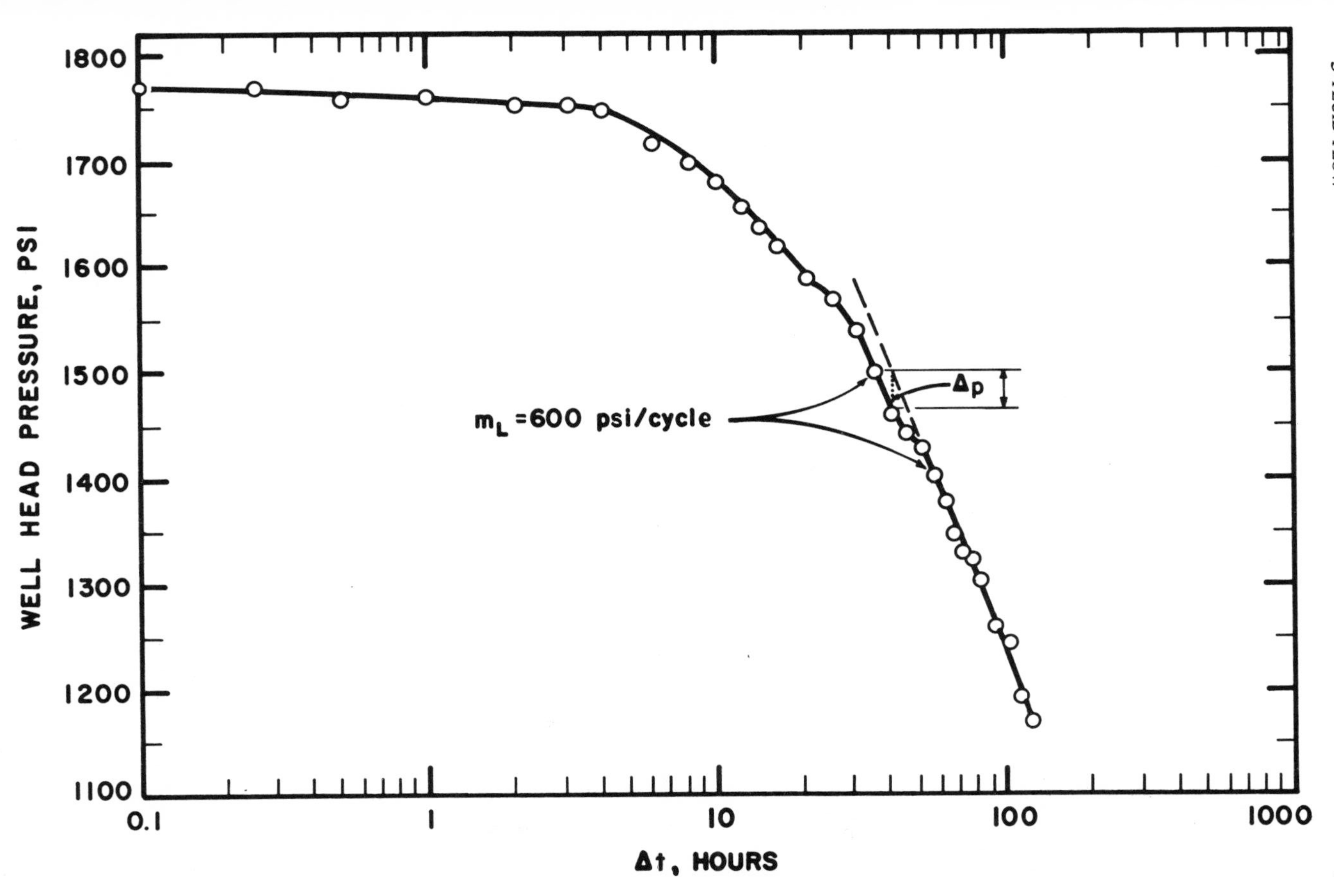

FIG. 10. Pressure falloff curve in a fracture-matrix system, Grayburg Formation, Lea County, New Mexico. (Courtesy of Continental Oil Co.)

In the few cases where the parallel slope sections appear in either buildup or falloff tests one may apply the following equation:

$$\omega = \phi_f c_f/(\phi_m c_m + \phi_f c_f) \tag{4}$$

where

$$\omega = e^{-2.303\,(\Delta p/m_L)} \tag{5}$$

and c = compressibility (psi^{-1}), ϕ = porosity, f refers to fracture and m to matrix, Δp = pressure difference (psi) between parallel slopes, and m_L = slope of pressure curve (psi/log cycle). The object of the calculation is to evaluate the proportion of fracture porosity, ϕ_f, to total porosity ($\phi_m + \phi_f$).

Comparison of Log and Core Analysis Data

Many more wells are logged than cored; and, as the quality and sophistication of logging techniques improve, this trend will continue. In an interesting study, Marchant and White[18] compared a large number of log and core analysis results for an extremely heterogeneous fine-grained oolitic limestone reservoir (Ratcliffe pay, Flat Lake Field, Montana). Contrary to most past comparisons, these investigations treated the log analysis as independently from the core analysis as was practical. Thus, the log analysts used only the rock grain densities and residual oil saturations measured in cores. Rock grain densities were used to give the applicable acoustic transit time, Δt_m, as shown in Fig. 13, p. 18. The sonic log porosity, ϕ_s, was then computed from the equation:

$$\phi_s = (\Delta t_{log} - \Delta t_m)/(\Delta t_f - \Delta t_m) \tag{6}$$

where Δt_{log} = sonic log deflection (μsec/ft), and Δt_f = interstitial fluid acoustic transit time, which was equal to 180 μsec/ft in this reservoir. The residual oil saturation, S_{or}, was determined from water-oil relative permeability tests on representative core samples. The resistivity log porosity, ϕ_r, was then calculated by using Eq. 7, which was unique for the carbonate reservoir:

$$\phi_r = \left(\frac{0.62 R_{mf}}{R_{xo}(1-S_{or})^2}\right)^{1/2.15} \tag{7}$$

where R_{mf} = mud-filtrate resistivity (ohm-meters), and R_{xo} = flushed zone resistivity (ohm-meters).

Figures 11 and 12 show comparisons of log and core analysis results for one well. The agreement, viewed as a whole, is reasonable. It would have been highly optimistic to expect a point-by-point agreement in a heterogeneous reservoir of this type. Wyllie[19] explains that formations are very seldom arranged like "layer cakes" and consequently the area surrounding the well should not be expected to conform exactly to the properties observed

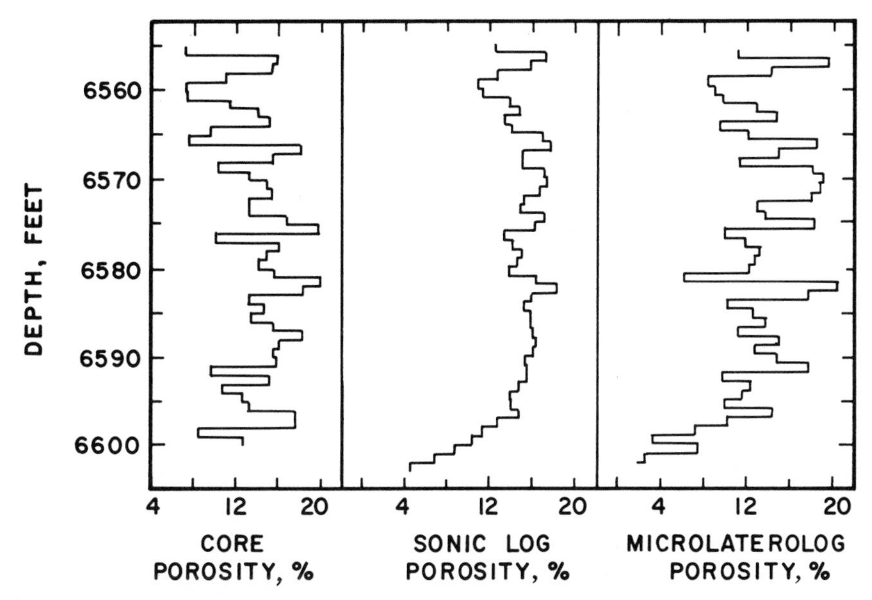

FIG. 11. Porosities from core analysis, sonic log, and microlaterolog. (After Marchant and White,[18] courtesy of SPWLA.)

in relatively small core samples taken from the center of the well. The logging tools actually examine a mass of rock surrounding the wellbore, which is larger and generally differs somewhat in average properties from the volume of rock represented by the core sample.

Comparison of Water Saturation Data

One of the most reliable methods for determining interstitial water saturation in a reservoir involves using oil-base drilling fluid during coring so that the interstitial brines will not be diluted with the water-base mud filtrate. Figure 13 shows a typical correlation between water saturations and air permeability determined from cores drilled with oil-base mud. There are generally lower water saturations in dolomite than in sandstone; the reason lies in the difference in pore geometry. One of the factors causing the difference is that there is generally less internal surface area per unit pore volume in carbonate rocks than in sandstones.

Capillary Pressure Data

Capillary pressure data are often used in determining the theoretical water saturation in a reservoir. The theory is based on the assumption that the

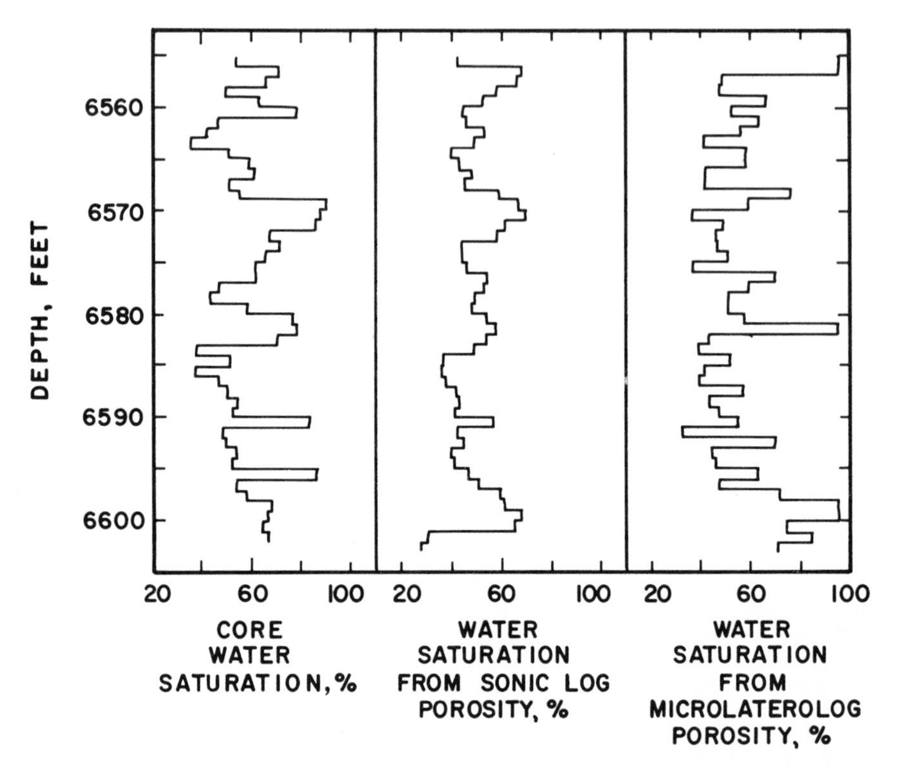

FIG. 12. Water saturations from core analysis, sonic log, and microlaterolog. (After Marchant and White.[18]) Core analysis water saturations were determined with the aid of capillary pressure tests. (Courtesy of SPWLA.)

formation was originally water filled and that oil later accumulated in it, or migrated into it, thereby displacing all of the initial "connate" water except for the amount held in the pore spaces by capillary forces. The force that is retaining the water (capillary pressure, P_c) is exactly balanced by the pull of gravity, which tends to drain the water out of the oil zone. Thus:

$$P_c = gh(\rho_w - \rho_o) \tag{8}$$

or

$$P_c = h(\rho_w - \rho_o) \times 0.433 \tag{9}$$

where P_c = capillary pressure in oil-water system (psig), h = height above free water level ($S_w = 100\%$) (ft), and ρ_w and ρ_o = specific gravity of water and oil, respectively. Figure 14 illustrates the interrelationship of oil-water relative permeability, capillary pressure, and liquid production in a reservoir. According to Knutson,[20] the water-oil contact (at the bottom of the transition zone), rather than the "free water level", is for practical purposes the

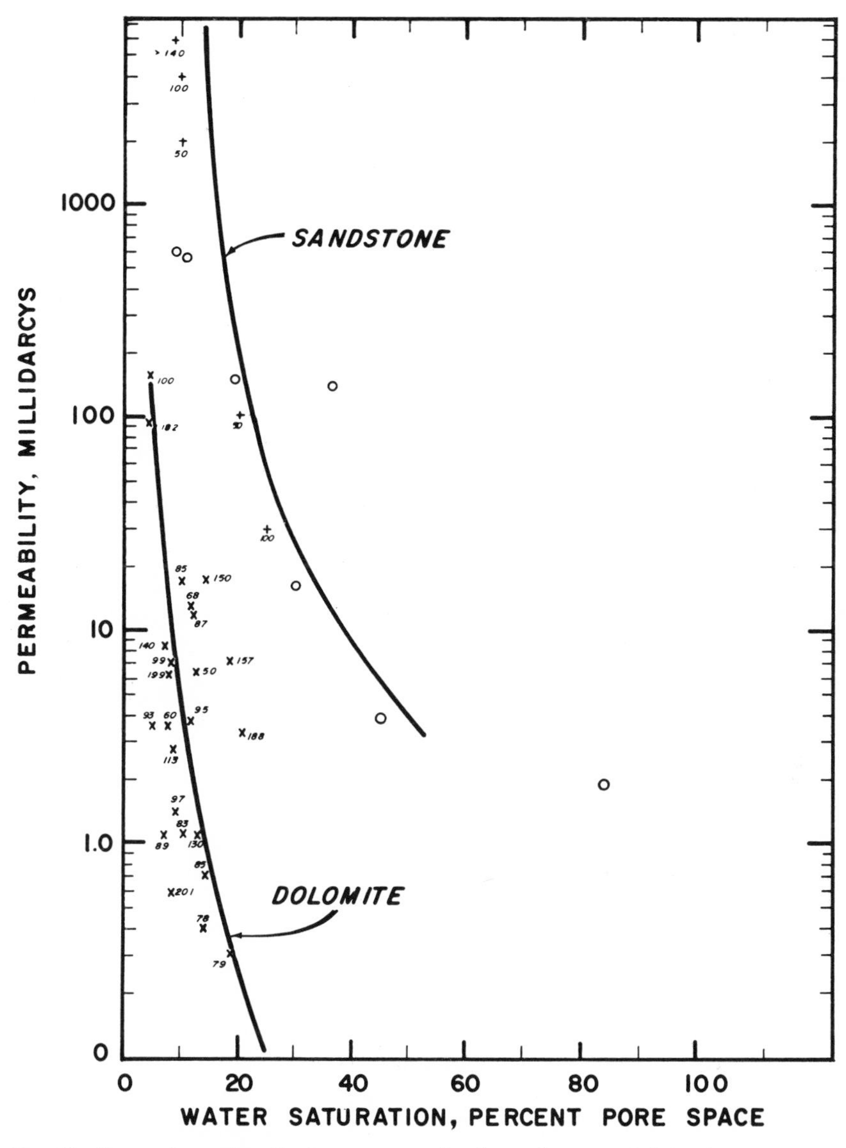

FIG. 13. Comparison of residual water saturation in sandstone and dolomite cores. (After Bulnes and Fitting.[3]) o East Texas sandstone; + Gulf Coast sandstone from 5 fields; × Wasson dolomite. Numbers represent approximate height of core above water table. (Courtesy of AIME.)

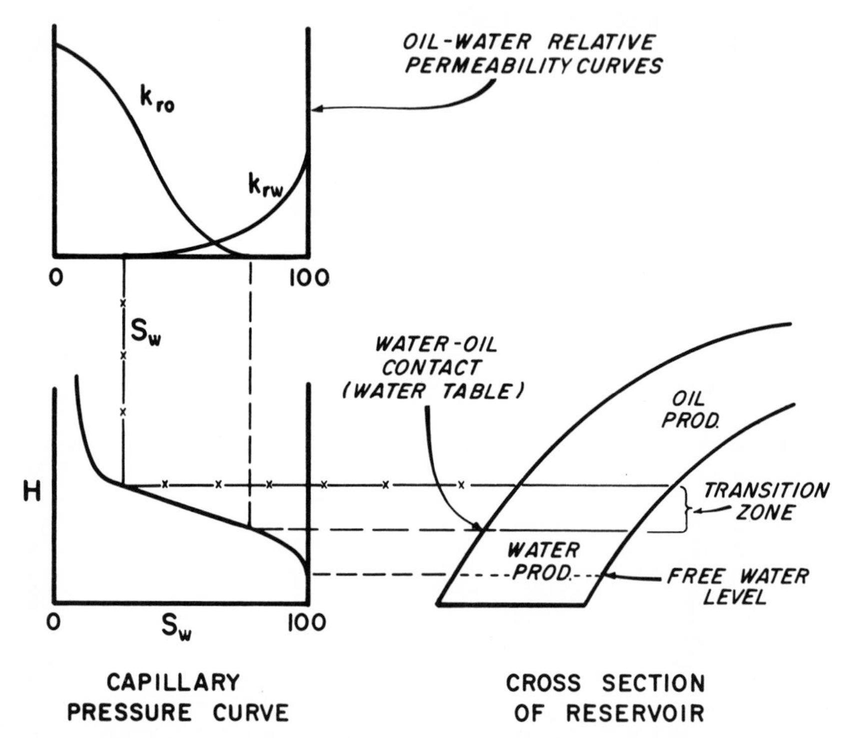

FIG. 14. Effect of relative permeability and capillary pressure on reservoir production.

water table of the reservoir. It is from this depth downward that only water is produced. Above this depth both oil and water are produced in the transition zone, and the following equation applies here:

$$\%\ \text{Water cut} = \frac{100}{1+(k_{ro}\mu_w B_w/k_{rw}\mu_o B_o)}. \tag{10}$$

For the sake of convenience, many capillary pressure tests are conducted in the laboratory using air-water or mercury-vacuum systems instead of the oil-water system that corresponds to the actual reservoir situation. In the resulting curves the nonwetting phases simulate the oil phase. The capillary pressures are proportional to each other as follows:

$$\frac{(P_c)_{o-w}}{(P_c)_{a-w}} = \frac{(\sigma \cos\theta)_{o-w}}{(\sigma \cos\theta)_{a-w}} \approx 0.33. \tag{11}$$

Brown[21] recommended a ratio of 6.4 for $(P_c)_{\text{Hg}}/(P_c)_{a-w}$ in limestones. In these proportionality relationships, *o* refers to oil, *w* to water, *a* to air, and Hg

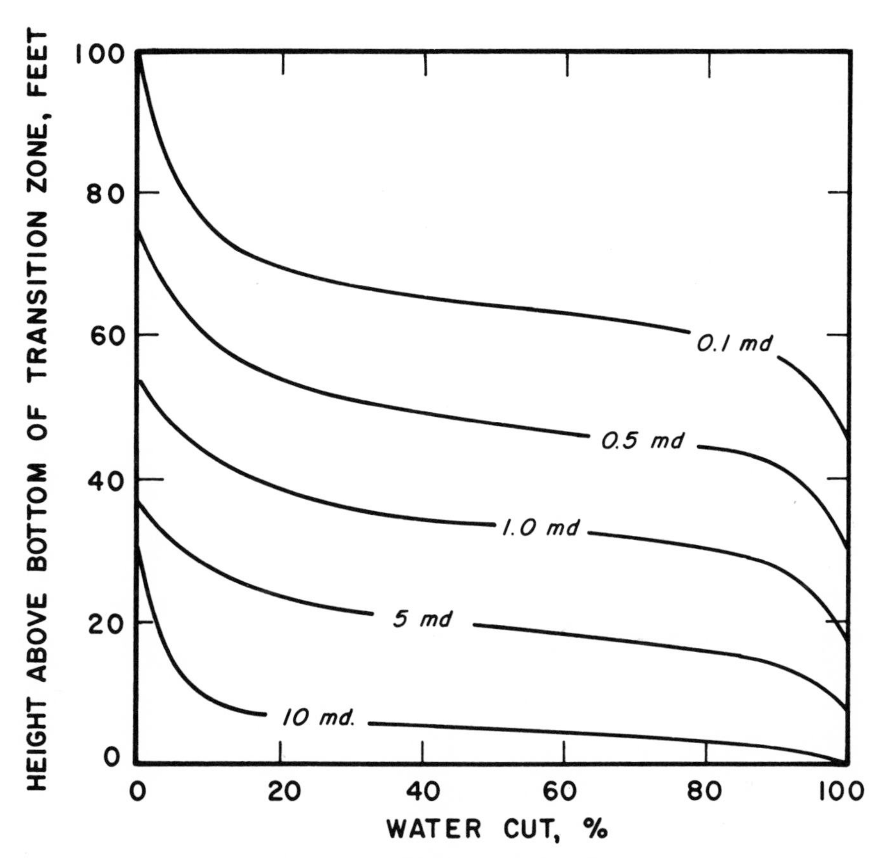

FIG. 15. Water cut curves, San Andres Dolomite. (After Aufricht and Koepf,[27] courtesy of AIME.)

to mercury; σ=interfacial tension (dynes/cm), and θ=contact angle of wetting phase with rock, measured through the liquid. (Also see Appendix B.)

The mercury-vacuum system is especially popular for carbonates because of the low permeabilities of many of the rock samples. To be useful in reservoir engineering evaluations, capillary data should show relationships between water saturations and height above bottom of transition zone (Eq. 8 or 9). Combining these data with appropriate water cut curves (derived from k_{rw}/k_{ro} test data and Eq. 10) gives data like those illustrated in Fig. 5, p. 232. Another illustration is presented in Fig. 15 to show water cuts as a function of height above the bottom of the transition zone and permeability.

Interstitial water saturation in the oil zone at a reasonable elevation above the free water level can often be correlated with either permeability or porosity or a combination of the two parameters, using the so-called J function proposed by Leverett.[22] The Leverett correlation is based on the work of Poiseuille,[23] Darcy,[24] and Kozeny[25] and in effect substitutes J for P_c, using the relationship:

$$J = \frac{P_c}{\sigma \cos \theta} \sqrt{\frac{k}{\phi}}. \tag{12}$$

The aim of correlating J versus S_w rather than P_c versus S_w is to reduce the scattering of data and to help in the averaging of capillary pressure data for a given lithological horizon. Ashford[26] related the J function also to relative permeabilities. (Also see Chapter 5, p. 240.)

Attempts to correlate various parameters with capillary pressure data for cores from wide geographic areas have been unsuccessful because of the wide range of pore geometries that exist in carbonates. Although no general relationship was obtained, Aufricht and Koepf[27] prepared a typical correlation of interstitial water with air permeability, based on a study of 205 core samples from 14 fields, as shown in Fig. 16.

Capillary pressure curves for carbonates often correlate better with porosity than with permeability. Figure 17 illustrates a correlation based on porosity. Similar correlations have been presented for various carbonate formations by Rockwood *et al.*[28]

Messer[29] described an evaporation method for determining the irreducible interstitial water saturation. This technique is rapid and gives results which agree reasonably well with capillary pressure data for about 100 ft above the bottom of the transition zone. In this method extracted core samples are resaturated with a liquid which is then allowed to evaporate. At first, the liquid flows to the core surfaces easily, and the evaporation rate is rapid. This rate declines abruptly, however, when the saturation becomes discontinuous and liquids move to the surface by alternate evaporation and condensation. The abrupt change of rate is noted, and the liquid volume retained at this point is related to the irreducible water saturation.

Primary Recovery

k_g/k_o Data

The relationship between the ratio of gas to oil permeability (k_g/k_o) and liquid saturation must be known in order to predict the primary recovery from a solution gas-drive reservoir. The k_g/k_o data derived from observed field data of solution gas-drive reservoirs and from laboratory measurements on core samples, which are believed to be representative of the pore geometry existing in the reservoir, are presented here.

Starting with the gas/oil ratio in the reservoir, which is simply the ratio

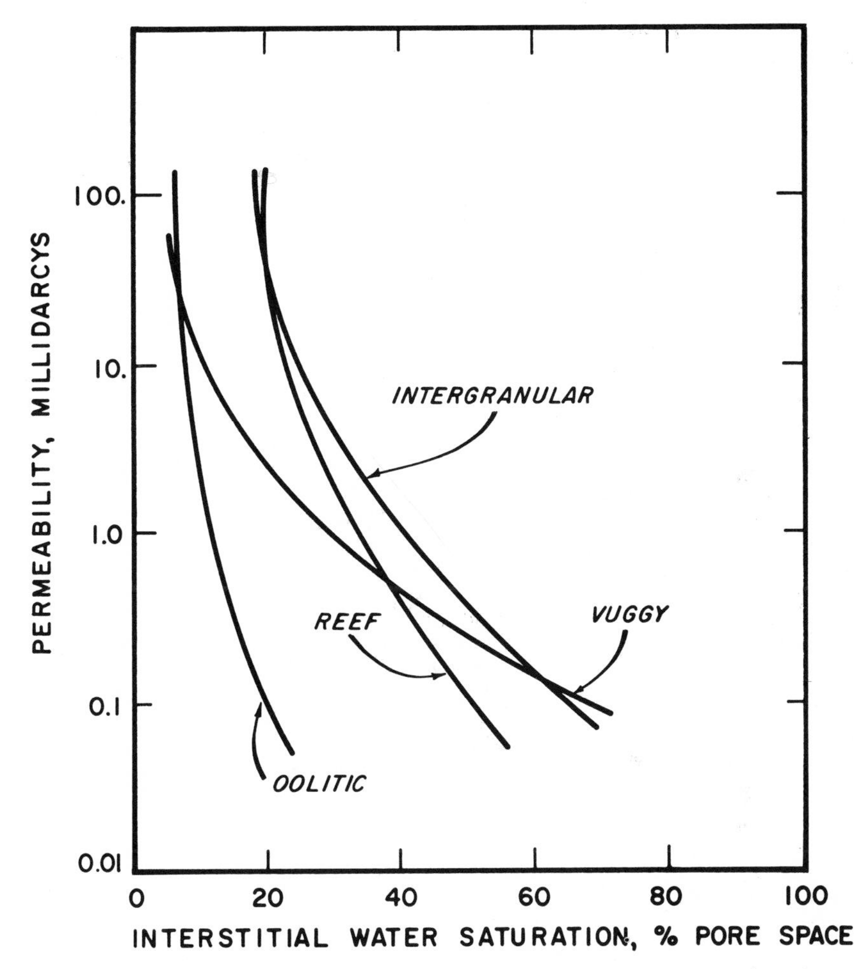

FIG. 16. Water distribution curves for typical carbonate formations at approximately 60 ft above free water. (After Aufricht and Koepf,[27] courtesy of AIME.)

of gas flow to oil flow, using Darcy's law for radial flow, and finally correcting reservoir flow data to surface conditions gives the equation:

$$R = \frac{k_g}{k_o} \frac{\mu_o B_o}{\mu_g B_g} + R_s \tag{13}$$

where R = surface gas/oil ratio (SCF/STB), R_s = solution gas/oil ratio (SCF/STB), k_g = effective gas permeability (md), k_o = effective oil permeability (md), μ = viscosity at reservoir conditions (cp), B_g = gas formation volume factor (reservoir barrels/SCF), and B_o = oil formation volume

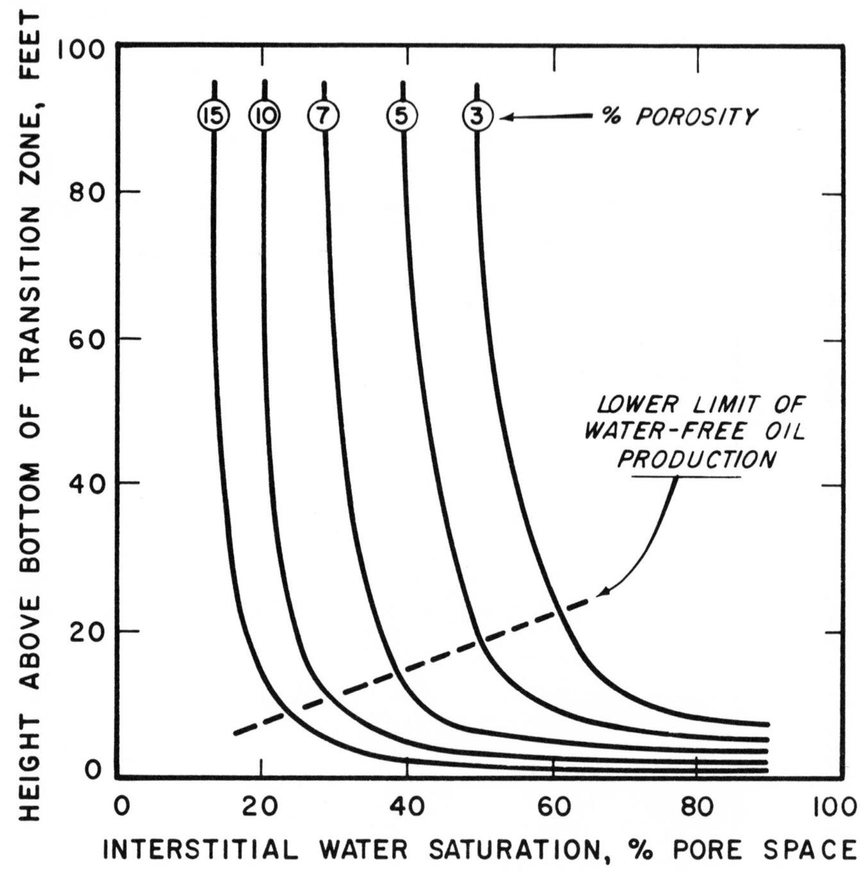

FIG. 17. Water distribution curves, Pennsylvanian dolomite. (After Aufricht and Koepf,[27] courtesy of AIME.)

factor (reservoir barrels/STB). The reservoir oil saturations corresponding to the field-derived k_g/k_o ratio are evaluated by the following equations, which are derived in reservoir engineering textbooks (e.g., Craft and Hawkins,[30] p. 116):

$$S_o = (1-S_w)\left(1-\frac{N_p}{N}\right)\frac{B_o}{B_{oi}} \tag{14}$$

and

$$S_L = S_o + S_w \tag{15}$$

where N_p = oil produced (STB), N = oil-in-place at bubble point pressure (STB), and B_{oi} = oil formation volume factor at bubble point pressure (reservoir barrels/STB). These equations were used to construct the field-derived k_g/k_o curves shown in Figs. 18–22.

Useful k_g/k_o data are presented in the literature by Elkins,[31] Patton,[32] Stewart *et al.*,[33] and Arps and Roberts.[34] These data are shown, together with other results obtained from unpublished sources, in Figs. 18–23.

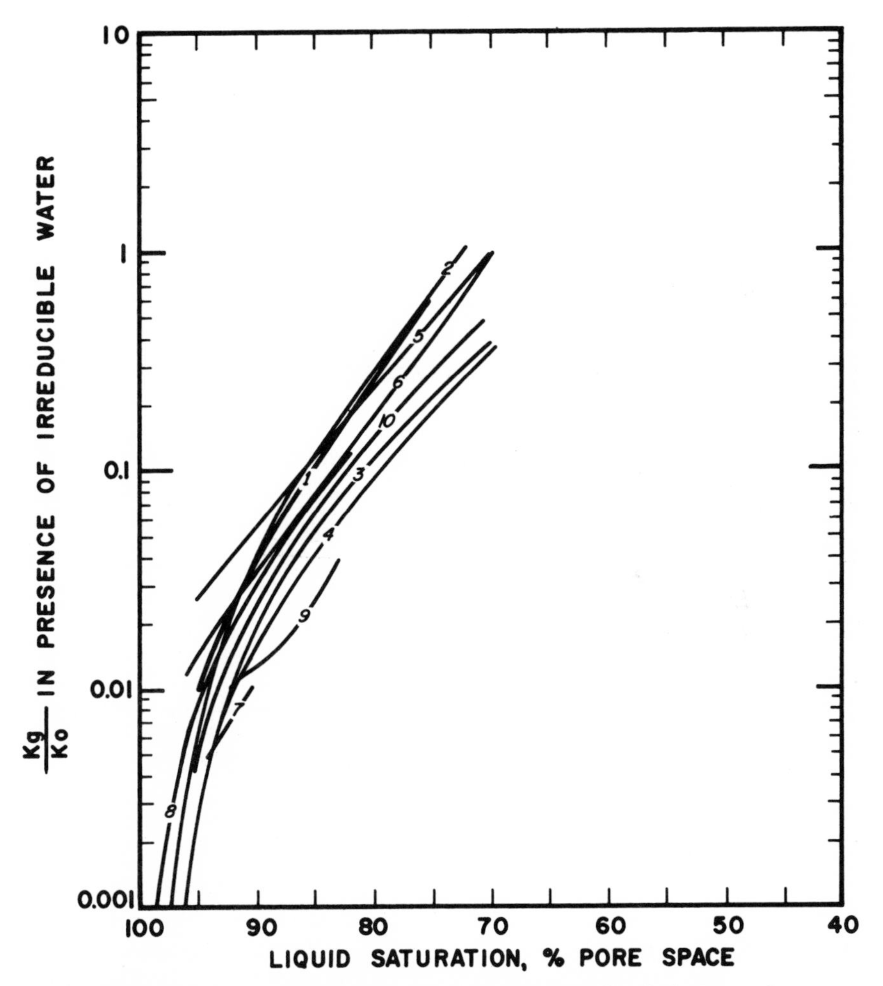

FIG. 18. Field-derived k_g/k_o data for dolomites. 1—Fullerton-Clearfork, Texas; 2—Skaggs-Grayburg, New Mexico; 3—S. Cowden-Foster (Grayburg), Texas; 4—N. Cowden (Grayburg), Texas; 5—Goldsmith, N. Dome (Grayburg-San Andres), Texas; 6—Goldsmith, S. Dome (Grayburg, San Andres), Texas; 7—Wasson (San Andres), Texas; 8—Harper (San Andres), Texas; 9—Penwell (San Andres), Texas; 10—Slaughter (San Andres), Texas. (After Arps and Roberts,[34] courtesy of AIME and unpublished sources.)

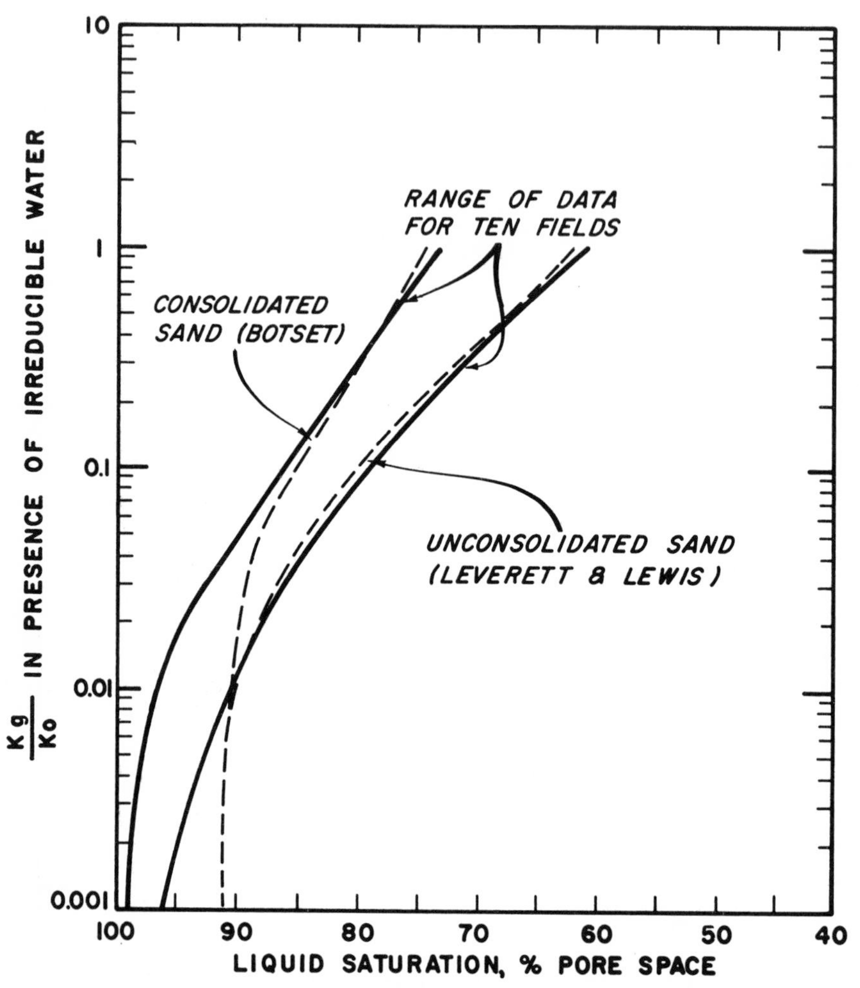

FIG. 19. Field-derived k_g/k_o data for dolomites compared with selected laboratory k_g/k_o data for sandstones. (Sandstone curves,[35, 36] courtesy of AIME.)

Figure 18 gives field-derived k_g/k_o curves for Permian dolomite fields in Texas and New Mexico. These data are compared in Fig. 19 with selected laboratory k_g/k_o data from the literature, namely, the curve for consolidated sandstone (k_{air} = 500 md, porosity = 22%) reported by Botset[35] and the curve for unconsolidated sand (k_{air} = 10,000 md, porosity = 42%) reported by Leverett and Lewis.[36] The range of the field-derived k_g/k_o curves for dolomite shows a remarkable and unexpectedly good agreement with the range of the literature data for consolidated and unconsolidated sand above a k_g/k_o ratio of 0.01.

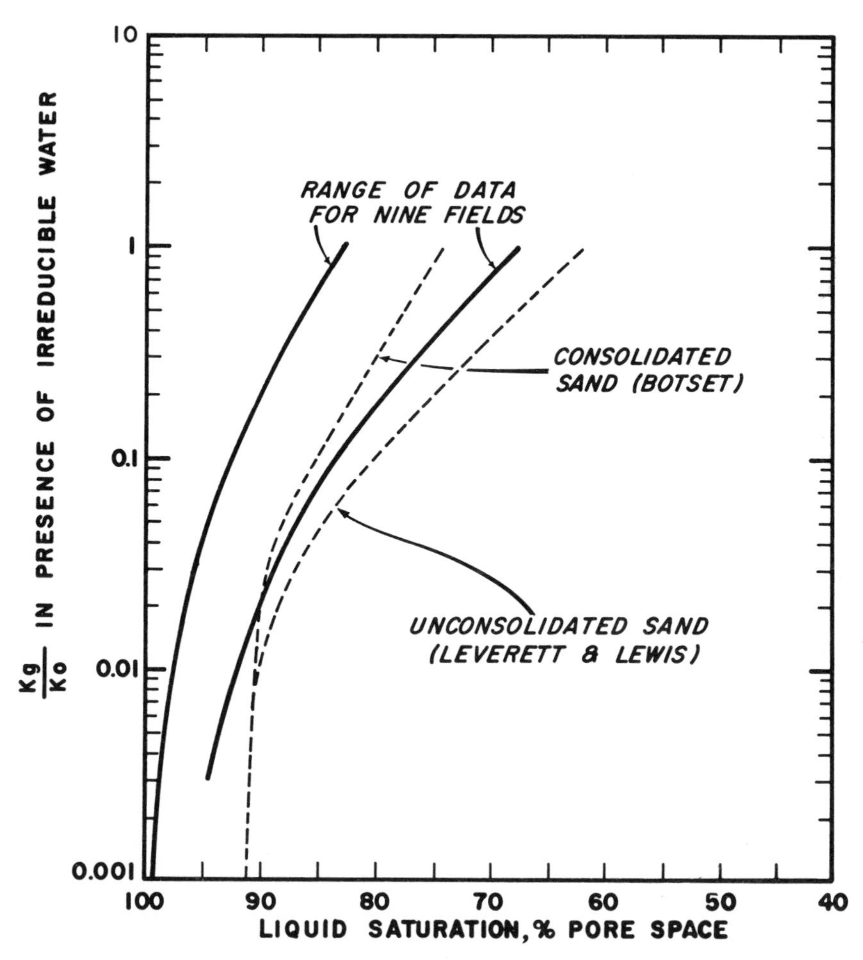

FIG. 20. Field-derived k_g/k_o data for sandstones compared with selected laboratory data. (Sandstone curves,[34–36, 39] courtesy of AIME.)

The discrepancy of field-derived and laboratory k_g/k_o curves below a k_g/k_o ratio of 0.01 in Fig. 19 is believed to be due to certain difficulties in the original laboratory tests in measuring extremely low gas rates with sufficient accuracy. Later tests conducted by Felsenthal and Conley[37] indicate that the k_g/k_o curves in the low k_g/k_o ratio region probably match the field-derived data much more closely than shown in Fig. 19.

In general, a steep k_g/k_o versus saturation curve indicates a low solution gas-drive efficiency, whereas a flat curve indicates a favorable efficiency. A comparison of field-derived data in Figs. 19 and 20 indicates that primary

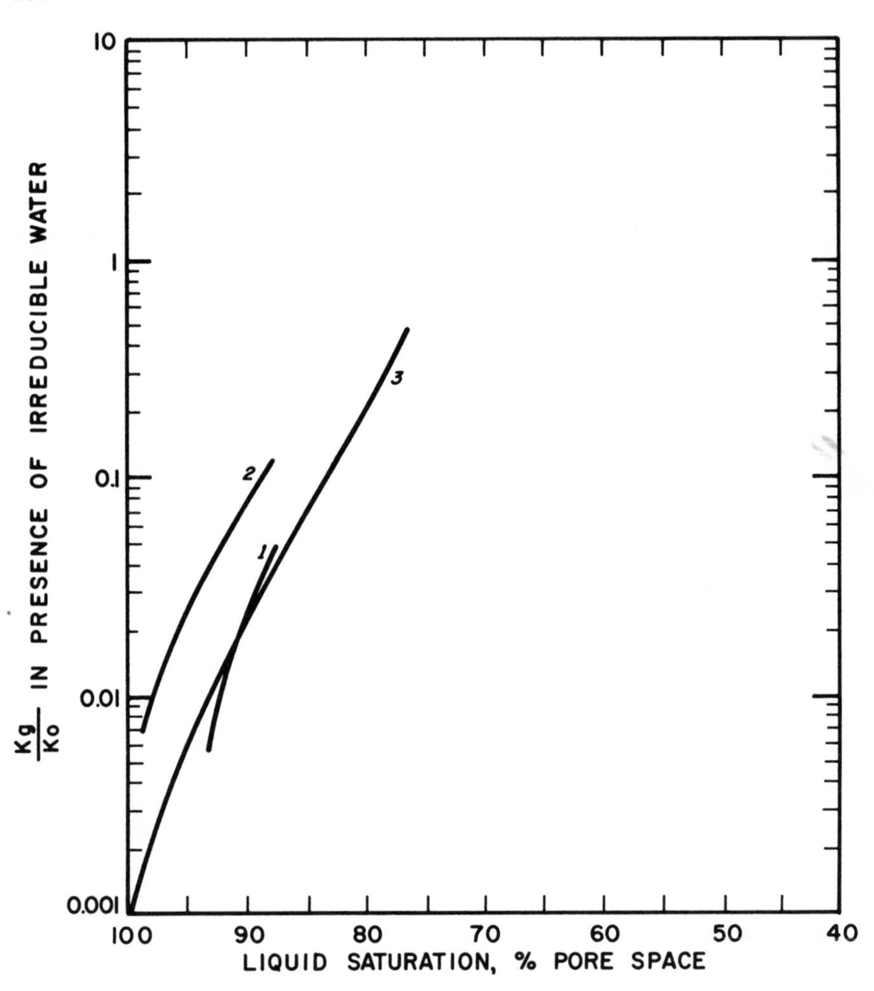

FIG. 21. Field-derived k_g/k_o data for limestones. 1—Haynesville-Pettit, Louisiana; 2—W. Edmond-Bois d'Arc (Hunton), Oklahoma; 3—Wheeler-Devonian (chert and limestone), Texas. (After Arps and Roberts,[34] courtesy of AIME.)

(solution gas-drive) efficiency may actually be as good on a percentage pore volume basis in dolomite reservoirs as in sandstone reservoirs or even better. The inference based on these limited data is startling, inasmuch as it suggests that the heterogeneous features of dolomites may help rather than hinder oil recovery by internal solution gas drive.

The sandstone field data shown for comparative purposes in Fig. 20 are based on those of Kaveler,[39] Elkins,[31] Arps and Roberts,[34] and various unpublished sources. The fields are all solution gas-drive reservoirs and cover a wide range of geologic ages (from Ordovician to Miocene) and various geographic locations (Arkansas, California, Louisiana, Oklahoma, Texas, and Wyoming).

Field-derived k_g/k_o data for limestones appear to span a wider range of values than was evident for dolomites. Figures 21–23 show data for ten limestone reservoirs, two of which also contained appreciable amounts of chert.

An interesting and extensive comparison of solution gas-drive recoveries in sandstone and carbonate reservoirs was reported by Arps *et al.*[38] Their results are summarized in Table 1. It may be seen that, on the basis of the

TABLE 1. Summary of Solution Gas-Drive Recovery Data Compiled by API Subcommittee on Recovery Efficiency[38]

(Courtesy of API.)

Distribution of Case Histories by Reservoir Mechanism and Rock Type

Class	Predominant reservoir mechanism	Reservoir rock type	
		Sand and sandstone	Limestone, dolomite, and other
A	Solution gas drive (without supplemental drive)	77	21
B	Solution gas drive (with supplemental drive)	60	21

Primary Oil Recoveries and Rock Parameters*

	Sand and sandstone			Limestone, dolomite, and other		
	Minimum	Median	Maximum	Minimum	Median	Maximum
Class A reservoirs						
Ultimate recovery						
S_{gr}, % pore space	13.0	22.9	38.2	16.9	26.7	44.7
bbl/acre-ft	47	154	534	20	88	187
Rock parameters						
k, md	6	51	940	1	16	252
ϕ, % bulk volume	11.5	18.8	29.9	4.2	13.5	20.0
Class B reservoirs						
Ultimate recovery						
S_{gr}, % pore space	7.7	25.5	43.5	11.2	26.0	42.6
bbl/acre-ft	109	227	820	32	120	464
Rock parameters						
k, md	10	216	2,500	2	19	867
ϕ, % bulk volume	12.0	21.0	35.9	3.3	13.3	24.8

* Minimum, median, and maximum values given for each separate parameter; values shown do not necessarily belong to same case history.

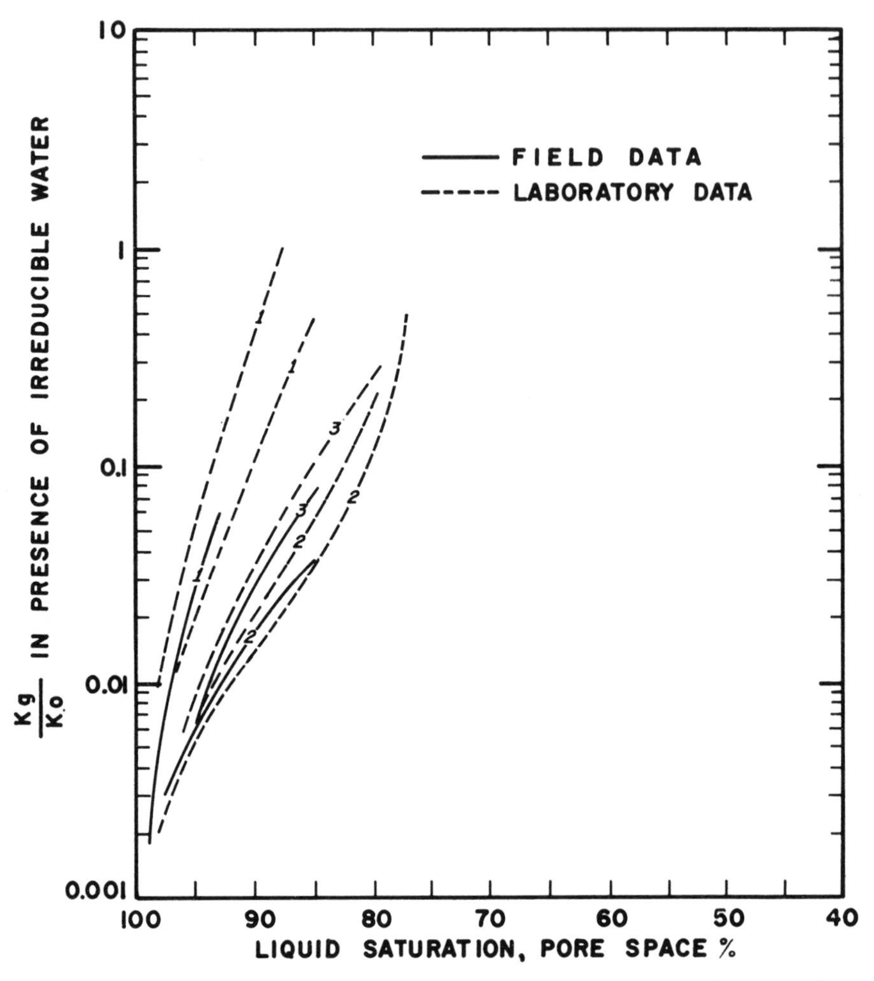

FIG. 22. Laboratory and field k_g/k_o data for limestones and chert. (After Stewart *et al.*[33]) 1—Fractured chert and limestone, Field F; 2—limestone with intergranular porosity, Field A; 3—limestone with intergranular porosity, Field B. (Courtesy of AIME.)

percentage of pore space voided as a result of primary production (expressed as residual gas saturation, S_{gr}), carbonate reservoirs compare favorably with sandstone reservoirs.It must be emphasized, however, that the carbonates produced significantly fewer barrels of oil per acre-foot than the sandstones. This difference stems simply from the fact that carbonates generally possess less porosity and consequently contain less oil initially in place per unit of bulk reservoir rock volume. (Also see Chapter 6, pp. 298–306, for additional discussion.)

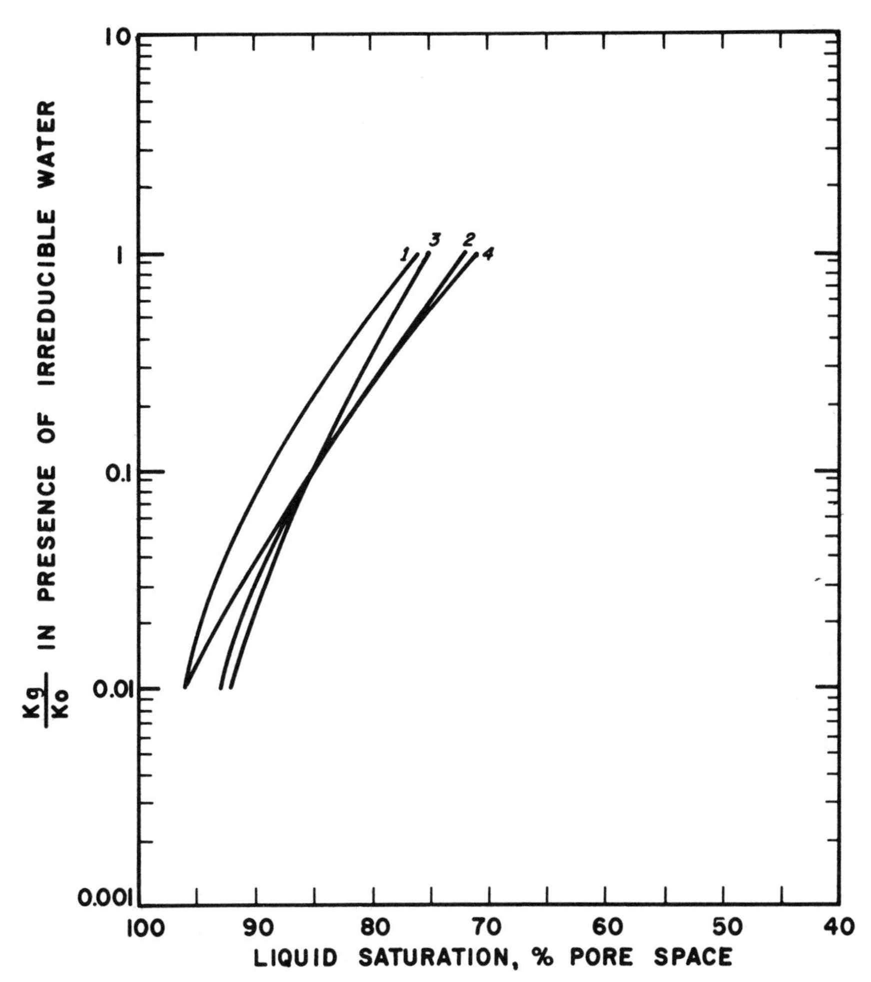

FIG. 23. Laboratory k_g/k_o data for reefs. (After Arps and Roberts.[34]) 1—Pennsylvanian reef A; 2—Pennsylvanian reef B; 3—Strawn Reef; 4—Palo Pinto Reef. (Courtesy of AIME.)

The laboratory data presented in Fig. 22 were obtained on full-diameter core samples, which apparently possessed pore geometries representative of the reservoir. The steepest (least favorable) k_g/k_o curve in Fig. 22 is for fractured chert and limestone (Field F). Microscopic examination indicated that the reservoir rock in Field F is made up of a large number of small, irregularly shaped, fractured chert pieces cemented together by limestone with intergranular porosity. The permeability of the test sample was 2.4 md, associated with an unusually high (for limestone) porosity of 30.3%. This high porosity may have been one reason for the steepness of the curve. A trend of increasing steepness of k_g/k_o curves with increasing porosity had been observed previously for sandstones.[40] (Also see Chapter 6, pp. 282–288.)

Calculating field-derived k_g/k_o curves for comparison with other fields in the area can be a strong diagnostic tool. It must be emphasized, however, that the calculations based on Eqs. 13–15 are not applicable to reservoirs with gravity drainage, strong gas-cap drive, or water influx.

Directional Permeability or Fracture Orientation

Most carbonate rock reservoirs are anisotropic; that is, permeabilities in one direction are often drastically different from those in another. This anisotropy is largely due to the geologic stresses existing in the reservoirs, as explained by Blanchet,[41] Martin,[42] and Alpay.[43]

Knowledge of this anisotropy may be of importance in the optimum location of wells for primary recovery. It becomes of crucial importance in the design of any subsequent fluid injection project. An insight into reservoir anisotropy can be gained during initial development of a field, if the reservoir oil is undersaturated.

Elkins and Skov[44] reported that, in the initial development of an undersaturated reservoir in the Spraberry-Driver area, an extra effort was made to measure the initial pressure in each of the 71 wells immediately after completion. At first it was assumed that the reservoir was isotropic and that production in a well caused circular drawdown isopotentials in the surrounding regions. Later, however, it was noted that the reservoir was anisotropic. The isopotentials actually had elliptical shapes, with the ratio of the major axis to the minor axis, a/b, proportional to the maximum/minimum permeability ratio, k_{max}/k_{min} as follows:

$$a/b = \sqrt{k_{max}/k_{min}}. \tag{16}$$

A typical isopotential graph in an anisotropic medium having $k_{max}/k_{min} = 8/1$ is shown in Fig. 24. In the Spraberry-Driver study, assumed values of k_{max}/k_{min} and azimuth of k_{max} were evaluated by a trial and error procedure, using an electronic computer. Seventy complete sets of calculations were performed until the "best fit" of assumed values and observed pressures was established. The equation used was as follows:

$$p_i - p = \frac{-q\mu_o B_o}{14.16\sqrt{k_x k_y}\, h}\; Ei - \left(\frac{(x-x_o)^2/k_x + (y-y_o)^2/k_y}{25.28t/\mu_o c\phi}\right) \tag{17}$$

where p_i = initial pressure (psi), p = pressure at x,y at time t (psi); q = production rate (B/D); μ_o = viscosity of oil (cp); B_o = oil formation volume factor (reservoir barrels/STB); h = thickness (ft); t = time (days); c = total compressibility of oil, water, and rock (psi^{-1}); ϕ = fractional porosity; k_x = effective permeability in x direction (d); k_y = effective permeability in y direction (d); $x - x_o$ = distance from producing well to pressure point in x direction (ft); and $y - y_o$ = distance from producing to pressure point in y direction (ft). Coordinates were rotated in each test run so that $k_y = k_{max}$ and $k_x = k_{min}$.

Arnold *et al.*[45] discussed equipressure lines created in an interference test in an undersaturated reservoir, as illustrated in Fig. 24. In this graph the center well was the producing well, and pressures in the outlying shut-in production wells were observed until approximate steady-state flow was established. In a test of this type, other producing wells in the field should ideally be far enough removed so that there is essentially no interference from them; otherwise, the elliptical isopotential lines shown will become distorted and will no longer remain elliptical. Multiple interference can be solved by making effects additive (principle of superposition).

It has been noted that fracture planes created by earth stresses in tectonically quiescent regions typical of many oil-producing basins are commonly in a vertical direction. Furthermore, as pointed out by Alpay[43] and others, fracture traces observed on the surface, for example, in the Permian Basin,

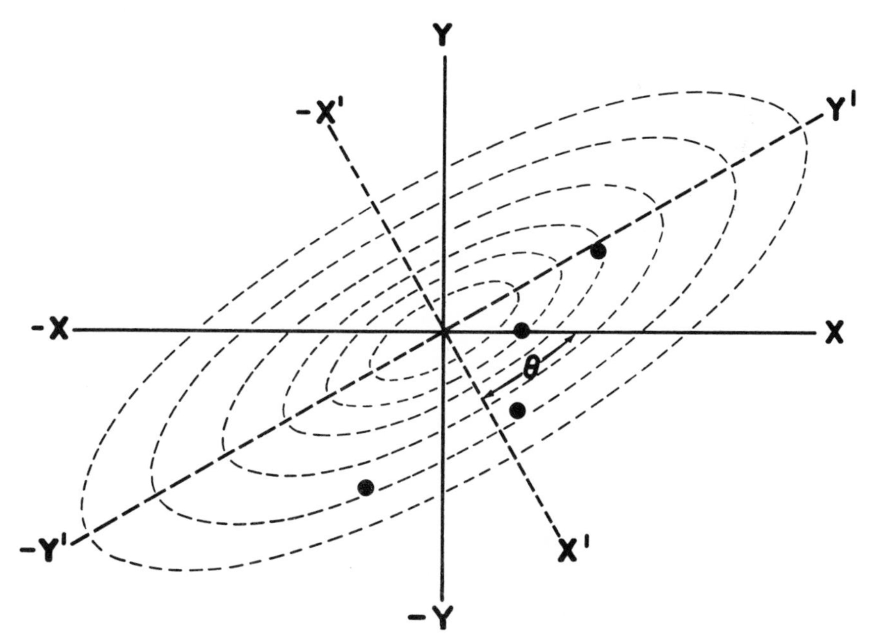

FIG. 24. Isopotential lines for $k_{max}/k_{min} = 8$. (After Arnold *et al.*,[45] courtesy of AIME.)

often bear a striking resemblance to the orientation of the major fracture directions in the reservoir. These surface traces may be investigated by aerial photography and the new techniques of airborne radar imagery (Dellwig *et al.*,[46] Miller,[47] Pasini and Overbey,[48] and MacDonald[49]). Thermal infrared imagery, as discussed by Sabins, [50] may also develop into a useful tool in this connection.

Reservoir Performance

After an extensive comparison of limestone and sandstone fields, Craze[4] concluded that the difference between the performances of limestone and of sandstone reservoirs does not appear to be fundamental. His studies included fields performing under solution gas drives, gas-cap drives, active water drives, and combination drives. One of several comparisons presented by Craze is reproduced as Fig. 25. One of the reservoirs shown, the Hendrick, produced from cavernous Permian Seven Rivers Limestone; the other reservoir in the Humble (Deep) Field was the Yegua Sand of Eocene age.

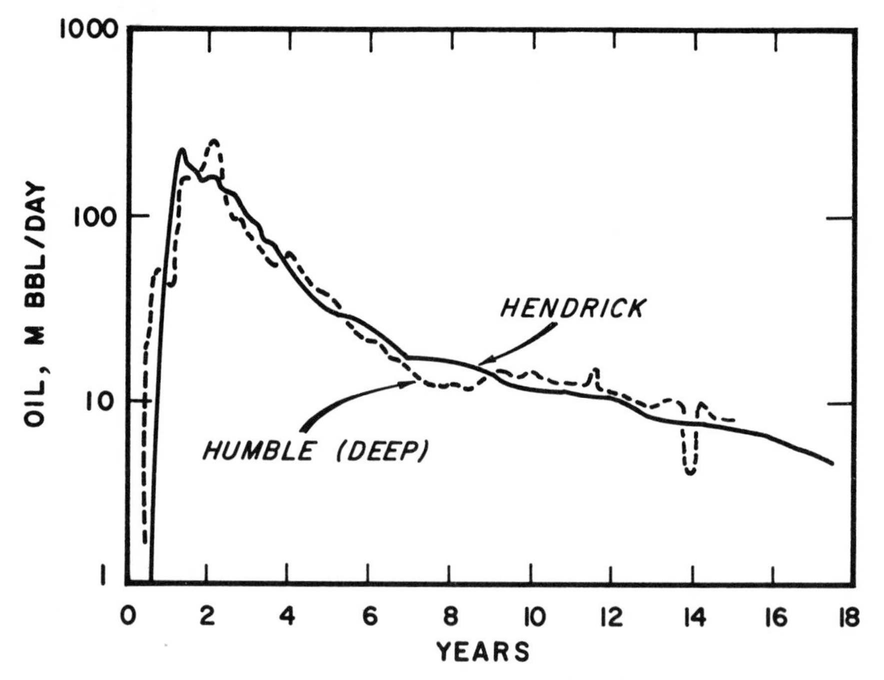

FIG. 25. Comparison of decline curves. Hendrick—produced from limestone, Humble (Deep)—produced from sandstone. Humble scale is one-fifteenth of Hendrick scale. (After Craze,[4] courtesy of AIME.)

Both fields were drilled up rapidly and started to produce as flush fields at peak open-flow rates. Production then declined; in the ensuing years, the two reservoirs yielded oil at low rates, accompanied by the production of large quantities of water. The noteworthy feature displayed in Fig. 25 is that, in spite of the differences of the porous media, the percentage rates of decline in oil production have been almost identical for the two fields.

These findings confirm the inference drawn from the k_g/k_o data, namely, that the primary recovery performance of carbonate reservoirs compares favorably with that of sandstone reservoirs. The favorable comparison applies, however, only on a percentage pore-space basis. As noted in the statistical API study,[38] carbonates generally have a smaller ultimate recovery in barrels per acre-foot than sands and sandstones (see Table 1). Craze noted that the volumetric material balance method, the unsteady-state radial flow equation, and other reservoir engineering techniques initially developed for sandstones are equally applicable to limestones; however, he added the following statement, which is similar in many respects to the conclusions drawn later by Elkins:[1] "Limestones require thorough understanding of the properties of the formation, of the fluids, their behavior during flow, and adequate production operating data." Craze particularly stressed the need for more complete coring and for comprehensive examination of core properties. Both Craze and Elkins also emphasized that proper use and interpretation of pressure transient tests in carbonates frequently require a longer time than is needed in the case of sandstones.

Matthies[51] presented results of interference tests between wells on 80-acre spacing in the Permian (Wolfcamp Age) limestone of the North Anderson Ranch Field of New Mexico. These results showed that the conventional unsteady-state flow equation was applicable to this carbonate reservoir. This equation, derived in reservoir engineering textbooks (e.g., Craft and Hawkins,[30] pp. 309–318), is as follows:

$$p_i - p = \frac{-q_o \mu_o B_o}{14.16\,kh}\, Ei - \left(\frac{r^2 \mu_o c \phi}{25.28\,kt}\right) \tag{18}$$

where p = formation pressure at distance r and time t (psi), r = distance between test wells (ft), k = effective permeability (d), and the other terms are the same as in Eq. 17.

In addition to some points of fundamental agreement with sandstone reservoir behavior during primary production, there are several unique features that are characteristic mainly of carbonate reservoirs. These features include the generally low porosity and the probable directionality of the natural fracture system, as discussed above. Another unique feature occurs during the invasion of a fractured reservoir by water because of a gradually rising water level. The rise of water in the fractures is accompanied by water imbibition and countercurrent oil production in the matrix blocks lying between the fractures. (See Aronofsky *et al.*[52] and Parsons and Chaney.[53])

A model study was presented by Jones-Parra and Reytor[54] for a fractured carbonate rock producing by solution gas drive. The porosities of the reservoir were divided into two broad types in accordance with their assumed effects on fluid distribution and flow. Results of the model study showed that under certain conditions a reservoir of this type would behave quite differently from a more homogeneous sandstone reservoir. Figure 26 shows the model

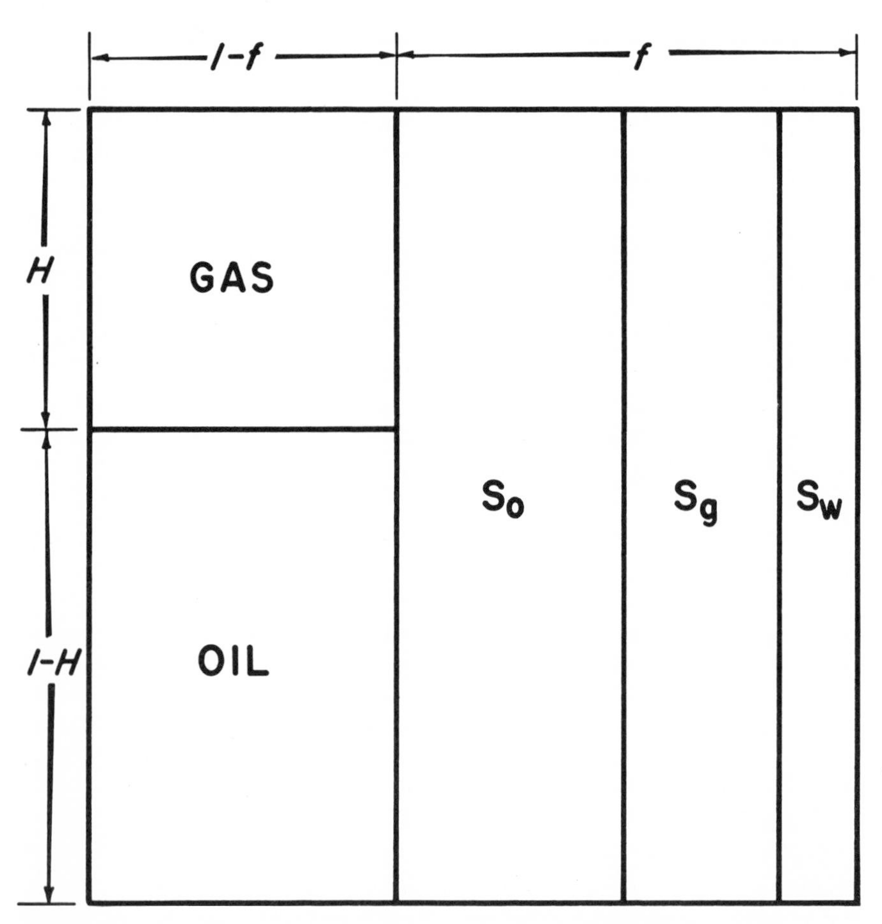

FIG. 26. Model of carbonate reservoir. (After Jones-Parra and Reytor,[54] courtesy of AIME.) f = fraction of total pore volume where gravity segregation does not take place; H = fraction of coarse porosity occupied by gas.

used. The coarse porosity is presented on the left side, where gravity segregation is believed to take place freely and the resistance to flow is very low. The fine porosity is presented on the right-hand side, where there is a high resistance to flow with relative permeability characteristics similar to those of tight sandstone. Gravity segregation does not occur here. Using the assumptions of the model, it is possible to recover more oil by producing at high rather than at low gas/oil ratios. In this manner the fine porosity is drained more effectively. Overall production declines less when producing at the higher gas/oil ratios in spite of the fact that at any given stage of depletion the pressures are lower. Needless to say, the validity of these conclusions depends as in all model studies, on how well the particular model used represents the actual reservoir situation. Inasmuch as the matrix porosity and fracture network may exist as an integral system, the model may be an oversimplification of the complex situation existing in many carbonate rock reservoirs.

The results of Jones-Parra and Reytor appear to contradict the inferences drawn by Elkins[31] on the basis of observations of a limited number of carbonate reservoirs. Elkins believed that the faster production rate in the Harper-San Andres Pool probably caused less favorable recovery characteristics than did the slower production rate in the Penwell-San Andres Field (Curves 8 and 9, Fig. 18).

Secondary Recovery

Initiation of Waterflooding

Fractures help the drainage of solution gas-drive reservoirs during primary production, when the natural energy of the reservoir is expelling oil from the rock. The situation changes decisively during conventional waterflooding operations, however, when these fractures become potential avenues for the driving fluid to bypass large volumes of the oil contained in the matrix of the reservoir. The two situations where imbibition is not effective are illustrated in Fig. 27. Tarr and Heuer[11] point out that the "presence of fractures should make one cautious about early water injection and even tend toward late water injection on the theory that we are certain of our primary production while secondary recovery may be speculative. We always favor pilot waterfloods in fractured reservoirs just to evaluate the effect of fractures."

In some carbonate reservoirs which contain large gas caps underlain by comparatively small oil zones or have good gravity drainage or strong natural water drives, the answer to the question "When should waterflooding be started?" may be "Never". There may be exceptions, but it would seem prudent to proceed with a great deal of caution under these circumstances.

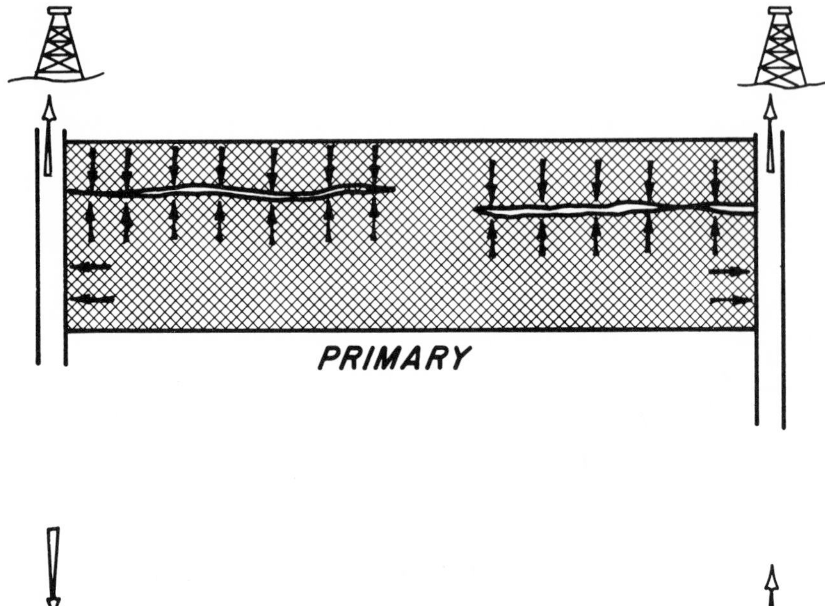

FIG. 27. Effect of fractures on primary and secondary recovery, in cases where imbibition is not effective.

k_w/k_o Data

Common to most waterflood prediction methods is the evaluation of residual oil saturation behind the waterflood front with the aid of the fractional flow equation (Buckley and Leverett[111] and Pirson,[59] pp. 555–606), which in its simplified form is as follows:

$$f_w = \frac{1}{1+(k_{ro}\mu_w/k_{rw}\mu_o)} \quad \text{or} \quad f_w = \frac{1}{1+(k_o\mu_w/k_w\mu_o)} \tag{19}$$

where f_w = fractional flow of water at the producing end of the system. The two equations presented above (19) are identical as long as relative permeabilities k_{ro} and k_{rw} are referred to the *same* base permeability, for example,

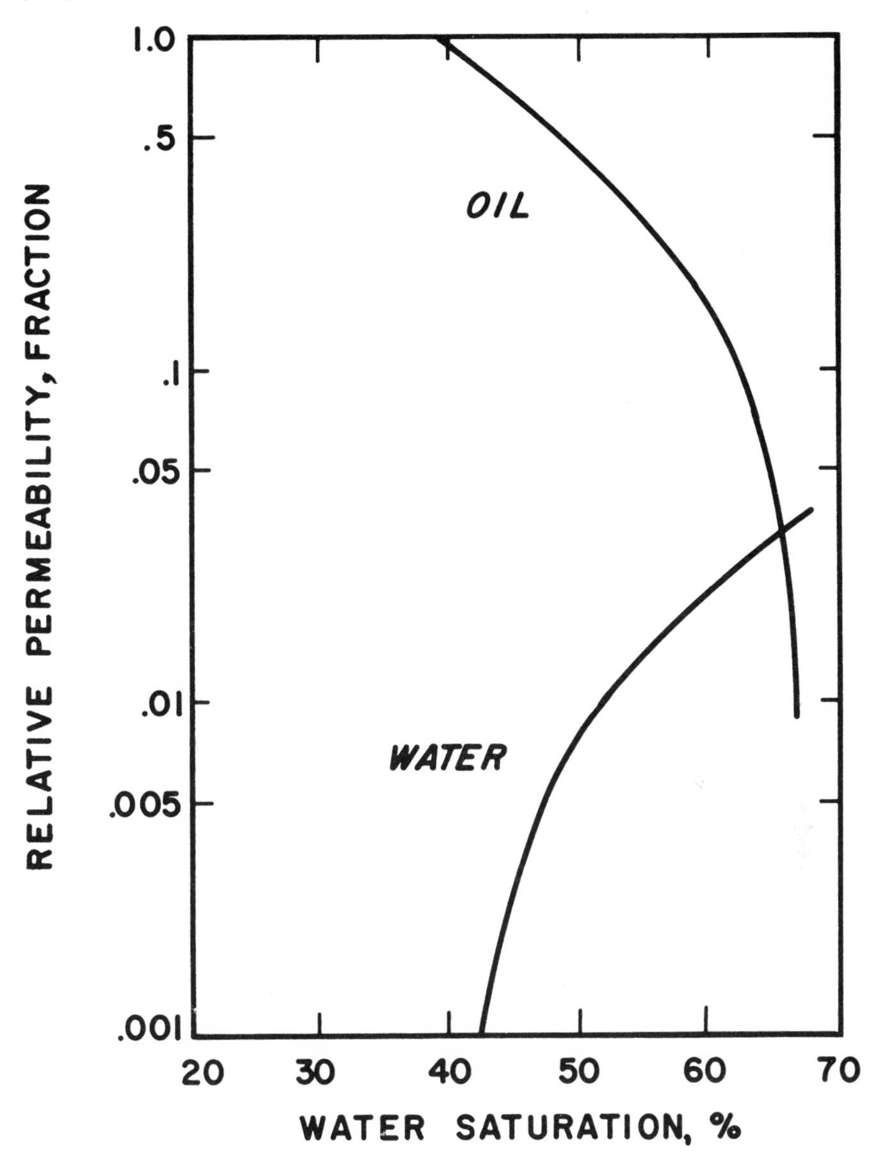

FIG. 28. Average oil and water relative permeabilities, Brown and White Dolomite, Panhandle Field, Texas. (After Abernathy,[94] courtesy of AIME.)

k_{air} or k_o in the presence of irreducible water saturation. Typical k_{ro} and k_{rw} versus saturation curves determined in the laboratory are shown in Figs. 28–30. There is a wide range of "end point" k_{rw} values, extending from 0.04 (Fig. 28) to 0.52 (Fig. 30). It is these "end point" values that govern

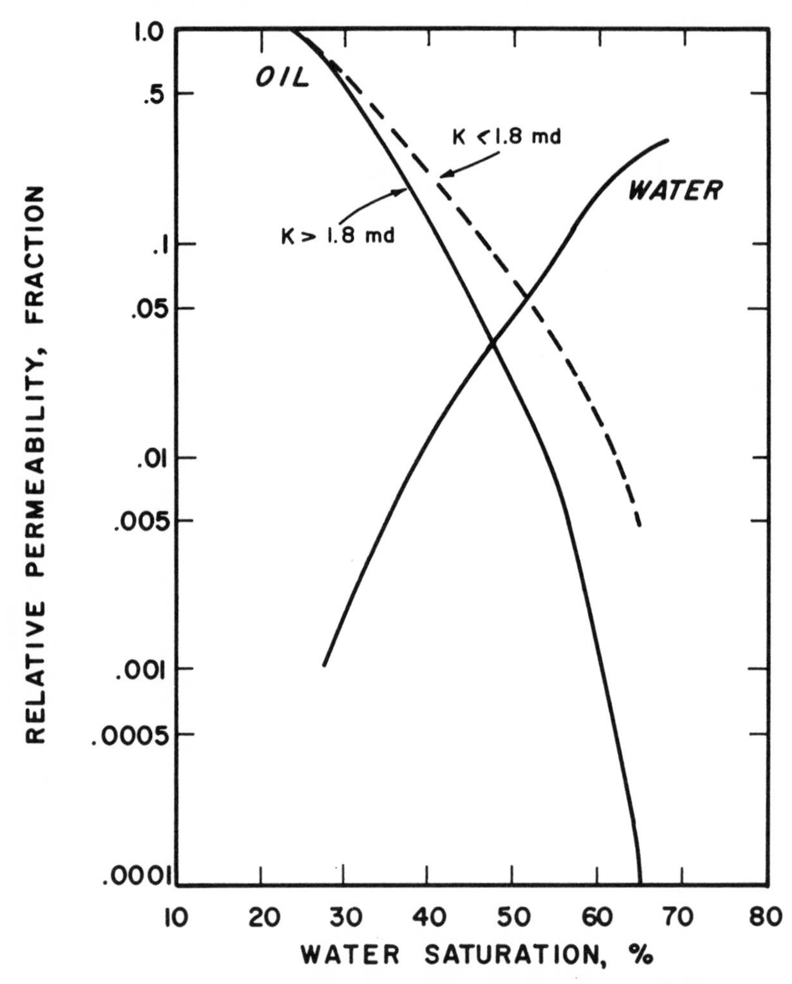

FIG. 29. Average oil and water relative permeabilities, Grayburg-Brown Dolomite, Foster Field, Texas. (After Abernathy,[94] courtesy of AIME.)

injectivities into the matrix of the rock. There is a shift of k_{ro} values with decreasing specific permeability in Fig. 29. In this specific case no corresponding change in the k_{rw} curve was noted.

A typical k_o/k_w ratio curve is presented in Fig. 31. Hendrickson[93] suggests that, if complete relative permeability curves are not available, one should let the k_o/k_w ratio equal 1,000 at connate water saturation and 0.01 at residual oil saturation, connect the two points with a straight line as shown by the dashed line in Fig. 31, and use the resulting data in the fractional flow Eq.

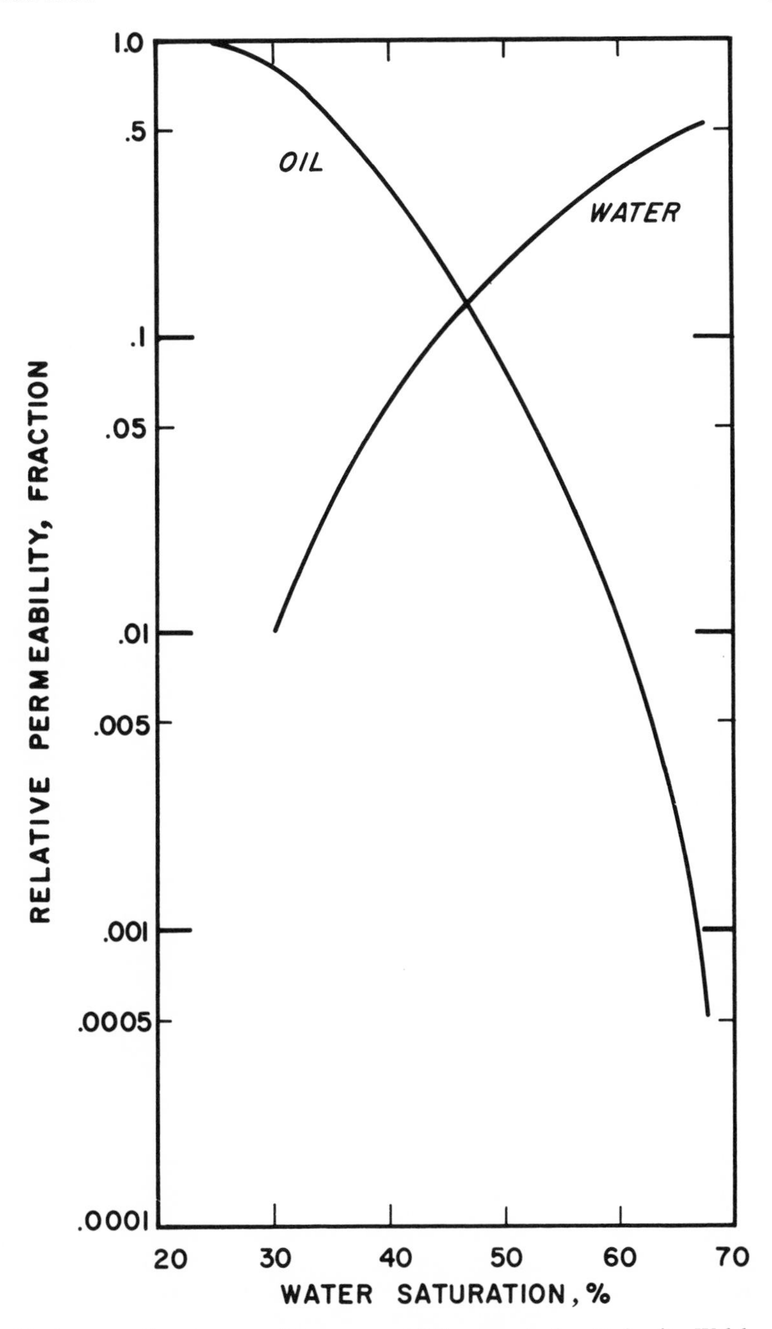

FIG. 30. Average oil and water relative permeabilities, San Andres Dolomite, Welch Field, Texas. (After Abernathy,[94] courtesy of AIME.)

19. It appears that very useful information can be gained from laboratory measurements that merely determine the "end points" of oil-water relative permeability curves.

Waterflood Recovery Predictions

The reader should not be misled by the apparent preponderance of evidence in the literature indicating good agreement between prediction and actual performance in carbonate rock waterfloods. Most authors are eager to present data that show good agreement between theory and performance but are less ready to report discrepancies. Although there are many cases in which prediction fails to match performance in carbonate waterfloods, comparatively few such instances have been reported in the literature.

The presence of fractures often causes total failures in conventional water injection projects. In some cases the resulting oil production is lower than the value extrapolated from the primary production trend. Willingham and McCaleb[55] reported a field case history for an area of the Permian Phosphoria Dolomite reservoir of the Cottonwood Creek Field, Wyoming, which contained wells that were very prolific during primary production because of a system of high-capacity fractures. The same area failed to yield secondary oil by conventional peripheral waterflooding because the injected water mostly flooded the fractures and displaced very little oil from the matrix portion of the reservoir rock. Elkins[1] described the failure of a conventional peripheral waterflood in the Bois d'Arc sequence of the Hunton Limestone of the West Edmond Field, Oklahoma. This failure was largely attributed to severe channeling of water through an extensive interconnected system of fractures and solution channels. Unpublished data for an area in a Pennsylvanian carbonate reservoir likewise indicated a waterflood failure under a combined peripheral and pattern flood. The area produced 11,000 bbl of oil per month and no water before water injection. Five months after the start of the flood, water broke through in the producing wells. Ten months later oil production declined to 1,100 bbl/month, which was equal to about 12% of the extrapolated primary trend, accompanied by a 97% water cut.

It is interesting to note that extensive waterflood predictions for this Pennsylvanian carbonate rock reservoir were based in part on permeability measurements made on several thousand feet of large (full-diameter) cores. These core analysis data were used in waterflood prediction calculations using the methods of Stiles[56] and of Dykstra and Parsons,[57] which were modified for resaturation of the gas space in each individual reservoir layer. With each method, the reservoir model used failed to represent the reservoir rock and its fracture system perfectly, but it was hoped that the data obtained by large core analyses would give at least an approximate representation of

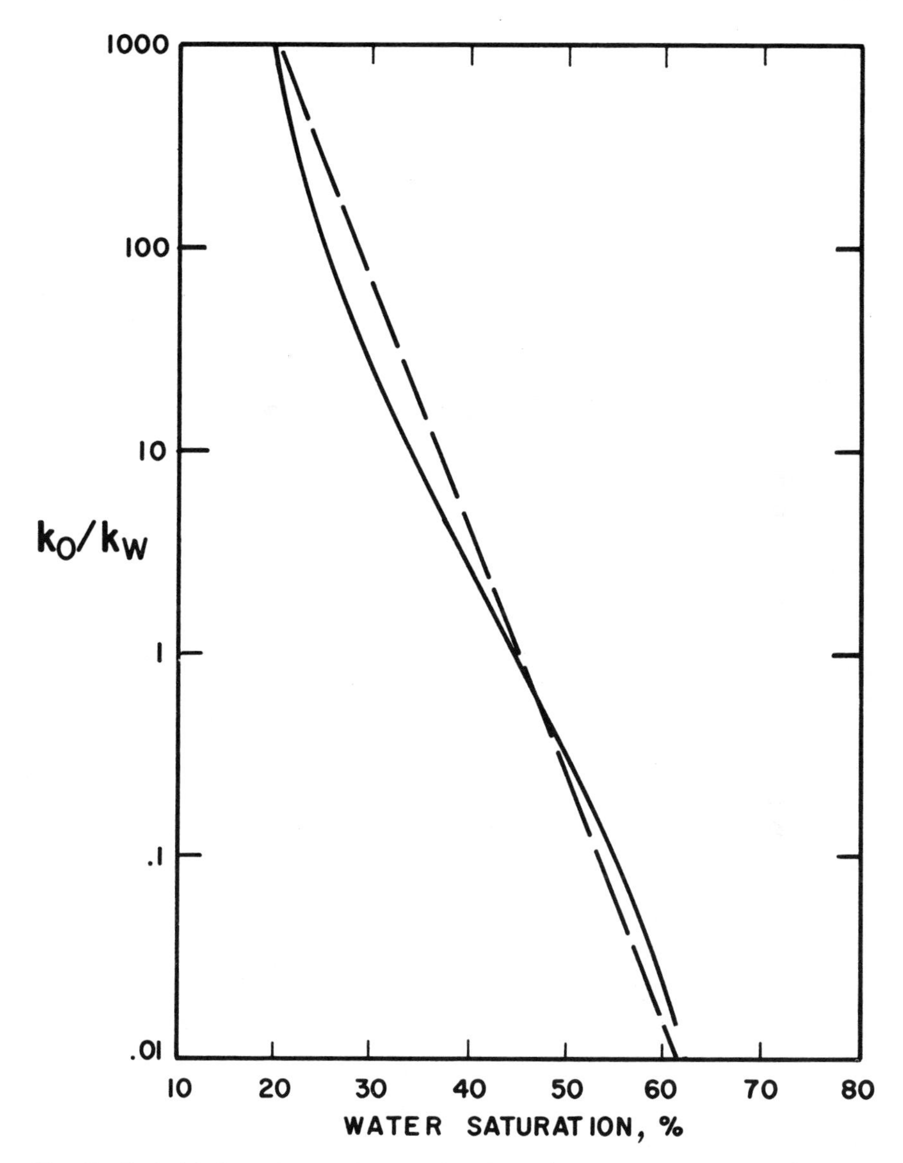

FIG. 31. Typical k_o/k_w ratios. Solid curve—based on laboratory measurements (after Tarr and Heuer[11]); dashed line—based on approximation (after Hendrickson,[93] courtesy of AIME).

the fractured rock system behavior. A large number of calculations were made with the aid of high-speed computers, and a wide range of predictive values was generated. Although it was hoped that field performance would follow at least one of the curves predicted, the actual field performance data failed to match *any* of the various predictions.

All waterflood prediction methods presented in the literature to date give overoptimistic results for reservoirs containing systems of fractures and solution channels. Most reservoirs containing such systems may not be amenable to conventional waterflooding, although there is some hope that the recognition and use of directional permeability, as well as cyclic injection schemes, may result in successful waterflood operations in reservoirs of this type. The presence of fracture systems, which are so troublesome in waterflooding, may be recognized from (1) visual examination of cores,[12] and (2) observations which indicate that interwell permeabilities determined from transient pressure data are several times greater than matrix rock permeabilities determined from cores.

Waterflood performance has matched prediction for several carbonate reservoirs. The methods of prediction were not greatly different from those of Stiles[56] and Dykstra and Parsons[57] and in a sense may be considered modifications of these methods. For instance, Hendrickson[93] described a close agreement between prediction and performance in a five-spot pilot waterflood in the San Andres Dolomite of the Welch Field (Texas). The results are illustrated in Fig. 32. The flood performance was predicted by

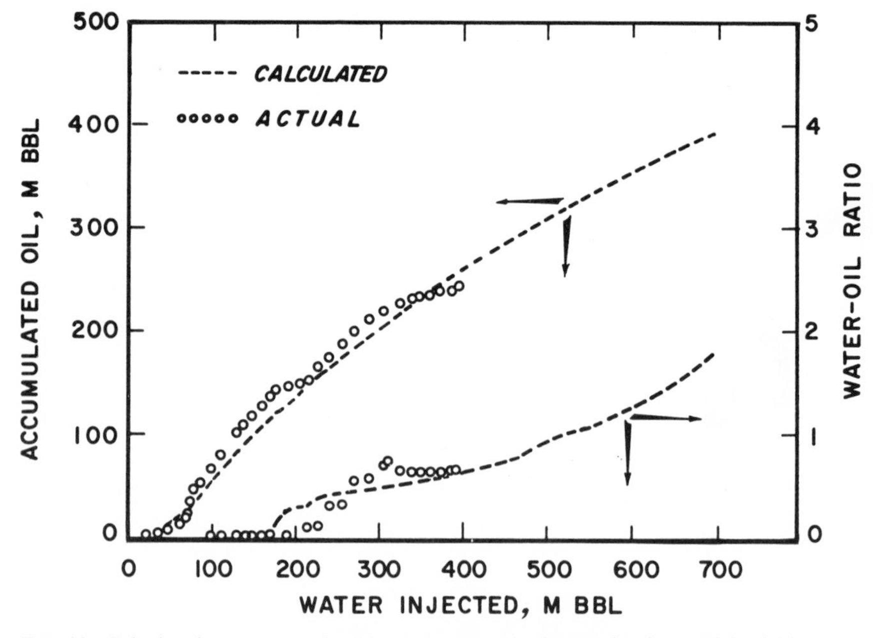

FIG. 32. Calculated versus actual performance, San Andres Dolomite, Welch Field pilot flood, Texas. (After Hendrickson,[93] courtesy of AIME.)

using the "band" concept, which combined the "layer cake" principle of Stiles with the fractional flow equation. Peculiar to the "band" concept was the fact that, after the core analysis data were arranged in descending order of permeability, the resulting array was subdivided into ten "bands" of *equal* pore volume. The performance of each "band" was calculated separately, and then the individual values were added to obtain the total predicted reservoir performance.

Abernathy[94] made additional calculations for the Welch Field and for two additional pilot floods in the reservoirs described in Table 2. Abernathy used

TABLE 2. Comparison of Waterflood Predictions and Performances in Carbonate Reservoirs

(After Abernathy,[59] courtesy of AIME.)

	Field		
	Welch	Panhandle	Foster
Formation	San Andres Dolomite	Brown and White Dolomite	Grayburg-Brown Dolomite
Average depth, ft	4,950	3,000	4,200
Net pay, ft	75	65	129
Average porosity, %	10	11.7	8.6
Average permeability (cores), md	6.3	7.2	2.6
Connate water saturation, %	23	39	23
B_o, res. bbl/STB	1.088 at 325 psi	1.044	1.061 at 244 psi
μ_o, cp	2.32 at 325 psi	2.33	2.91 at 244 psi
μ_w, cp	0.82	0.83	0.80
°API gravity	34.4	40	35
BHP at start of flood, psi	325	—	244
N_p/N at start of flood	0.03	0.256*	0.12
Current actual recovery, % oil-in-place	21.7	34.2	21
Calculated ultimate recovery under flood, % oil-in-place	38	35	28

* Includes gas injection.

three prediction methods: (1) that of Stiles,[56] (2) that of Craig *et al.*,[112] and (3) a combination of the two methods, called the Craig-Stiles method. The results shown in Figs. 33–35 indicate that the Craig-Stiles method gave the best predictions for the three fields.

The Craig-Stiles method provides for growth of the swept zone after initial water breakthrough, reduction of residual oil in the swept zone after

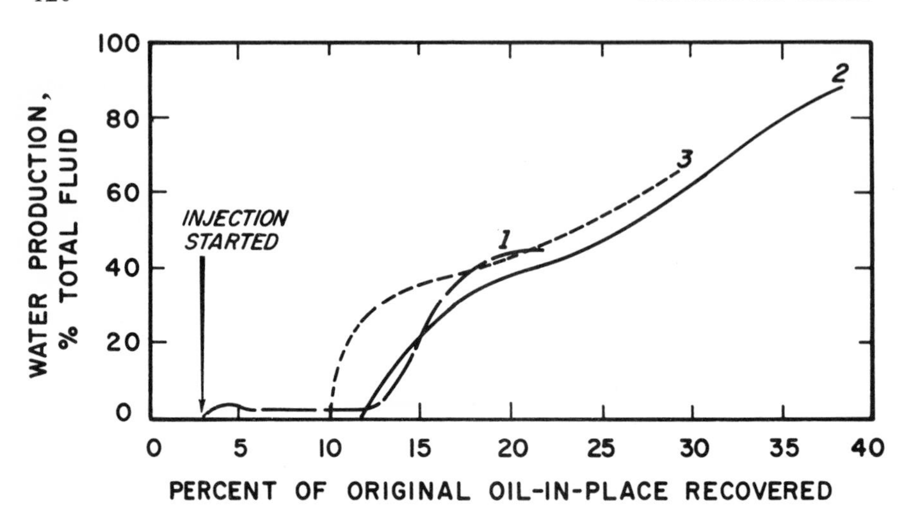

FIG. 33. Comparison of actual performance with two prediction calculations, San Andres Dolomite, Welch Field pilot flood, Texas. (After Abernathy.[94]) Legend: 1—actual, 2—calculated by Craig-Stiles method, 3—calculated by Band method. (Courtesy of AIME.)

water breakthrough, and liquid resaturation of the initial gas space (before flooding) in each layer. Oil recovery from a given layer is equal to:

$$N_p = V_p \left[(S_{wr} - S_{wi})\, E_{as} - S_{gi} \right] \frac{1}{B_o} \tag{20}$$

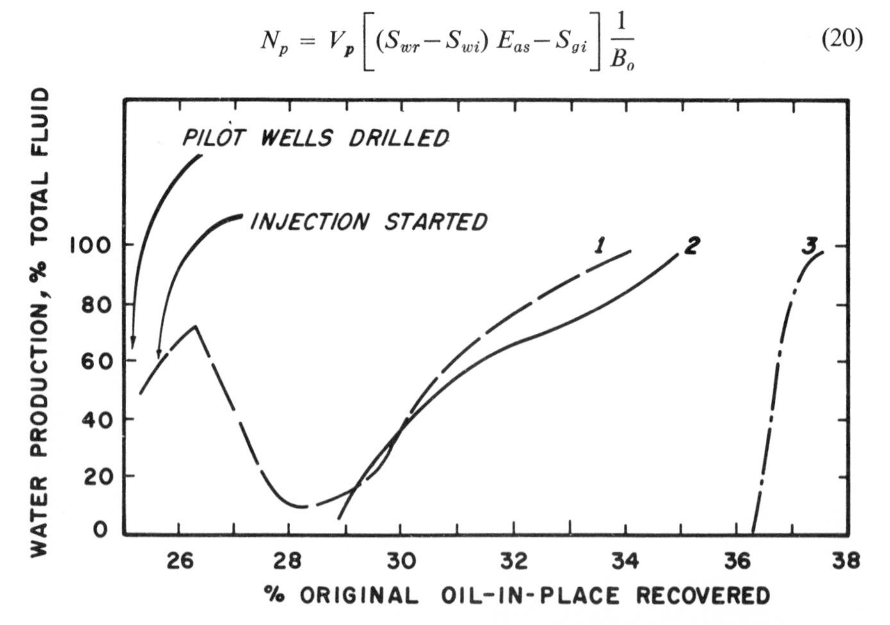

FIG. 34. Calculated versus actual performance, Brown and White Dolomite, Panhandle Field pilot flood, Texas. (After Abernathy.[94]) Legend: 1—actual, 2—calculated by Craig-Stiles method, 3—calculated by Craig *et al.* method. (Courtesy of AIME.)

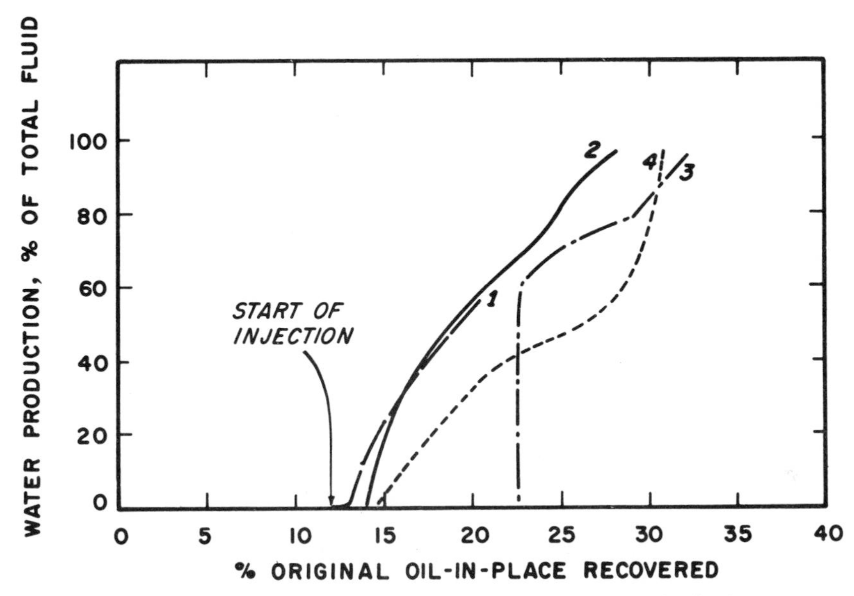

FIG. 35. Calculated versus actual performance, Grayburg-Brown Dolomite, Foster Field pilot flood, Texas. (After Abernathy.[94]) Legend: 1—actual, 2—calculated by Craig-Stiles method, 3—calculated by Craig *et al.* method, 4—calculated by Stiles method. (Courtesy of AIME.)

where V_p = pore volume (bbl), S_{wr} = final average water saturation in swept area (fraction), S_{wi} = initial water saturation at start of flood (fraction), S_{gi} = initial gas saturation at start of flood (fraction), and E_{as} = aerial sweep efficiency (fraction). After resaturation was complete (i.e., fillup had occurred), water injection was set equal to total liquid production.

In using the Craig-Stiles method, special care was taken to break the reservoirs into only a few layers so that the combined performance of the layers was representative of the performance of the whole reservoir. The layering technique of Miller and Lents,[113] which requires that sufficient core analysis data be available to describe the reservoir, was used. In this technique the reservoir is subdivided only into obviously separable subdivisions or layers (12 were used for the Welch Field and 3 each for the Foster and Panhandle Fields). The permeabilities and associated reservoir properties were evaluated for each layer from the core analysis data obtained from all wells at the same relative position in the reservoir. As in all of the methods used by Abernathy and Hendrickson, no cross flow was assumed to occur between layers. An improved method for subdividing a reservoir by statistical procedures has been presented by Testerman,[107] and a similar method for the zonation of a carbonate reservoir has recently been described by Gill.[99]

Natural water drives are generally considerably less effective in carbonate rocks than in sandstones. The reason for this difference stems largely from the greater heterogeneity of the carbonates. A look at natural water-drive recoveries is pertinent here, because these drives are believed to result in as high recoveries as those achieved through the best combination of normal primary production and waterflooding. The results of a statistical study by the API[38] of natural water drives in 72 sand and sandstone reservoirs and 39 limestone, dolomite, and other rock types are summarized in Table 3.

TABLE 3. Range of Parameters of Reservoirs with Water Drive as Predominant Recovery Mechanism*

(After Arps *et al.*,[38] courtesy of API.)

	Sand and sandstone			Limestone, dolomite, and other		
	Worst	Median	Best	Worst	Median	Best
Ultimate recovery						
Residual oil, % pore space	63.5	32.7	11.4	90.8	42.1	24.7
bbl/acre-ft	155	571	1,641	6	172	1,422
Rock parameter						
k, md	11	568	4,000	10	127	1,600
ϕ, % bulk volume	11.1	25.6	35.0	2.2	15.4	30.0
S_{wi}, % pore space	47	25	5.2	50	18	3.3

* Values shown for each separate parameter; values do not necessarily belong to same case history.

Injection Rate Predictions

All injection rate equations given in standard reservoir engineering textbooks, such as those by Muskat,[8] Calhoun,[58] Pirson,[59] and Craft and Hawkins,[30] contain permeability capacity (kh) terms, which to a very large extent control the predicted injection rates. In other words, injection calculations require knowledge of either $k_g h$ or $k_w h$. This information can, in many cases, be derived from $k_o h$ data obtained from pressure buildup tests, when the reservoir was undersaturated, and from appropriate relative permeability test data. For water injection predictions the $k_w h$ term can be evaluated as follows:

$$k_w h = k_o h \times (k'_{rw}/k'_{ro}) \tag{21}$$

where k'_{rw} = relative permeability to water at residual oil saturation, and k'_{ro} = relative permeability to oil at irreducible water saturation. Typical

k'_{rw} and k'_{ro} data are represented by the "end points" of the oil-water relative permeability curves shown in Figs. 28–30.

Gas well tests yield $k_g h$ data, but these tests are often affected by turbulence in the wellbore region. A factor of turbulence can be evaluated either from well or from laboratory tests. Gewers and Nichols[98] measured turbulence factors on cores from microvugular carbonate gas reservoirs in the Turner Valley Member of the Rundle Formation (Mississippian) of western Canada and obtained values that were as much as one order of magnitude higher than previously reported data for nonvuggy rocks.

The best approach for obtaining $k_w h$ data is to use results of injectivity tests and/or subsequent pressure falloff tests. Care must be taken during the injectivity tests that the formation is not being fractured or "parted", thereby producing an erroneously high permeability capacity value. Creation of new fractures in the formation can be evaluated with the aid of "step rate injectivity" tests.[60] In these tests injection rates are increased stepwise and plotted against $(p_w - p_e)$, where p_w = bottom hole injection pressure and p_e = shut-in formation pressure. It can be assumed that no new fractures are being opened as long as the curve is a straight line *and* goes through the origin of a Cartesian plot.

Fracturing or "pressure parting" pressures can also be predicted and evaluated by means of the fracturing equation presented by Hubbert and Willis[61] and Eaton[62]:

$$\frac{p_f}{D} = \left(\frac{p_o}{D} - \frac{p_e}{D}\right)\left(\frac{\nu}{1-\nu}\right) + \frac{p_e}{D} \tag{22}$$

where p_f = bottom hole pressure that causes fracture propagation (psi), p_o = overburden pressure (psi), D = depth (ft), and ν = Poisson's ratio. A plot indicating observed fracture gradients and Poisson's ratios for a Mississippian limestone reef and a Permian dolomite reservoir is shown in in Fig. 36.

Analysis of pressure falloff tests is described in detail by Matthews and Russell[13] (pp. 72–83). Of the methods described, the one developed by Hazebroek *et al.*[63] appears to apply well to carbonate reservoirs. This method yields a $k_w h$ value and also a dimensionless skin factor, s, which is defined as follows:

$$s + \ln r_e/r_w = 0.00708(p_w - p_e)k_w h/i\mu_w \tag{23}$$

where r_e = external boundary radius (ft), r_w = wellbore radius (ft), and i = stabilized injection rate (B/D) before pressure falloff testing. If the skin factor is negative, one may assume that r_w is in effect larger than the originally assumed value based on hole size. The effective well-bore radius, r_{we}, which

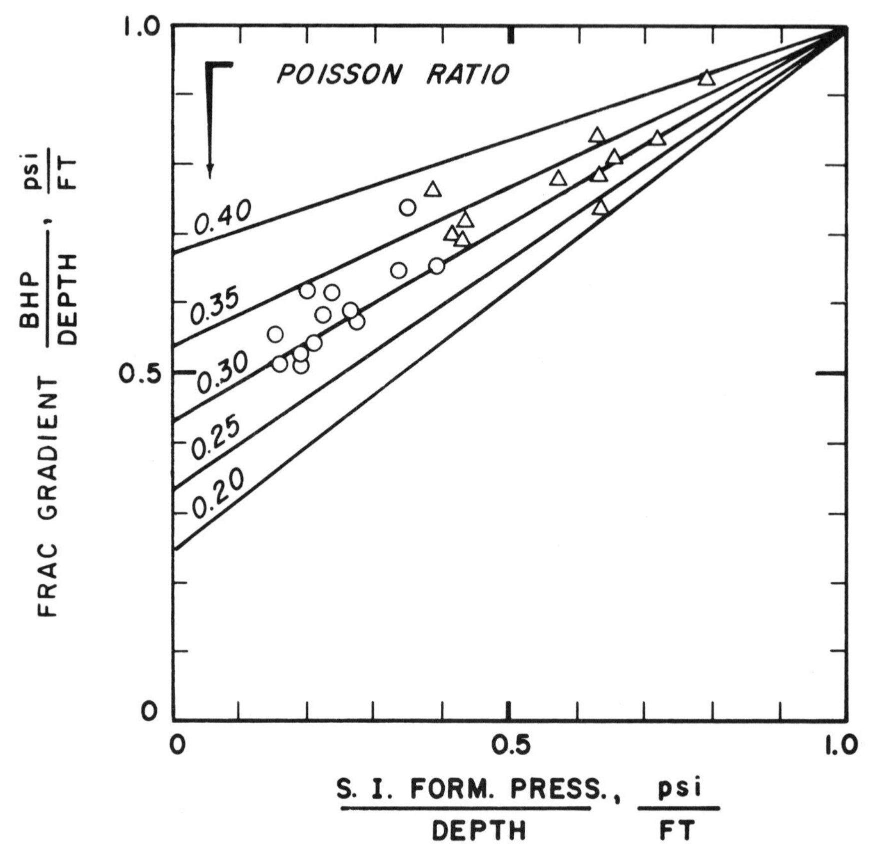

FIG. 36. Observed fracture pressure gradients and Poisson's ratios as a function of shut-in formation pressure/depth ratios. o Mississippian limestone reef, at 6,500-ft depth; Δ Permian dolomite reservoir, at 3,900-ft depth. (Courtesy of Continental Oil Co.)

later must be substituted into all injectivity prediction equations, can then be evaluated by setting $s = 0$. This results in the following equation:

$$r_{we} = r_e e^{-a} \tag{24}$$

where $a = 0.00708(p_w - p_e)k_w h / i\mu_w$.

A practical method for evaluating the true matrix permeability of a formation has been described by Clark.[64] This method has been found to be useful for evaluating the properties of carbonate rocks surrounding water injection wells. Millheim and Cichowicz[65] described a similar method for low-permeability fractured gas wells. Both methods first evaluate the effects of possible induced fractures and then separate this effect from the contribution of the rock matrix. Adams *et al.*[66] described transient pressure tests in a complex gas reservoir, indicating a tight dolomite matrix near the wellbore and a higher matrix permeability at some distance from the well.

The Importance of Geology

One of the keys to successful secondary operations is a thorough understanding of reservoir geology. As explained earlier, directional permeability trends and fracture orientation can be determined during initial field development and primary production. Such data are of great value in the design ot injection-well patterns.

Figure 37 shows a reservoir that had an east-west preferential fracture orientation, which was not recognized during primary production. Infill wells were drilled in the original well pattern to accommodate a pilot flood. This pilot flood gave excellent results (high sweep efficiencies) because

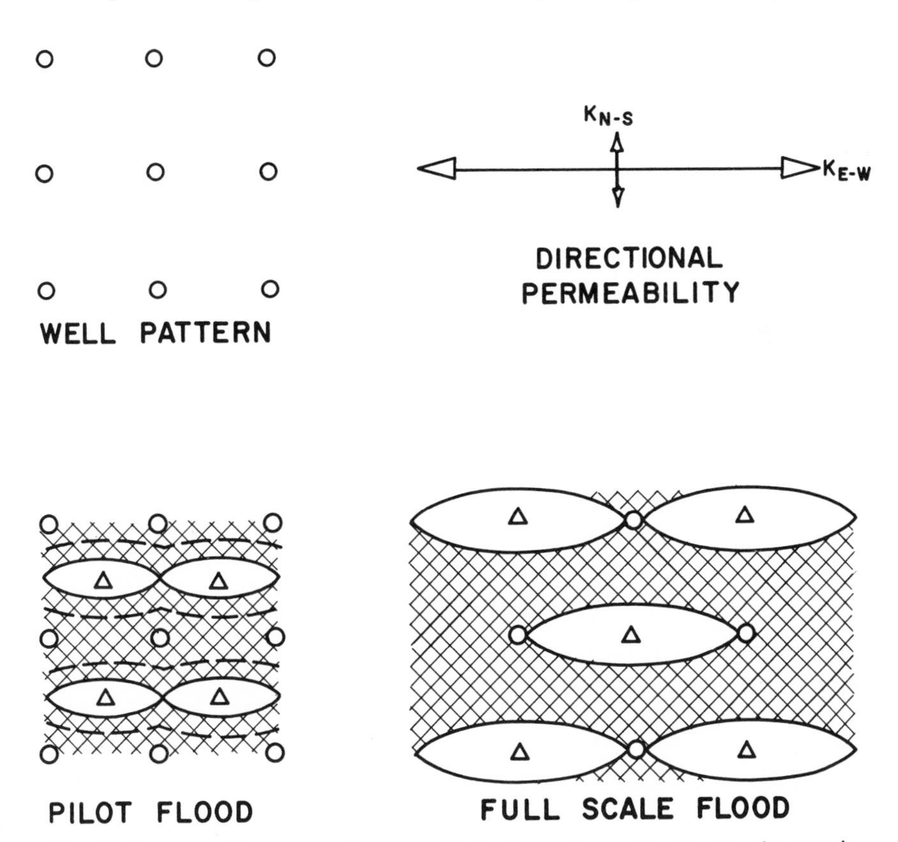

FIG. 37. Proper and improper alignment of injection pattern with directional reservoir trends.

injection wells happened to be lined up parallel to the preferential fracture orientation. When a full-scale flood was instituted, using the wider initial well spacing, however, it was discovered that the injection wells were aligned improperly with the fracture trend. The pattern resulted in an early water breakthrough and a poor sweep efficiency. As soon as the preferential fracture orientation was recognized, the injection well pattern in the full-scale flood was realigned. This saved the project from what appeared to be certain failure and turned it into a commercial success.

A *moderate* degree of induced fracturing is usually needed in low-permeability carbonate reservoirs in order to complete a waterflood in a reasonable length of time. The degree of fracturing should be kept under control and analyzed periodically with the aid of pressure falloff analyses such as the one described by Clark.[64] If fracturing or pressure parting during water injection is allowed to proceed without restraint and control, excessively long fractures may form, which, in turn, can lead to early water breakthrough. Fracturing of the reservoir cap rock could also conceivably result, causing loss of oil and injection fluid.

Induced fractures in tectonically relaxed areas, characterized by normal faulting, generally have a preferred orientation. As pointed out earlier, it is extremely important to recognize this directionality and to make the flood pattern take advantage of it. Fracture and permeability directionality can be ascertained by specially designed pilot waterflood tests (Carnes[67] and Lane[68]); evaluation of gas injection tests (Armstrong *et al.*,[69] Crawford,[70] and Alpay[43]); interference tests (Elkins and Skov[44] and Kunkel and Bagley[71]); tracer flow tests (Baldwin[72]); impression packers (Fraser and Pettitt[73] and Anderson and Stahl[74]); and/or downhole acoustical measurements (Zemanek *et al.*[75]). A refinement of the interference testing technique, called "pulse testing", has also been applied successfully by McKinley *et al.*[76] to a highly permeable (k_{air} = 1,200 md) dolomitic limestone reservoir in order to define interwell properties. This method shows promise of making it possible to locate directional permeabilities and flow barriers in some fields.

The use of statistical methods for analyzing porosity trends in a carbonate reservoir has been described by Campa and Romero.[114] Knowledge of directionally oriented reservoir trends of this type facilitate the optimum design of a secondary recovery injection well pattern.

It is a common experience that the performance of individual wells varies widely in a carbonate reservoir. Figure 38 shows the directions of pressure interference and the locations of barriers noted in a low-permeability ($k_{air} \approx 5$ md) limestone reef reservoir. The interwell relationships indicated in this graph were established in a series of interference tests lasting several months. Obviously, the chances of successful secondary recovery operations in complex situations of this type will be enhanced by close cooperation

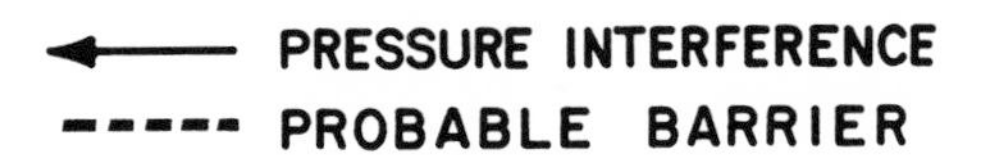

FIG. 38. Pressure interference and flow barriers observed from interference tests in a portion of a limestone reef reservoir. (Courtesy of Continental Oil Co.)

between engineers and geologists. Excellent examples of such cooperation have been reported. For instance, Willingham and McCaleb[55] noted extensive lithofacies changes and variations in fracture frequency in the Permian Phosphoria Reservoir of the Cottonwood Creek Field, Wyoming. Their review of the engineering and geologic data led to beneficial changes in secondary recovery operations, causing an increase from 2,100 to 3,600 B/D of oil during the first year of revised operating practices.

In another example, Wayhan and McCaleb[77] reported a review of the geology and secondary recovery performance of the Mississippian Madison Limestone in the Elk Basin Field of Wyoming and Montana. Included in the geological portion of the study were geochemical, isotopic, and X-ray analyses, paleontological studies, and a study of sedimentary petrology, all designed to determine the depositional environment of the reservoir rock and its subsequent diagenesis. These were supplemented by an expanded suite of logs (run on new wells) that included gamma-ray, neutron, sonic, density, laterolog, and microlog. As a result of these studies, a new concept of the reservoir was developed, which after changes in the injection scheme and additional drilling produced an extra 5,400,000 bbl of oil.

Dowling[118] reported geologic studies conducted to improve secondary recovery operations in a multi-zone Permian dolomite field (Monahans Clearfork Field, Ward and Winkler counties, Texas). Application of carbonate environmental concepts led to better definition of field limits, quality and continuity in each zone, including delineation of potential productive zones that were either cased off or not penetrated as a result of early evaluation techniques.

Miscible Flooding

Unique opportunities for very efficient vertical miscible displacement drives occur in pinnacle reef pools, for example, in western Canada.[115–117] A solvent bank rich in LPG (ethane through butane) is generally spotted at the existing gas-oil contact and then driven downward by dry gas. Recoveries ranging as high as 96% of the oil initially in place are claimed for these processes.

Cyclic Flooding

The theory behind cyclic flooding is to induce oil to move out from the rock matrix by forcing injection fluid, either water or gas and water, into the rock and thus raising the reservoir pressure, followed by a depressuring phase and production from the matrix. Cycles can be repeated several times. During the depressuring phase of the "cycle", water is retained by capillary attraction, and oil is released into natural fractures or high-permeability zones, whence it can flow into producing wells. The cyclic process is intended to apply to reservoirs which are difficult to waterflood by conventional methods.

Laboratory data and theoretical work supporting the concept of cyclic flooding have been presented by Owens and Archer,[78] Bokserman and Shalimov,[79] Tatashev,[80] Felsenthal and Ferrell,[81] and Felsenthal *et al.*[82] The process is closely related to the imbibition process (Brownscombe and

Dyes,[83] Aronofsky *et al.*,[52] Graham and Richardson,[84] Mattax and Kyte,[85] Raimondi and Torcasso,[86] Blair,[87] and Parsons and Chaney[53]).

Field data on the use of cyclic flooding in the Spraberry Field, Texas, were presented by Elkins and Skov.[88] In this field, injection wells had been lined up carefully with the fracture trend. The spacing of fractures ranged from a few inches to a few feet. Matrix permeability was less than 1 md, and porosity was in the 8–15% range. During secondary recovery, high injection rates were used to repressure the reservoir to above the bubble point pressure. Then injection was stopped. Within a few days oil production rose substantially and then declined gradually. The pressure phase lasted 1 month, and the depressuring phase about 7 months. This process was repeated three times, causing substantial oil rate increases, but each cycle produced less oil than the preceding one.

Only one pressure cycle was successful in extremely tight Cretaceous carbonate formations, the Austin Chalk and the Buda Limestone in the Salt Flat Field, Texas. These formations are characterized by a tight (k_{air} = 0.1 md, ϕ = 12%) matrix. The low permeability shown by cores was confirmed by well performance. Hester *et al.*[89] described a fairly successful test in one well that yielded 10,000 bbl of oil after water had been injected into it at pressures just below the fracturing pressure. Only one third of the injection water was produced back. A second cycle essentially failed to produce any oil, probably because of the low formation permeability and the lack of expulsive energy to propel oil to the wellbore during the depressuring phase of the cycle.

Willingham and McCaleb[55] and Elkins[1] reported somewhat more successful applications of the injection-production scheme just described. The first instance occurred in a highly fractured portion of the Permian Phosphoria Formation at Cottonwood Creek, Wyoming. Twelve former water injection wells, after a cumulative injection of 25.4 million bbls of water, were put on production and started to yield oil after 2–3 weeks. After 1 year they were producing 1,850 B/D of oil with a 67% water cut. The second instance occurred in the heterogeneous Bois d'Arc Limestone of the West Edmond Field, Oklahoma, where a former injection well, after receiving 1 million bbl of water, produced 131,500 bbl of "cyclic" oil. In both instances, conventional waterfloods had been tried unsuccessfully in the areas surrounding these wells.

Cyclic water injection projects have also been used successfully in sandstone reservoirs for which inefficient sweep was expected by conventional waterflooding (Bleakley[90] and Maslyantsev *et al.*[91]). Successful cyclic flooding in a highly permeable fractured carbonate rock was described by Vasilechko *et al.*[92] The injection-production scheme in the latter project was similar to the one reported for the Spraberry Field in Texas. During the water injection phase, which lasted 1.5–2 months, formation pressure was increased. Injection was then stopped for 1–1.5 months to allow depressuring of the reservoir. Contrary to the Spraberry, the reservoir of the A_4 Formation in the Pokrovsk Field, USSR, had a high permeability and porosity (k = 1,065 md, ϕ = 23%). Cyclic flooding was started after conventional waterflooding produced oil at a 75% water cut. During 8 cycles, about 1.5 million extra bbl of oil were recovered. Production declined with each successive cycle. Vasilechko *et al.* concluded that cyclic waterflooding should be effective in other reservoirs as well.

Summary

The information contained in this chapter, as well as supplemental literature data,[93–108] indicates that carbonate rocks are considerably more heterogeneous than the generally homogeneous sandstones for which most of our classical fluid flow concepts were developed.[109, 110] This heterogeneity stems largely from a combination of a low-permeability matrix and openings of supercapillary size, such as vugs, fractures, fissures, and solution channels. Large void spaces may also owe their origin to the presence of oolites, pisolites, and stylolites.

The heterogeneity of carbonates often causes great differences in the performance of wells in the same reservoir. One of the keys that help to unlock the mystery of these reservoirs is a thorough understanding of the geology, with respect to both depositional and postdepositional (diagenetic) events. A good understanding of the distribution of the oil initially in place requires a knowledge of the manner in which oil entered the reservoir during geologic time. Furthermore, a knowledge of regional geologic stresses is important for delineating preferred fracture orientations, which in turn play a crucial role in the selection of optimum patterns for secondary recovery operations.

Primary recovery by solution gas drive appears to be aided by the presence of natural fractures. These fractures act as drainage channels and are conducive to favorably high oil recoveries on a percentage of pore space basis. It should be emphasized, however, that primary recoveries on a *barrels per acre-foot* basis are generally significantly lower in carbonates than in sandstones. This difference can be attributed to the fact that carbonates tend to have lower porosities and fewer barrels of oil initially in place per bulk reservoir volume than do sandstones.

The fractures and other supercapillary openings that appear to be so helpful in primary recovery are a source of potential trouble during secondary recovery. Fractures and fracture-like features frequently act as channels for the early breakthrough of injected fluids, making secondary recovery operations more hazardous in carbonates than in sandstones.

In some carbonate reservoirs secondary recovery operations have performed as predicted by techniques based largely on sandstone reservoir models, but in many cases actual recoveries have fallen far short of the values predicted. Secondary recovery operations in these difficult-to-flood reservoirs can be optimized by (1) paying close attention to geologic features in the design of the injection pattern, (2) providing optimum stimulation of producing wells, and (3) keeping injection pressures under careful control so as to avoid the creation of excessively long induced fractures. There is also a possibility that recovery can be enhanced by using intermittent injection "cycles", a technique that has been reported to be effective in a number of fields in the United States and other countries.

References

1. Elkins, L. F.: "Internal Anatomy of a Tight, Fractured Hunton Lime Reservoir Revealed by Performance—West Edmond Field", *J. Pet. Tech.* (1969) Vol. 21, 221–232.
2. Littlefield, M., Gray, L. L. and Godbold, A. C.: "A Reservoir Study of the West Edmond Hunton Pool, Oklahoma", *Trans.*, AIME (1948) Vol. 174, 131–164.
3. Bulnes, A. C. and Fitting, R. U., Jr.: "An Introductory Discussion of the Reservoir Performance of Limestone Formations", *Trans.*, AIME (1945) Vol. 160, 179–201.
4. Craze, R. C.: "Performance of Limestone Reservoirs", *Trans.*, AIME (1950) Vol. 189, 287–294.
5. Daniel, E. J.: "Fractured Reservoirs of Middle East", *Bull.*, AAPG (1954) Vol. 38, 774–815.
6. Huitt, J. L.: "Fluid Flow in Simulated Fractures", *AIChE J.* (1956) Vol. 2 (2), 259–264.
7. Parsons, R. W.: "Permeability of Idealized Fractured Rock", *Trans.*, AIME (1966) Vol. 237, II, 126–136.
8. Muskat, M.: *Physical Principles of Oil Production*, McGraw-Hill Book Co., New York (1949) 922 pp.
9. Kelton, F. C.: "Analysis of Fractured Limestone Cores", *Trans.*, AIME (1950) Vol. 189, 225–234.
10. Warren, J. E., Skiba, F. F. and Price, H. S.: "An Evaluation of the Significance of Permeability Measurements", *J. Pet. Tech.* (1961) Vol. 13, 739–744.
11. Tarr, C. M. and Heuer, G. J.: "Factors Influencing the Optimum Time to Start Water Injection", paper SPE 340 presented at SPE Fifth Biennial Secondary Recovery Symp., Wichita Falls, Tex. (1962) 10 pp.
12. Sangree, J. B.: "What You Should Know to Analyze Core Fractures", *World Oil* (1969) Vol. 168 (5), 69–72.
13. Matthews, C. S. and Russell, D. G.: *Pressure Buildup and Flow Tests in Wells*, Soc. Pet. Engs. AIME Monograph 1, Dallas (1967) 167 pp.
14. Pollard, P.: "Evaluation of Acid Treatments from Pressure Build-up Analysis", *Trans.*, AIME (1959) Vol. 216, 38–43.
15. Warren, J. E. and Root, P. J.: "The Behavior of Naturally Fractured Reservoirs", *Trans.*, AIME (1963) Vol. 228, II, 245–255.
16. Russell, D. G. and Truitt, N. E.: "Transient Pressure Behavior in Vertically Fractured Reservoirs", *Trans.*, AIME (1964) Vol. 231, 1159–1170.
17. Odeh, A. S.: "Unsteady-State Behavior of Naturally Fractured Reservoirs", *Trans.*, AIME (1965) Vol. 234, II, 60–66.

18. Marchant, L. C. and White, E. J.: "Comparison of Log and Core Analysis Results for an Extremely Heterogeneous Carbonate Reservoir", paper presented at SPWLA Ninth Annual Logging Symp., New Orleans, La. (1968) 16 pp.
19. Wyllie, M. R. J.: "Reservoir Mechanics—Stylized Myth or Potential Science", *J. Pet. Tech.* (1962) Vol. 14, 583–588.
20. Knutson, C. F.: "Definition of Water Table", *Bull.*, AAPG (1954) Vol. 38, 2020–2027.
21. Brown, H. W.: "Capillary Pressure Investigations", *Trans.*, AIME (1951) Vol. 192, 67–74.
22. Leverett, M. C. "Capillary Behavior in Porous Solids", *Trans.*, AIME (1941) Vol. 142, 152–169.
23. Poiseuille, J.: "Recherches Expérimentales sur le Mouvement des Liquides dans les Tubes de Très Petits Diamètres", *Mém. Savants Etrangers* (1846) Vol. 9, 543.
24. Darcy, H.: *Les Fountaines Publiques de la Ville de Dijon*, Victor Dalmont, Paris (1856).
25. Kozeny, J.: "Über Kapillare Leitung des Wassers im Boden", *Sitzber. Akad. Wiss, Wien, Math.-naturw. kl.* (1927) Vol. 136–2A, 271.
26. Ashford, F. E.: "Computed Relative Permeability, Drainage and Imbibition", paper SPE 2582 presented at SPE 44th Annual Fall Meeting, Denver, Colo. (1969.)
27. Aufricht, W. R. and Koepf, E. H.: "The Interpretation of Capillary Pressure Data from Carbonate Reservoirs", *Trans.*, AIME (1957) Vol. 210, 402–405.
28. Rockwood, S. H., Lair, G. H. and Langford, B. J.: "Reservoir Volumetric Parameters Defined by Capillary Pressure Studies", *Trans.*, AIME (1957) Vol. 210, 252–259.
29. Messer, E. S.: "Interstitial Water Determination by an Evaporation Method", *Trans.*, AIME (1951) Vol. 192, 269–274.
30. Craft, B. C. and Hawkins, M. F.: *Applied Petroleum Reservoir Engineering*, Prentice-Hall, Englewood Cliffs, N.J. (1959) 437 pp.
31. Elkins, L. E.: "The Importance of Injected Gas as a Driving Medium in Limestone Reservoirs as Indicated by Recent Gas-Injection Experiments and Reservoir-Performance History", *Dril. and Prod. Prac.* (1946) 160–174.
32. Patton, E. C., Jr.: "Evaluation of Pressure Maintenance by Gas Injection in Volumetrically Controlled Reservoirs", *Trans.*, AIME (1947) Vol. 170, 112–155.
33. Stewart, C. R., Craig, F. F., Jr. and Morse, R. A.: "Determination of Limestone Performance Characteristics by Model Flow Tests", *Trans.*, AIME (1953) Vol. 198, 93–102.
34. Arps, J. J. and Roberts, T. G.: "The Effect of the Relative Permeability Ratio, the Oil Gravity, and the Solution Gas-Oil Ratio on the Primary Recovery from a Depletion Type Reservoir", *Trans.*, AIME (1955) Vol. 204, 120–127.
35. Botset, H. G.: "Flow of Gas-Liquid Mixtures through Consolidated Sand", *Trans.*, AIME (1940) Vol. 136, 91–105.
36. Leverett, M. C. and Lewis, W. B.: "Steady Flow of Gas-Oil-Water Mixtures through Unconsolidated Sands", *Trans.*, AIME (1941) Vol. 142, 107–116.
37. Felsenthal, M. and Conley, F. R.: "Discussion on Analysis of Reservoir Performance k_g/k_o Curves and a Laboratory k_g/k_o Curve Measured on a Core Sample", *Trans.*, AIME (1955) Vol. 204, 240.
38. Arps, J. J., Brons, F., van Everdingen, A. F., Buchwald, R. W. and Smith, A. E.: *A Statistical Study of Recovery Efficiency*, API *Bull.* D14, New York (1967) 33 pp.
39. Kaveler, H. H.: "Engineering Features of the Schuler Field and Unit Operation", *Trans.*, AIME (1944) Vol. 155, 58–87.
40. Felsenthal, M.: "Correlation of k_g/k_o Data with Sandstone Core Characteristics", *Trans.*, AIME (1959) Vol. 216, 258–261.
41. Blanchet, P. H.: "Development of Fracture Analysis as Exploration Method", *Bull.*, AAPG (1957) Vol. 41 (8), 1748–1759.
42. Martin, F. G.: "Mechanics and Control in Hydraulic Fracturing", *Pet. Eng.* (1967) Vol. 39 (13), 63–72.

43. Alpay, O. A.: "Application of Aerial Photographic Interpretation to the Study of Reservoir Natural Fracture Systems", paper SPE 2567 presented at SPE 44th Annual Fall Meeting, Denver, Colo. (1969) 12 pp.
44. Elkins, L. F. and Skov, A. M.: "Determination of Fracture Orientation from Pressure Interference", *Trans.*, AIME (1960) Vol. 219, 301–304.
45. Arnold, M. D., Gonzales, H. T. and Crawford, P. B.: "Estimation of Reservoir Anisotropy from Production Data", *Trans.*, AIME (1962) Vol. 225, I, 909–912.
46. Dellwig, L. F., Kirk, J. N. and Walters, R. L.: "The Potential of Low-Resolution Radar Imagery in Regional Geologic Studies", *J. Geophys. Res.* (1966) Vol. 71, 4995–4998.
47. Miller, V. C.: "Current Trends in Photogeology and in the Use of Other Remote Sensing Methods in Geological Interpretation", *Earth-Sci. Revs.* (1968) Vol. 4, 135–152.
48. Pasini, J., III and Overbey, W. K., Jr.: "Natural and Induced Fracture Systems and Their Application to Petroleum Production", paper SPE 2565 presented at SPE 44th Annual Fall Meeting, Denver, Colo. (1969) 7 pp.
49. McDonald, H. C.: "Sensing Techniques in Exploration", *Oil and Gas J.* (1969) Vol. 67 (22), 110–112, and (23), 116–117.
50. Sabins, F. S.: "Thermal Infrared Imagery and Its Application to Structural Mapping in Southern California", *Bull.*, GSA (1969) Vol. 80, 397–404.
51. Matthies, E. P.: "Practical Application of Interference Tests", *J. Pet. Tech.* (1964) Vol. 16, 249–252.
52. Aronofsky, J. S., Massé, L. and Natanson, S. G.: "A Model for the Mechanism of Oil Recovery from the Porous Matrix Due to Water Invasion in Fractured Reservoirs", *Trans.*, AIME (1958) Vol. 213, 17–19.
53. Parsons, R. W. and Chaney, P. R.: "Imbibition Model Studies on Water-Wet Carbonate Rocks", *Trans.*, AIME (1966) Vol. 237, II, 26–34.
54. Jones-Parra, J. and Reytor, R. S.: "Effect of Gas-Oil Ratio on the Behavior of Fractured Limestone Reservoirs", *Trans.*, AIME (1959) Vol. 216, 395–397.
55. Willingham, R. W. and McCaleb, J. A.: "The Influence of Geologic Heterogeneities on Secondary Recovery from the Permian Phosphoria Reservoir, Cottonwood Creek, Wyoming", SPE paper 1770 presented at SPE Rocky Mountain Regional Meeting, Casper, Wyo. (1967) 12 pp.
56. Stiles, W. E.: "Use of Permeability Distribution in Waterflood Calculations", *Trans.*, AIME (1949) Vol. 186, 9–13.
57. Dykstra, H. and Parsons, R. L.: "The Prediction of Oil Recovery by Waterflood", *Secondary Recovery of Oil in the United States*, API, New York (1950) 160–174.
58. Calhoun, J. C., Jr.: *Fundamentals of Reservoir Engineering*, Univ. of Oklahoma Press, Norman (1953) 417 pp.
59. Pirson, S. J.: *Oil Reservoir Engineering*, McGraw-Hill Book Co., New York (1958) 735 pp.
60. Yuster, S. T. and Calhoun, J. C., Jr.: "Pressure Parting", *Producers Monthly* (1945) Vol. 9 (4), 16–26.
61. Hubbert, M. K. and Willis, D. G.: "Mechanics of Hydraulic Fracturing", *Trans.*, AIME (1957) Vol. 210, 153–166.
62. Eaton, B. A.: "Fracture Gradient Prediction and Its Application in Oilfield Operations", *Trans.*, AIME (1969) Vol. 246, 1353–1360.
63. Hazebroek, P., Rainbow, H. and Matthews, C. S.: "Pressure Fall-off in Water Injection Wells", *Trans.*, AIME (1958) Vol. 213, 250–260.
64. Clark, K. K.: "Transient Pressure Testing of Fractured Water Injection Wells", *J. Pet. Tech.* (1968) Vol. 20, 639–643.
65. Millheim, K. K. and Chichowicz, L.: "Testing and Analyzing Low-Permeability Fractured Gas Wells", *Trans.*, AIME (1968) Vol. 243, 193–198.
66. Adams, A. R., Ramey, H. J., Jr. and Burgess, R. J.: "Gas Well Testing in a Fractured Carbonate Reservoir", *J. Pet. Tech.* (1968) Vol. 20, 1187–1194.

67. Carnes, P. S.: "Effects of Natural Fractures or Directional Permeability in Water Flooding", SPE paper 1423 presented at SPE Seventh Biennial Secondary Recovery Symp., Wichita Falls, Tex. (1966) 8 pp.
68. Lane, B. B.: "A Skeleton Pilot Flood, an Aid to Proper Flood Pattern Selection", SPE paper 2525 presented at SPE 44th Annual Fall Meeting, Denver, Colo. (1969) 7 pp.
69. Armstrong, F. E., Howell, W. D. and Watkins, J. W.: "Radioactive Inert Gases as Tracers for Petroleum Reservoir Studies", *USBM Rept. Invest.* 5733 (1961).
70. Crawford, P. B.: "Method Proposed for Determining Reservoir Heterogeneity Prior to Waterflooding", *Producers Monthly* (1962) Vol. 26 (10), 6–7, and 26 (11), 20–21.
71. Kunkel, G. C. and Bagley, J. W., Jr.: "Controlled Waterflooding, Means Queen Reservoir", *J. Pet. Tech.* (1965) Vol. 17, 1385–1390.
72. Baldwin, D. E., Jr.: "Prediction of Tracer Performance in a Five-Spot Pattern", *Trans.*, AIME (1966) Vol. 237, I, 513–517.
73. Fraser, C. D. and Pettitt, B. E.: "Results of a Field Test to Determine the Type and Orientation of a Hydraulically Induced Formation Fracture", *J. Pet. Tech.* (1962) Vol. 14, 463–466.
74. Anderson, T. O. and Stahl, E. J.: "A Study of Induced Fracturing Using an Instrumental Approach", *J. Pet. Tech.* (1967) Vol. 19, 261–267.
75. Zemanek, J., Caldwell, R. L., Glenn, E. E., Holcomb, S. V., Norton, L. J. and Straus, A. J. D.: "The Borehole Televiewer—a New Logging Concept for Fracture Location and Other Types of Borehole Inspection", *J. Pet. Tech.* (1969) Vol. 21, 762–774.
76. McKinley, R. M., Vela, S. and Carlton, L. A.: "A Field Application of Pulse Testing for Detailed Reservoir Description", *J. Pet. Tech.* (1968) Vol. 20, 313–321.
77. Wayhan, D. A. and McCaleb, J. A.: "Elk Basin Madison Heterogeneity—Its Influence on Performance", *J. Pet. Tech.* (1969) Vol. 21, 153–159.
78. Owens, W. W. and Archer, D. L.: "Waterflood Pressure Pulsing for Fractured Reservoirs", *Trans.*, AIME (1966) Vol. 237, I, 745–752.
79. Bokserman, A. A. and Shalimov, B. V.: "Cyclic Input to Strata with Dual Porosity During Water Displacement of Oil", *Izv. Akad. Nauk SSSR, Mekh. Zhidkostey i Gazov* (1967) (2), 168–174. (In Russian.)
80. Tatashev, K. K.: "Effect of the Change in Layer Pressure on the Oil Displacement from Blocks of Fissured-Porous Oil Reservoirs by Water Imbibition", *Geol. Nefti i Gaza* (1968) Vol. 1, 55–59. (In Russian.)
81. Felsenthal, M. and Ferrell, H. H.: "Oil Recovery from Fracture Blocks by Cyclic Injection", *J. Pet. Tech.* (1969) Vol. 21, 141–142.
82. Felsenthal, M., Ferrell, H. H. and Matthews, R. R.: "Pressure Pulsing Oil Production Process", U.S. Patent No. 3,480,081 (1969) 4 pp.
83. Brownscombe, R. E. and Dyes, A. B.: "Water-Imbibition Displacement—a Possibility for the Spraberry", *Drill. and Prod. Prac.* (1952) 383–390.
84. Graham, J. W. and Richardson, J. G.: "Theory and Application of Imbibition Phenomena in Recovery of Oil", *Trans.*, AIME (1959) Vol. 216, 377–381.
85. Mattax, C. C. and Kyte, J. R.: "Imbibition Oil Recovery from Fractured Water-Drive Reservoirs", *Trans.*, AIME (1962) Vol. 225, II, 177–184.
86. Raimondi, P. and Torcasso, M. A.: "Distribution of the Oil Phase Obtained upon Imbibition of Water", *Soc. Pet. Eng. J.* (1964) Vol. 4 (1), 49–55.
87. Blair, P. M.: "Calculation of Oil Displacement by Counter-Current Water Imbibition", *Trans.*, AIME (1964) Vol. 231, II, 195–202.
88. Elkins, L. F. and Skov, A. M.: "Cyclic Waterflooding the Spraberry Utilizes 'End Effects' to Increase Oil Production Rates", *Trans.*, AIME (1963) Vol. 228, I, 877–884.
89. Hester, C. T., Walker, J. W. and Sawyer, G. H.: "Oil Recovery by Imbibition Water Flooding in the Austin and Buda Formations", *J. Pet. Tech.* (1965) Vol. 17, 919–925.

90. Bleakley, W. B.: "Cyclic Water-Injection Scheme Doubles Output of Reservoir", *Oil and Gas J.* (1969) Vol. 67 (29), 61–63.
91. Maslyantsev, Y. V., Ogandzhanyants, V. G., Surguchev, M. L., Gavura, V. E. and Ivanovskii, G. I.: "An Experiment with a Cyclic Method of Water Injection into A_4 Formation in Pokrovsk Field", *Nauch.-Tekh. Sb. Ser. Neftepromyslovoe Delo* (1969) (1), 7–10. (In Russian.)
92. Vasilechko, V. P., Gnatyuk, R. A. and Petrash, I. N.: "Effectiveness of Cyclic Water-flooding in Oil Formations of Pre-Carpathian Fields", *Nauch.-Tekh. Sb. Ser. Neftepromyslovoe Delo* (1969) (1), 10–12.
93. Hendrickson, G. E.: "History of the Welch Field San Andres Pilot Water Flood", *J. Pet. Tech.* (1961) Vol. 13, 745–748.
94. Abernathy, B. F.: "Waterflood Prediction Methods Compared to Pilot Performance in Carbonate Reservoirs", *J. Pet. Tech.* (1964) Vol. 16, 276–282.
95. Chekhovskaya, G. Y., Kharchenko, B. S. and Postash, M. F.: "The Problem of Predicting Waterflood Behavior of Fractured Reservoirs", *Neft. Khoz.* (1969) (3), 27–30. (In Russian; English abstract No. 117,933, T.U. Pet. Abstracts.)
96. Dobrynin, V. M.: "Evaluating Oil Reserves in Fractured and Vuggy Reservoirs by the Elastic Material Balance Technique", *Geol. Nefti i Gaza* (1968) (5), 50–56. (In Russian; English abstract No. 115,705, T.U. Pet. Abstracts.)
97. Dyes, A. B., Kemp, C. E. and Caudle, B. H.: "Effect of Fractures on Sweep-Out Pattern", *Trans.*, AIME (1958) Vol. 213, 245–249.
98. Gewers, C. W. and Nichols, L. R.: "Gas Turbulence Factor in a Microvugular Carbonate", *J. Can. Pet. Tech.* (1969) Vol. 8 (2), 51–56.
99. Gill, D.: "A Computer Oriented Method for the Zonation of Ordered Data Sets and Its Application to Reservoir Evaluation and Digitized Log Analysis", *Abstracts*, Annual Geol. Soc. Am. and Assoc. Soc. Meeting, Atlantic City, N.J. (1969) Part 7, 78.
100. Greenkorn, R. A., Johnson, C. R. and Schallenberger, L. K.: "Directional Permeability of Heterogeneous Anisotropic Porous Media", *Trans.*, AIME (1964) Vol. 231, II, 124–132, and II, 363–364.
101. Kazemi, H.: "Pressure Transient Analysis of Naturally Fractured Reservoirs with Uniform Fracture Distribution", *Soc. Pet. Eng. J.* (1969) Vol. 9, 451–462.
102. Kazemi, H., Seth, M. S. and Thomas, G. W.: "The Interpretation of Interference Tests in Naturally Fractured Reservoirs with Uniform Fracture Distribution", *Soc. Pet. Eng. J.* (1969) Vol. 9, 463–472.
103. Medvedskii, R. I.: "A Method of Solving by Successive Approximations Problems of Unsteady-State Flow of a Liquid in a Fractured Porous Medium", *Izv. Akad. Nauk SSSR, Mekh. Zhidkostey i Gazov* (1969) (2), 162–168. (In Russian; English abstract No. 117,940, T.U. Pet. Abstracts.)
104. Overbey, W. K., Jr. and Rough, R. L.: "Surface Joint Patterns Predict Wellbore Fracture Orientation", *Producers Monthly* (June, 1968) 16–19.
105. Schneider, F. N. and Owens, W. W.: "Sandstone and Carbonate Three-Phase Relative Permeability Characteristics", SPE paper 2445 presented at SPE Rocky Mountain Regional Meeting, Denver, Colo. (1969) 10 pp.
106. Vieira, L. P.: "Fractured Reservoirs", *Proc.*, Seventh World Pet. Cong., Dril. and Prod. (1967) Vol. 3, 219–227.
107. Testerman, J. D.: "A Statistical Reservoir-Zonation Technique", *Trans.*, AIME (1962) Vol. 225, I, 889–893, 1394.
108. Bedcher, A. Z., Okun, M. I. and Shcherbakov, E. L.: "Relationship Between Saturation Parameter and the Coefficient of Oil or Gas Saturation in Fractured Reservoirs", *Neft. Geol. Geofiz.* (1968) (1), 33–35. (In Russian; English abstract No. 115,279, T.U. Pet. Abstracts.)
109. Collins, R. E.: *Flow of Fluids through Porous Materials*, Reinhold Publ. Corp., New York (1961) 270 pp.
110. Scheidegger, A. E.: *The Physics of Flow through Porous Media*, The Macmillan Co., New York (1960) 313 pp.

111. Buckley, S. E. and Leverett, M. C.: "Mechanism of Fluid Displacement in Sands", *Trans.*, AIME (1942) Vol. 146, 107–111; *SPE Reprint Series* (1942) (2), 91–95.
112. Craig, F. F., Jr., Geffen, T. M. and Morse, R. A.: "Oil Recovery Performance of Pattern Gas or Water Injection Operations from Model Tests", *Trans.*, AIME (1955) Vol. 204, 7–15.
113. Miller, M. G. and Lents, M. R.: "Performance of Bodcaw Reservoir, Cotton Valley Field Cycling Project", *Dril. and Prod. Prac.* (1946) 128–149.
114. Campa, M. F. and Romero, R. M.: "Statistical Zonation Technique and Its Application to the San Andres Reservoir in the Poza Rica Area, Vera Cruz, Mexico", *Rev. Inst. Mex. Pet.* (1969) Vol. 1 (1), 56–60. (In Spanish; English abstract No. 113,958, T.U. Pet. Abstracts.)
115. Anonymous: "Miscible Flood May Capture 95% of Crude in Place", *Oil and Gas J.* (1965) Vol. 63 (30), 164–166.
116. Willman, G. J.: "Vertical Miscible Flood to Hike Recovery by 70 Million Barrels", *World Oil* (1966) Vol. 162 (1), 75–78.
117. Anonymous: "Rainbow Project May Hit 96% Recovery", *Oil and Gas J.* (1968) Vol. 66 (51), 48–49.
118. Dowling P. L., "Application of Carbonate Environmental Concepts to Secondary Recovery Projects", SPE paper 2987, presented at 45th Annual SPE Fall Meeting, Houston, Texas (1970) 16 pp.

CHAPTER 4

Formation Evaluation

LEENDERT DE WITTE

Introduction

This chapter deals with the evaluation, from geophysical well surveys, of the formation characteristics and fluid contents of porous strata.

The most important formation characteristic is the effective porosity, ϕ, defined as the fraction of bulk volume occupied by interconnected void space.

The parameter most commonly used in specifying fluid content is the water saturation, S_w, which is the fraction of interstitial space occupied by formation water. If S_w is less than unity, it is assumed that the remaining space is occupied by hydrocarbons. The quantity $S_{o,\,g} = 1 - S_w$, which is the fraction of pore space occupied by oil or gas, is, of course, of primary interest in estimating the producing potential of a particular stratum.

With few exceptions, all methods of calculating S_w are based on determination of the electrical resistivity of the undisturbed formation. This so-called true resistivity, although strongly dependent on S_w, is, however, a function of several other parameters. The most important of these are (1) a matrix cell constant, called the formation factor, F, which is related to porosity; (2) the resistivity of the formation water, R_w, and (3) the colloidal or clay content of the formation. The latter factor, while very important in sandstone-shale series, usually plays not more than a minor role in carbonate reservoirs.

The general approach in the evaluation of well surveys is to delineate porous strata and determine the porosity and the formation factor, to find the true resistivity and the formation water resistivity, and from a combination of the last three factors to calculate the hydrocarbon saturation. Whereas the true resistivity is always determined from electrical surveys, the porosity may be found by acoustical methods, radioactivity measurements, or formation factor determinations from electrical resistivity measurements on the zone surrounding the wellbore and flushed by invading mud filtrate.

Table 1 summarizes the starting parameters in formation evaluation and the techniques employed in obtaining them. The principal logging methods mentioned in the table are discussed in separate sections, after presentation

References p. 210.

TABLE 1. Techniques for Evaluating Formation Parameters

Method \ Parameter	Porosity		Formation water resistivity	True Resistivity	Clay Content
	Direct	Formation factor from invaded zone			
Electrical		Microlog Microlaterolog Proximity log Short normal limestone curve	Self-potential	Long lateral Long normal Two normals Deep induction Dual induction-laterolog Laterolog (salt mud)	Self-potential
Radioactivity	Neutron-gamma Neutron-neutron (thermal) Neutron-neutron (epithermal) Density log (gamma-gamma)		Salinity log		Gamma ray Neutron vs. density log
Acoustical	Sonic log				
Direct sampling	Core analysis Side wall samples		Produced water analysis		Differential thermal analysis Electrochemical (diffusion cell)

of a brief outline of the petrophysical relations between formation characteristics and the quantities measured in electrical logging.

At the end of the chapter special techniques for the evaluation of other properties, such as vuggy porosity, fracture patterns, and boreface characteristics are reviewed.

Objectives in Electrical Logging

All of the numerous forms of electrical logging record in their right-hand traces measurements of an average or apparent resistivity of the borehole and its surrounding formations. The characteristic volume for the particular measurement and the weighting for the regions comprised in this volume, however, differ for the various devices.

The left-hand trace normally records a so-called spontaneous potential curve. This is sometimes replaced by a gamma-ray or a hole caliper survey.

In order to appreciate the intent and methodology of electrical surveys, a general schematic of the down-hole environment of a porous formation, as shown in Fig. 1, will be considered.

The term *resistivity* denotes an electrical property of matter, which is the inverse of conductivity and is defined as the resistance of a cube of the material to current flow parallel to one of its sides. The unit of resistivity commonly used in well logging is the ohm-meter, which is the resistance of a cube the sides of which are 1 m long. In Fig. 1, R_t denotes the true or undisturbed formation resistivity, R_m is the resistivity of the drilling fluid in the borehole traversing the formation, and R_i is the average resistivity of the portion of the stratum surrounding the borehole, which has been invaded by the filtrate of the drilling fluid and is referred to as the invaded zone resistivity. The outward boundary of the zone of invasion is normally not a sharply defined cylindrical surface, as indicated in the schematic, but rather a region of transition. Nor are the fluid content and the resistivity constant through the remainder of the zone. Immediately adjacent to the boreface, the flushing by mud filtrate is usually more complete than at greater distances. The resistivity of the thoroughly flushed part is frequently treated separately and denoted by R_{xo}. Whereas in water-bearing formations R_{xo} will reflect 100% saturation by mud filtrate, in the presence of hydrocarbons there will be residual oil or gas saturation even in the flushed zone. The portion of the hydrocarbons that is displaced by the invading fluids is referred to as movable oil or gas. Inasmuch as the rates of displacement of interstitial water and movable oil by the mud filtrate are different, one may find near the outer boundary of the invaded zone a low-resistivity annulus where most of the movable oil, but little of the formation water, has been displaced. This annulus can be especially pronounced in very porous oil-bearing formations

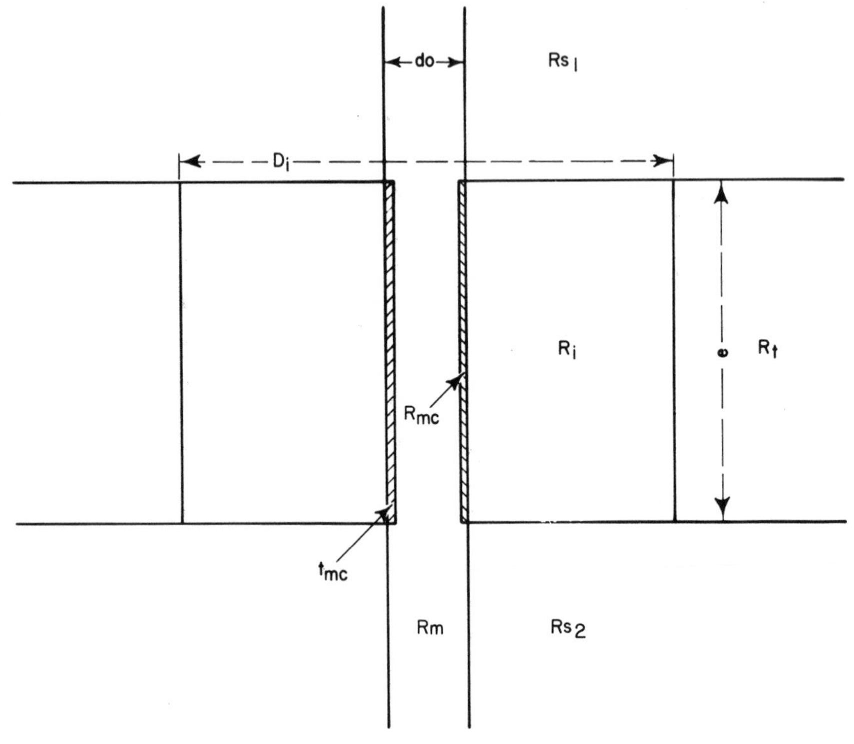

FIG. 1. Schematic diagram of resistivity distributions in a porous formation.

invaded by fresh mud filtrate. For purposes of expediency the invaded zone is treated in most interpretation techniques as a homogeneous region concentric with the drill hole and having constant resistivity R_i.

The invasion of permeable strata by mud filtrate usually leaves a colloidal filter cake on the boreface. The thickness of the "mud cake," denoted by t_{mc}, varies with the water loss, colloidal content, and salinity of the mud and will range from nearly negligible for very saline muds to more than 1 in. in extreme cases for fresh muds. Calcium-rich muds normally have thinner mud cakes than ordinary bentonite muds. The mud cake resistivity, R_{mc}, is usually higher than the mud resistivity, R_m. Its value is of importance in the interpretation of contact-type resistivity logs such as the microlog, the microlaterolog, and their equivalents.

The quantitities R_{s1} and R_{s2} in Fig. 1 denote the resistivities of the overlying and underlying adjacent beds. When the porous beds under investigation are thin, corrections to their recorded apparent resistivities may have to be made for adjacent bed effects. These corrections, which involve R_{s1} and

R_{s2}, are often quite large and very complex for the conventional resistivity logs in hard rock environments. Therefore, the tendency in the last two decades has been to replace the conventional logs in limestone areas by focused logging devices such as induction logs, laterologs, and guard electrode surveys.

The principal function of the macroresistivity logs (those devices other than contact-micro types) is to give measurements from which R_t and/or R_i can be ascertained. The purpose of the micro-devices is the detailed delineation of permeable zones and the determination of R_{xo} (or R_i). The spontaneous potential (SP) or natural potential (NP) curve registered in the left-hand trace of electrical surveys records the variations of the electrical potential occurring spontaneously in the drill hole. These potentials are of electrochemical nature and are caused by the contrast in ionic content between the drilling mud and the formation waters, and by the difference in permeability to ionic diffusion of successive strata. Qualitatively the SP curve gives a ready record of the shaliness of formations. With proper caution it can be used for the calculation of R_w, the formation water resistivity.

Quantitative Relations in Electric Log Interpretation

(References 1–22)

In the absence of conductive solids, the resistivity of water-saturated rock formations is directly proportional to the resistivity of the formation water. The constant of proportionality is called the formation factor, F. The relationship is expressed by:

$$R_o = FR_w \tag{1}$$

where R_o is the resistivity of the rock when 100% saturated with water having resistivity R_w.

Although certain native metals and metal sulfides are recognized as conductive rock constituents, in practice conductive solids in sedimentary rocks are represented solely by clay minerals. Carbonate rocks with an appreciable disseminated clay content are termed marls. The instances where marly formations have sufficient porosity to be of interest as reservoir rocks are so rare that little attention is devoted to their treatment.

The formation factor has an inverse relationship to porosity: the lower the porosity, the higher the formation factor. Empirically, the relation is given by:

$$F = \phi^{-m} \tag{2}$$

where the exponent, m, is called the cementation factor.

For certain oolitic limestones the value of m may be as low as 1.6. For sucrose dolomites it is of the order of 1.8–2; for dolomitic limestone and other forms of carbonates with homogeneous matrix porosity the value is 2–2.2. In cases of extreme irregularity or tortuosity of the interstices m can be as high as 2.4; certain oolicastic limestones have m values around 2.8. In the absence of reasonable lithological knowledge about a horizon it is common to use the simplified relation:

$$F = 1/\phi^2. \tag{3}$$

In general, the following equations are used for water-bearing carbonate formations:

$$R_t = R_o = FR_w \approx R_w/\phi^2 \tag{4}$$

and

$$R_i = R_{io} = FR_{mf} \approx R_{mf}/\phi^2. \tag{5}$$

Here R_{io} denotes the resistivity of the invaded zone, 100% saturated with mud filtrate of resistivity R_{mf}.

The R_w is often referred to as connate water resistivity. This usage, however, is not always correct, because formation waters are not necessarily connate with the formation. Especially at the edges of basins, the original waters dating from the time of deposition of the formation are frequently replaced by later artesian waters.

The presence of hydrocarbons in any given formation increases its resistivity because part of the conductive formation water is replaced by non-conductive oil or gas. As the hydrocarbon content increases, the water saturation, S_w, decreases; but even in reservoirs producing clean oil or gas an irreducible amount of water remains, coating the pore surfaces. The water wetting the interstitial surfaces, referred to also as capillary bound water, provides a continuous path for electrical current conduction. In very porous and clay-free formations the irreducible water saturation may be as low as 5%. In most cases, however, it constitutes at least 10% of pore space, and for very shaly reservoir rocks it may reach 60%. In carbonate reservoirs the usual range for irreducible water saturation is 5–35% of pore space. When S_w exceeds the irreducible value, the presence of free or movable water is indicated and the formation will produce with a water cut.

The quantitative relation between formation resistivity and S_w has been found empirically to be of the form:

$$S_w = \left(\frac{R_o}{R_t}\right)^{1/n} \tag{6}$$

where $1.8 \leqslant n \leqslant 2.2$.

The precise value of the saturation exponent, n, is rarely known in practice. Unless it has been experimentally established for the particular facies of the formation under investigation, the value 2 is generally assumed for n.

Relations 1, 2, and 6 were first derived by Archie,[2, 3] and Eq. 6 is commonly referred to as Archie's formula.

As has been mentioned previously, the invaded zone of hydrocarbon-bearing rocks will have a residual oil or gas saturation so that the mud filtrate saturation in the invaded zone, S_{wi}, is less than unity. Thus, Archie's formula gives:

$$S_{wi} \approx \left(\frac{R_{io}}{R_i}\right)^{1/2} = \left(\frac{FR_{mf}}{R_i}\right)^{1/2} \tag{7}$$

This may be rewritten as:

$$F = \frac{R_i}{R_{mf}} \cdot S_{wi}{}^2 \tag{8}$$

Relation 8 is frequently used to calculate the formation factor from a measured or derived value of the invaded zone resistivity. After F is found, Eq. 2 or 3 gives the porosity, and knowledge of R_t and R_w then allows computation of S_w from Eqs. 1 and 6. Whereas the apparent resistivities recorded on electric logs have absolute scales from zero to a specified number of ohm-meters, the spontaneous potential is a relative measurement. The position of the curve and the scale are adjusted by the logging operator so that the maximum and minimum deflections for most of the run will be accommodated within the ten-division field for the trace.

For interpretation purposes, the value of the SP deflection is taken with respect to the readings opposite homogeneous compacted shale bodies or the so-called shale base line. Values to the left of the shale base line are taken as negative, deflections to the right as positive. When the mud column is homogeneous, the shale base line formed by the most positive shale deflections is a fairly vertical trace. When there are significant salinity gradients in the mud column, the base line will drift. In such cases it is helpful to draw a slanted line on the log, representing the trend of the base line, in order to determine particular formation deflections.

The predominant components of the spontaneous potential differences encountered in boreholes are due to electrochemical diffusion-type potentials. In clay-free formations the salinity contrast between the formation water and the mud filtrate will cause salt ions to diffuse from the more concentrated to the less saline solution. The negative ions are less hydrated and diffuse faster, causing a buildup of negative charges on the less concentrated side and a relative excess of positive ions on the more saline side. The buildup of electrical charge will slow down the negative ions and accelerate the positive ions until a dynamic equilibrium is reached, at which both ions, on the average, migrate at the same rate. The electrical equilibrium potential difference across the boundary between the solutions is called a simple diffusion or liquid-liquid junction potential.

For solutions of univalent ions (e.g., NaCl solutions) the value of this potential difference is given by:

$$E_d = \frac{RT}{F} \cdot \frac{u-v}{u+v} \ln \frac{a_1}{a_2} \tag{9}$$

where R is the gas constant; F is the Faraday; T is the absolute temperature (°K); u and v are the positive and the negative ionic mobilities, respectively; and a_1 and a_2 are the ionic activities of the two solutions.

Across clays or shales, the diffusion process between the solutions is altered by the fixed negative charges on the clay mineral lattices, and the clouds of positive counter ions in the bound water layers of the clays. Through pure endurated shales only the positive ions are able to diffuse, so that opposite the shales the weak solution acquires a positive charge. Again a dynamic equilibrium is reached when the diffusion tendency is balanced by the electrical back conduction. The resulting potential is called a perfect membrane potential; in the case of NaCl solutions it has the value:

$$E_m = \frac{RT}{F} \ln \frac{a_1}{a_2} \tag{10}$$

The total potential difference in the drill hole between a portion opposite the shale and a region facing an adjacent inert or clean formation of finite porosity is found by combining Eqs. 9 and 10:

$$E_t = \frac{RT}{F}\left(1 - \frac{u-v}{u+v}\right) \ln \frac{a_1}{a_2} \tag{11}$$

Using the approximate inverse proportionality between activities and resistivities of aqueous solutions and inserting the values of the various electrochemical constants, one obtains:

$$(\mathrm{SP})_s = -K_t \log_{10} \frac{R_{mf}}{R_w} \tag{12}$$

For NaCl solutions, K_t has the value of 70.6 at 77°F. At 100°F, 200°F, and 300°F, the K_t values are roughly 75, 85, and 95, respectively.

The subscript s in $(\mathrm{SP})_s$ indicates that Eq. 12 refers to the so-called static SP, which is the value of the electrochemical potential when not diminished by ohmic voltage drops caused by the currents resulting from the shunting of the diffusion cells at the formation boundaries. Effects of these SP currents are considered in the next section.

Equation 12 clearly provides a means for calculating the formation water resistivity when the conditions for its applicability are fulfilled. Combination of Eq. 12 with the resistivity formulas gives the useful relation:

$$(SP)_s = -K_t \left(\log \frac{R_i}{R_t} + 2 \log \frac{S_{wi}}{S_w} \right) \quad (13)$$

As indicated before, most carbonate reservoir rocks are sufficiently free of interstitial clay so that normally the quantitative relations given above for clean formations apply without modification. In formations where the interstitial clay content is not negligible, the resistivity relations are complicated by the fact that part of the electrical current is conducted by the counter ions associated with the clay lattices. This effect becomes especially noticeable for the invaded zones with fresh mud filtrate. The ratio R_{io}/R_{mf} then no longer gives the true formation factor, but rather yields an apparent formation factor, F_a, which is smaller than F. The SP behavior changes correspondingly in that the reservoir rock no longer acts as an inert cell in which a simple diffusion potential is established. Instead, the shaly formation acts as a leaky membrane, giving rise to a potential value somewhere between the diffusion potential and the shale membrane potential. As a consequence, the magnitude of the SP deflection is decreased. Application of Eq. 12 no longer yields the true R_w, but rather gives an apparent water resistivity, R_{wa}, which is higher than R_w. It was suggested by Tixier[16] that because of the opposite effects of shale on F and R_w the relation

$$R_o = F_a R_{wa} \quad (14)$$

might still hold. This relation, later confirmed theoretically for water-bearing strata, is referred to as the Tixier relation for shaly formations.

In oil-bearing shaly rocks the presence of oil accentuates the shaliness, because it increases the ratio of interstitial clay to water-filled pore space. Both the resistivity and SP relations become quite complex in this case. In the limit of fresh muds and very shaly reservoir rocks Eq. 13 changes to:

$$(SP)_s = -K_t \left(\log \frac{R_i}{R_t} + \log \frac{S_{wi}}{S_w} \right) \quad (15)$$

In carbonate reservoirs the degree of shaliness will rarely be sufficient to justify the use of Eq. 15. It is suggested that for shaly carbonate reservoirs an intermediate relation be used of the form:

$$(SP)_s = K_t \left(\log \frac{R_i}{R_t} + \gamma \log \frac{S_{wi}}{S_w} \right) \quad (16)$$

where $1.5 \leqslant \gamma < 2.0$. The parameter γ will depend on the degree of shaliness and the value of R_{mf}.

Determination of Formation Water Resistivity from the SP Curve—The SP Plot

(References 23–28)

Relation 12, expressing the dependence of the SP deflection on the mud filtrate and formation water resistivities for clay-free formations, can be used directly for calculation of R_w only over the following limited range of intermediate water resistivity values: $0.05 < R_w < 0.3$. Outside this range, either the inverse proportionality of solution activity and resistivity or the assumption that an electrolyte approximates an NaCl solution breaks down. An empirical method for correcting the average deviations outside this range was introduced by Gondouin *et al.*[23] The method consists of finding an effective water resistivity, $(R_w)_e$, from the SP equation and subsequently making a correction to $(R_w)_e$ to find the true R_w. The correction chart can be found in the article by Gondouin *et al.*, and in the Schlumberger *Log Interpretation Charts.*[38] The latter charts also give a graphical interpretation of Eq. 12 for different temperatures or K_t values.

For the evaluation of a single, remote wildcat well in an area where formation water salinities have not been mapped, the above procedure is all one can follow unless a special salinity log has been run on the well. In cases of development wells or relatively closely spaced wildcats, that is, where a number of neighboring wells penetrate the same formations, one can usually enhance the analysis by combining the available information from these wells into a so-called SP plot. For each of the wells the mud filtrate resistivity at formation temperature is plotted on semilog paper against the SP deflection for the formation in question. Charts or tables to convert mud resistivities to filtrate resistivities and to make the appropriate temperature corrections are available in any of the service companies' collections of log interpretation charts. The SP deflection is entered on the linear scale, the R_{mf} value on the log scale. If the formation water composition is locally constant, the mud or filtrate resistivity measurements and the temperature conversions are made correctly, and the formation is free of colloidal matter, the points of the SP plot should fall roughly on a straight line. The intersection of this line with the SP = 0 ordinate then gives the value of $(R_w)_e$.

In practice, however, one seldom obtains a linear plot. Although some of the scatter is due to erroneous measurements or incomplete corrections, the most significant departures are caused either by lithological facies changes or by differences in the value of R_w. The points corresponding to the lowest R_w value in the region and to the most clay-free facies will form the lower right-hand margin of the plot, and will normally represent a good approximation to clean rock behavior. A line drawn through these points, with a slope corresponding to K_t, will intersect the ordinate at the lowest $(R_w)_e$

value for the area or the field. To establish this slope a line is drawn with an SP interval equal to K_t for one cycle (or factor of 10) on the R_{mf} scale and moved parallel to itself until it passes through the outermost points of the plot. If there is no geographical pattern to the departure of the scattered

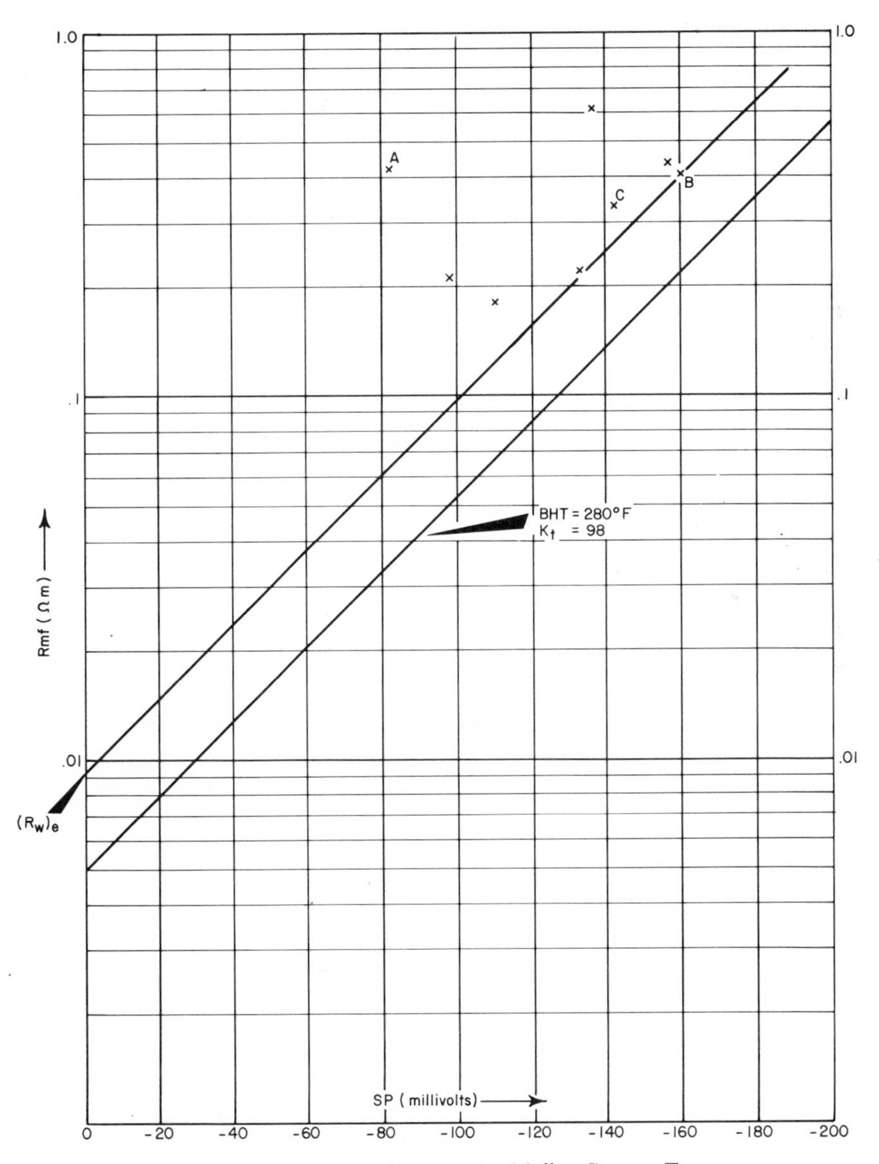

FIG. 2. SP plot, Edwards Lime, Dilworth Dome, McMullen County, Texas.

points from this clean facies line, the differences are probably due to measurement errors and variations in the ratio of clay content to water-filled pore space. In low-porosity formations even very small amounts of interstitial clay can cause significant decreases in the SP deflection. By the same token a formation completely free of clay-type material can have a fully-developed static SP, no matter how low its porosity.

Figure 2 shows an example of an SP plot for the Edwards Lime gas field at Dilworth Dome, McMullen County, Texas. The formation temperature is around 280°F with a corresponding K_t value of 98. The line with a slope of $K_t = 98$ indicates an $(R_w)_e$ value of 0.0092. This corresponds to a true R_w of 0.017 Ωm, indicating that the formation water is an almost completely saturated brine. Inasmuch as the formation is truncated by a piercement-type salt column, this is a logical result. Four of the points fall very closely to the boundary line of the plot. Three other points show sufficient departures to indicate some colloid content. The departure of point *A* is too large to be caused by a clay content variation in a limestone reservoir. More probably the well in question occurs in a separate fault block where the formation water has a different (higher) resistivity.

For thin resistive limestone horizons interbedded with shale, the measured SP can be substantially lower than the static SP because of the shunting effect of the SP currents. Theoretical correction curves for this effect were first introduced by Doll.[28] More recently they have been replaced by empirically determined SP bed thickness correction charts. For $R_i/R_m = 200$, corrections are negligible when the bed thickness exceeds 50 ft. For $R_i/R_m = 5$, the corrections can be ignored unless the bed is less than 15 ft thick.

In saline muds, the SP deflection will be reversed if the resistivity of the muds is lower than that of the formation waters. The SP current or bed thickness effects will be more severe to the extent that the SP curve loses definition. Departures from the inverse proportionality of activity and resistivity must be taken into account. Furthermore, saline mud columns are frequently inhomogeneous, causing appreciable base line shifts and difficulties in specifying the mud resistivity at a given depth. Needless to say, quantitative interpretation of SP curves for saline muds in carbonate rocks requires extreme caution.

Conventional Resistivity Logs

(References 28–37)

It has been indicated that in present practice conventional electric logs are rarely used for the surveying of holes drilled through carbonate formations. During preceding decades, however, hundreds of thousands of holes were logged with these devices. The records of all fields developed before the late nineteen fifties contain almost exclusively the standard E-logs. For this reason it is necessary to devote some attention to the use of these logs in formation evaluation. The conventional resistivity logs usually display curves recorded with three different devices: short normal, short lateral, and long lateral curves. A long normal may take the place of the short lateral. The normal devices emit current from a current electrode, *A*, and record the value of potential at a distance *AM* from this electrode, with respect to a reference electrode, *N*, at the surface (infinity). The current from *A* returns through a remote electrode, *B*. The distance *AM* is called the spacing of the device. The most common spacing for the short normal is 16 in., but values of 10 in. and 18 in. are also used. The usual long normal spacing is 64 in. The lateral curves use current flow between a pair of electrodes, *A* and *B*, and measure potential at an electrode *M*, again with respect to a surface reference electrode. The distance *AB* is small compared to *AM*. The spacing of a lateral device is the distance between *M* and the midpoint, *O*, between *A* and *B*. An equivalent lateral arrangement is to have one down-hole current electrode, *A*, and two pickup electrodes, *M* and *N*. The spacing is then measured between *A* and the midpoint, *O*, between *MN*. Common lateral spacings, usually denoted by *AO* in log headings, are 18 ft 8 in. and 24 ft. An additional shorter lateral of 10 ft was used for a long time.

Inasmuch as the emitted current for the devices is kept constant, the measured potential differences would be directly proportional to the resistivity if the electrodes were embedded in a homogeneous medium, and the recordings could be calibrated in ohm-meters of resistivity. The required calibration constants are a function of the value of the current and of the electrode spacings. In actual practice, where the current passes through the drill hole and the surrounding formations, the same calibration or conversion constants are used and the readings on the resistivity scales are termed apparent resistivities.

The apparent resistivities recorded with each device at any point opposite a formation depend on all of the variables shown in Fig. 1, as well as on the distance of the pickup and current electrodes from the formation boundaries. An excellent description of the details of this dependence is given in the Schlumberger *Interpretation Handbook for Resistivity Logs.*[37] In general, the smaller the spacing of a device, the smaller is its radius of investigation, that is, the apparent resistivity depends primarily on the region close to the electrodes. Thus, the short normal apparent resistivities are influenced strongly by R_i and less by R_t. They are very sensitive to variations in borehole diameter, whereas adjacent bed effects are less pronounced than for devices with larger spacings. The long normal is more sensitive to R_t, and the long laterals have apparent resistivities that approximate R_t in thick beds unless invasion is very deep.

If a formation is permeable and infiltrated by drilling fluids, the rate of invasion is governed largely by the hydraulic pressure of the mud column and the mud cake characteristics. The lower the porosity, the deeper a given amount of invasion fluid will penetrate. Inasmuch as in most carbonate reservoirs matrix porosities are fairly low (as compared to those for sandstones), generally moderate to deep invasion is encountered in carbonates with invaded zone diameters $> 3d_o$, where d_o = hole diameter (d_h is used also).

Another distinguishing characteristic of carbonate reservoirs is that potential producing zones usually occur as porous intervals within a larger body of dense rock. Thus, the beds adjacent to horizons of interest frequently have very high resistivity. This gives rise to strong adjacent bed effects, which are difficult to evaluate or correct for. The adjacent bed effects increase with a decrease in mud resistivity. In saline muds the quantitative evaluation of conventional resistivity curves becomes rather hopeless in all but the most homogeneous carbonate formations. Most published bed thickness correction charts for the normal devices are for cases where the adjacent beds are low-resistivity shales.

When the beds are sufficiently thick, the short normal may be used after a borehole correction to yield an approximation of R_i. Similarly, the long lateral apparent resistivities can give an approximate value for R_t. The long normals are usually more affected by invasion than the long laterals. Under favorable conditions the combination of the two normals can be used to obtain a reasonable estimate of the ratio R_i/R_t.

Charts depicting the apparent resistivities as functions of the invasion parameters and R_t are referred to as resistivity departure curves. Various publications of such charts are available from the service companies. When the charts combine several apparent resistivities for devices of given spacings and yield directly R_i, R_t, their ratio, or values corrected for hole effects, they are called simplified departure curves.

Figure 3 shows an example of a conventional electric log through a section of Kansas City Lime in the Admussen Pool, Butler County, Kansas. Qualitatively it is noted that the formation is divided into various beds separated by shaly laminations. At 2,228–40 and 2,244–56 ft intervals the short normal reading is appreciably higher than the long normal, which in turn is much higher than the long lateral. These are typical characteristics of a water-bearing formation with saline formation water, invaded by fresh mud filtrate. Toward the top of the formation the resistivity markedly increases, whereas the SP deflection decreases. This could be caused either by a decrease in porosity or by an oil-water transition. Although the individual beds are not very thick, the resistivity contrast of the intervening layers is small enough so that a quantitative approach appears feasible. In the lower water-bearing

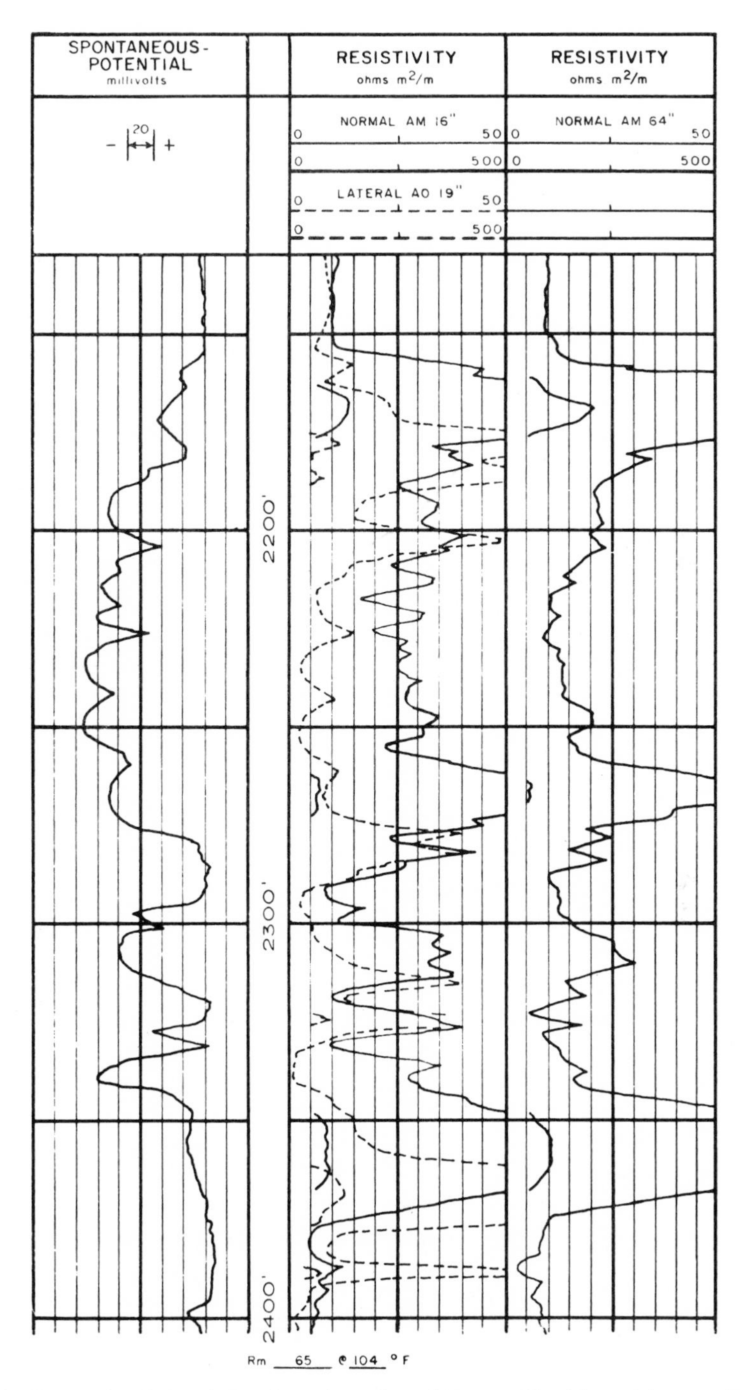

FIG. 3. Conventional electric log through section of Kansas City Lime, Admussen Pool, Butler County, Kansas. (Courtesy of McCulloch Oil Corp. of California.) $R_m = .65$ at 104°F.

beds the behavior of all three resistivity curves appears fairly regular. In the upper zones the long lateral curve is complicated by a number of boundary effects and the influence of some thin resistive streaks.

Before tackling the more complicated upper horizons, a quantitative interpretation of one of the lower zones will be made. For the interval at 2,228–40 ft the following data are obtained:

Long normal apparent resistivity $(R_a)_{64} = 13.2\ \Omega\text{m}$.
Short normal apparent resistivity $(R_a)_{16} = 26\ \Omega\text{m}$.

$R_m = 2.4\ \Omega\text{m}$ at 57°F $= 1.3\ \Omega\text{m}$ at 104°F (BHT—i.e., bottom hole temperature) $= 1.4\ \Omega\text{m}$ at 97°F (interpolated formation temperature).

$(R_a)_{64}/R_m = 9.44$; and $(R_a)_{16}/R_m = 18.6$.

Hole diameter $d_o \approx 7\frac{3}{8}$ in.

By using two-normal simplified departure curves of the type shown in Fig. 4, a value of R_i/R_m and a value of R_t/R_m for each of a number of assumed depths of invasion are found as follows (d = hole diameter corrected for presence of logging sonde, or effective hole diameter):

D_i/d	R_i/R_m	R_t/R_m
1.5	80	5
2.5	45	4.8
5.0	28	4.2
8.0	23	3.3
12.0	21	1.5

(A full set of simplified departure curves was published in 1962 by the Lane-Wells Company, now merged with Dresser-Atlas, under the title "R_i–R_t Conversion Charts.") From the normal curves alone one cannot decide which of these combinations actually prevails.

The information from the long lateral is used next. In this environment of low resistivity it is possible to assume that $(R_a)_{lat} \approx R_t$; thus,

$$(R_a)_{19\ ft} = 2.8\ \Omega\text{m} \approx R_t \text{ or } R_t/R_m \approx 2.0$$

Interpolating in the above listing based on the two normals, one obtains $D_i/d_o \approx 10$, $R_i/R_m \approx 22$, and $R_t/R_m \approx 2.0$.

From a R_m versus R_{mf} chart, $R_{mf} = 1.07\ \Omega\text{m}$ at 97°F, so that $F = R_i/R_{mf} = 28.8$; and from Eq. 3, the porosity $\phi = 0.185$ or 18.5% of bulk volume. A consistency check with the use of the SP deflection can be made next. A reading with respect to the shale base line at 2,150 ft gives SP $= -112$ mV. At 97°F the value of $K_t = 73$. Equation 12 yields:

$$-112 = -73 \log \frac{1.07}{(R_w)_e} \quad \text{or} \quad (R_w)_e = 0.07\ \Omega\text{m}.$$

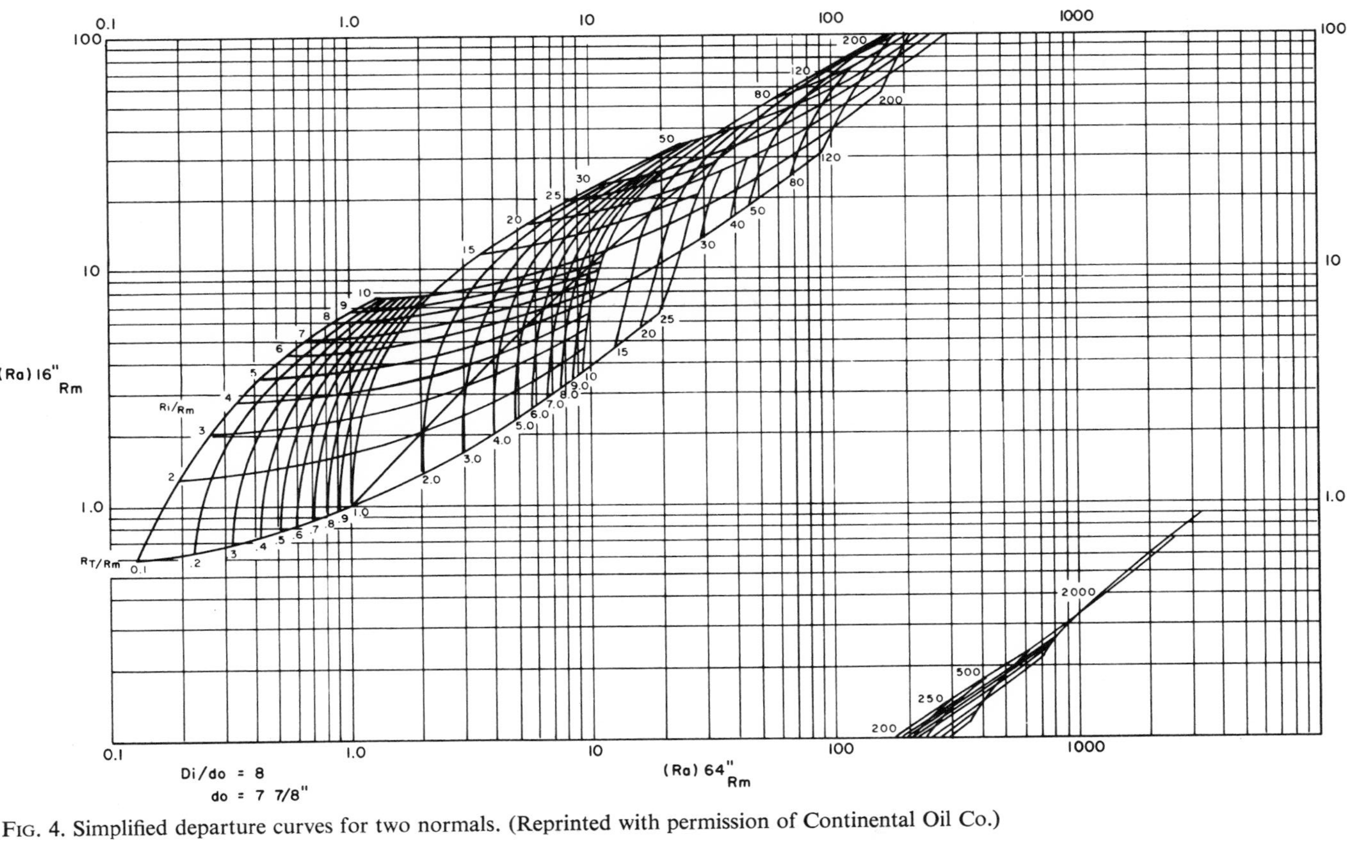

FIG. 4. Simplified departure curves for two normals. (Reprinted with permission of Continental Oil Co.)

The Gondouin chart[23] gives $R_w = 0.086\ \Omega\text{m}$. For a water-bearing formation:

$$R_t = R_o = FR_w = 28.8 \times 0.086 = 2.48\ \Omega\text{m}.$$

This is in very reasonable agreement with the value of $R_t = 2.8\ \Omega\text{m}$, as obtained from the long lateral curve. The slightly higher value given by the lateral device is readily explained by the influence of the fairly deep invasion ($D_i/d \approx 10$) on the lateral apparent resistivity.

Having established a degree of confidence in the interpretation approach for this formation, one can now apply the two-normal analysis to one of the beds exhibiting the resistivity gradient. The interval at 2,188–203 ft is selected as being the thickest and most homogeneous, even though the short normal shows that the resistivity is not constant throughout the zone.

Using average values, one has:

$$(R_a)_{64} = 22\ \Omega\text{m} \quad \text{and} \quad (R_a)_{64}/R_m = 15.7$$

$$(R_a)_{16} = 32\ \Omega\text{m} \quad \text{and} \quad (R_a)_{16}/R_m = 22.9$$

If the depth of invasion is assumed to be at least as deep in the upper zones as in the lower beds, the simplified two-normal departure curves yield:

D_i/d	R_i/R_m	R_t/R_m
10	25.5	6.75

A comparison with the results for the lower zone shows that, whereas R_i/R_m has increased by less than 20%, the true resistivity has increased by more than a factor of 3. This appears to be indicative of an oil-water transition with some residual oil in the invaded zone rather than a lithological (porosity) change, because the latter would affect R_i and R_t equally.

No quantitative analysis is presented for the uppermost zones at the 2,155–78 ft interval. As can be readily seen from the simplified departure curves in Fig. 4, the resolution of these graphs decreases very rapidly for high R_a/R_m values. Moreover, the readings are complicated by the presence of thin dense streaks at depths of 2,160 and 2,183 ft and shaly breaks at 2,164 and 2,178 ft.

The increased resistivities for the interval at 2,260–77 ft below the water-bearing zones are caused, in all probability, by a marked decrease in porosity. The distinction between resistivity increases caused by the presence of hydrocarbons and those owing to low porosity constitutes the essence of the problem of log interpretation in carbonate reservoirs.

As an example of the use of the short normal for the determination of R_i, one could have entered the value $(R_a)_{16}/R_m = 18.6$ for the interval at 2,228–40 ft directly on the simplified short normal departure curves shown by the solid lines in Fig. 5. For $d_o = 7\frac{3}{8}$ in.,

$$\left(\frac{R_{16}}{R_m}\right)_{\text{corr.}} = 16 \approx \frac{R_i}{R_m}$$

This in turn gives $R_i/R_{mf} = F = 20.9$ and $\phi = 0.22$. The lower R_i value and higher porosity found by this procedure result from neglecting the influence of R_t on the short normal apparent resistivity. In this case, inasmuch as invasion was fairly deep, the effect is rather small and the discrepancy in porosity of 22% versus 18.5% is not too serious.

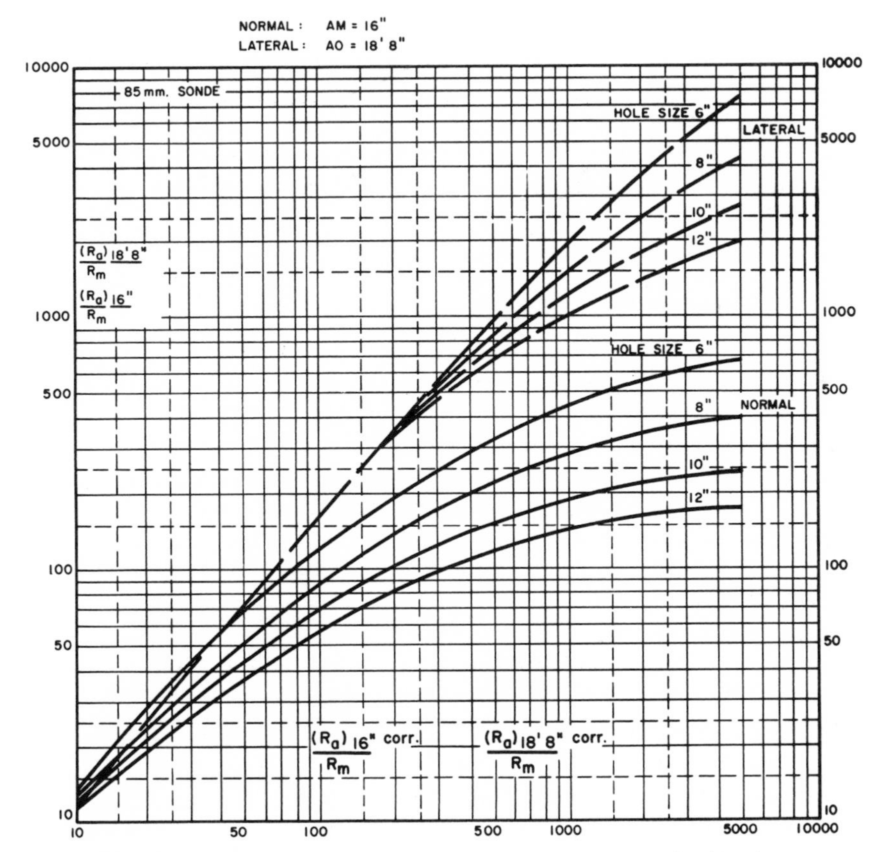

Fig. 5. Electric log simplified resistivity departure curves. (Courtesy of Schlumberger Technology Corp.)

The Limestone Curve

(Reference 54)

The example of porosity determination from the short normal curve in the Kansas City Lime showed that even for fairly deep invasion the apparent resistivity is influenced by parts of the formation beyond the invaded zone. For formation factor and porosity evaluation it is preferable to have a device that has a smaller depth of investigation and gives results less influenced by adjacent beds. These characteristics are provided by the so-called limestone curve, which is obtained with a short lateral device. In order to avoid the complicated asymmetrical formation boundary effects associated with lateral devices, this device has a symmetrical arrangement of pickup electrodes, consisting of two measuring electrodes above and two below the central current electrode. The electrodes nearest the current electrode are interconnected, as are the two outside electrodes. The spacing between the current electrode and the center of either set of pickup electrodes is normally 32 in., although other spacings are used occasionally. In general, the limestone curve represents a depth of investigation corresponding to a 32-in. lateral, and its boundary response is symmetrical. The short range of investigation also decreases adjacent bed effects. A 32-in. limestone curve will reflect fewer adjacent bed effects and, therefore, will show more detail than a 10-in. short normal. In some areas the limestone curve arrangement is run in conjunction with the laterolog, a focused resisitivity log, which is discussed later. The limestone curve run on a laterolog-7 sonde has a spacing, AO, of 37 in.

A drawback of the shallow radius of investigation is greater sensitivity to hole diameter and mud resistivity variations. The essence of limestone curve interpretation is the assumption that the apparent resistivity when corrected for hole effects gives the value of R_i, with negligible influence of R_t.

Figure 6 shows an example of an electric log, run with a 10-in. normal, a 32-in. limestone curve, and a 19-ft long lateral, through the Permian Wolfcamp Limestone in the Anderson Ranch Field, Lea County, New Mexico. There is a significant difference in detail on the limestone curve and on the 10-in. normal curve.

Pertinent data for the interpretation are as follows: $d_o = 8\frac{3}{4}$ in., $R_m = 0.4\ \Omega$m at 150°F, $R_{mf} = 0.25\ \Omega$m at 150°F. Simplified departure curves for the 32-in. limestone curve are shown in Fig. 7. The intervals 9,667–74 ft, 9,811–30 ft, and 9,879–91 ft have been analyzed.

Interval 9,667–74 *ft*: $(R_a)_{32} = 25\ \Omega$m, $(R_a)_{32}/R_m = 62.5$. From Fig. 7, $R_i/R_m = 102$ and $R_i = 40.8\ \Omega$m.

Assuming 30% residual oil saturation, $S_{wi} = 0.7$, $R_{io} = 20\ \Omega$m, and $R_{io}/R_{mf} = 80 \approx F$.

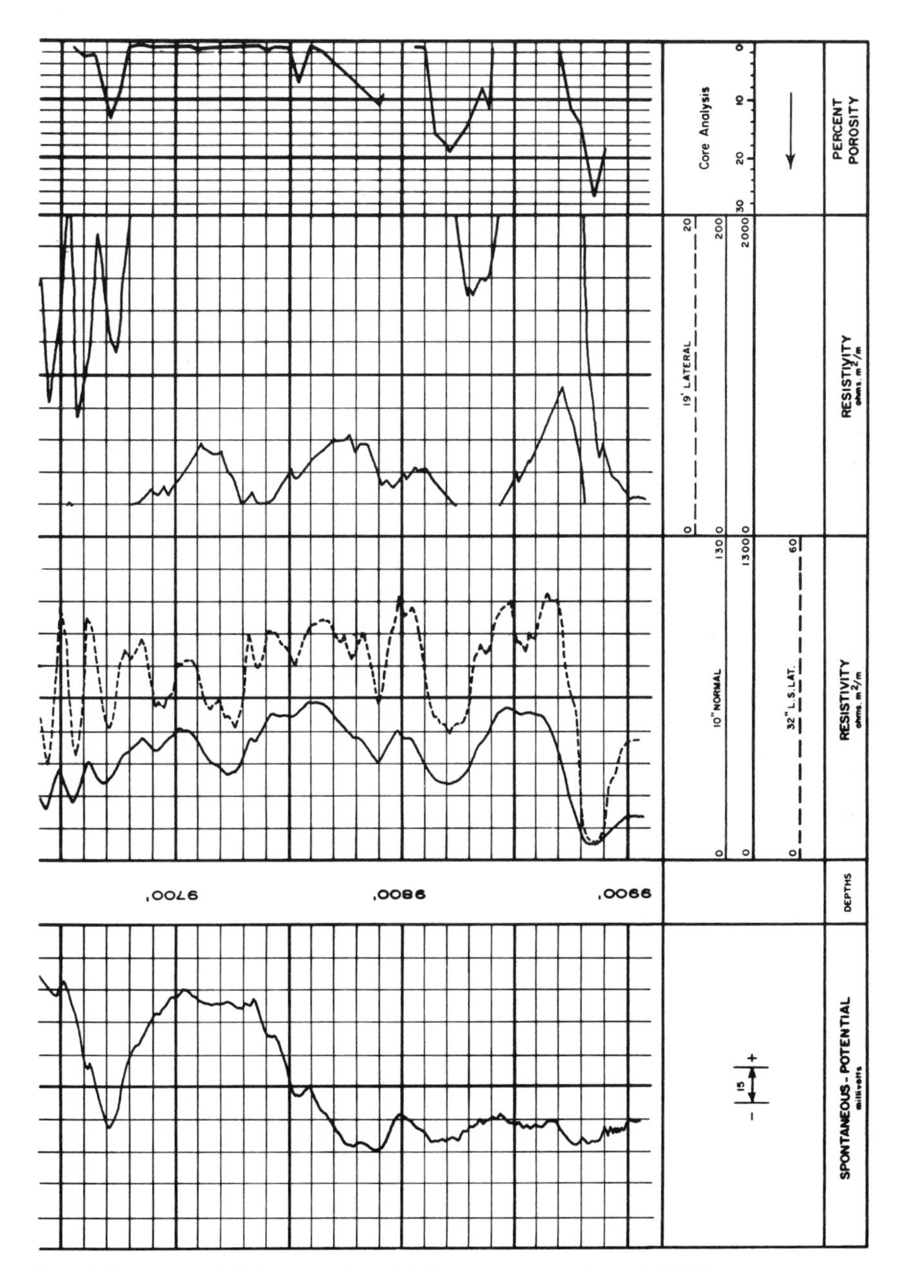

FIG. 6. Conventional E-log, together with limestone curve through Wolfcamp Limestone, Anderson Ranch Field, Lea County, New Mexico. Porosities determined by core analysis are presented on the right-hand side. (Courtesy of Continental Oil Co.)

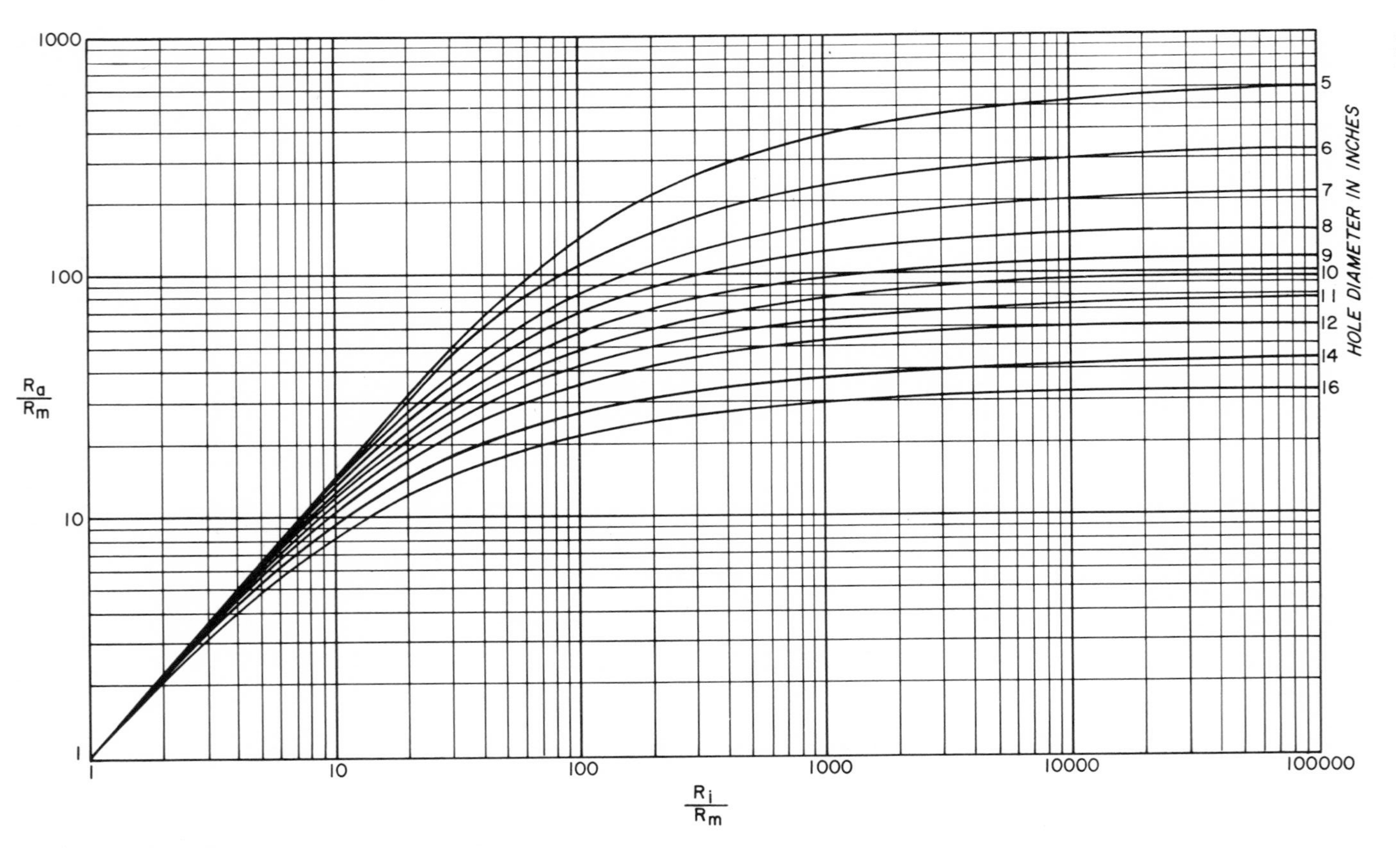

FIG. 7. Simplified departure curves for the 32-in. limestone curve.

Porosities can be found for different values of the cementation factor, m: $m = 2$, $\phi = 11.2\%$; $m = 2.2$, $\phi = 13.7\%$; $m = 2.5$, $\phi = 17.3\%$.

The highest SP value for the Wolfcamp is -81 mV, and K_t at $150° = 79$.

If the formation is assumed to be clean at the points of the highest SP reading, the following calculations can be made:

$$-81 = -79 \log \frac{0.25}{(R_w)_e}, \ (R_w)_e = 0.0234\,\Omega\text{m}, \ R_w = 0.036\,\Omega\text{m},$$

$$R_{lat} = 113\,\Omega\text{m}, \ R_{lat}/R_m = 282.$$

From Fig. 5, $(R_{a\,lat}/R_m)_{corr} \approx R_t/R_m = 170$ and $R_t = 68\,\Omega$m.

$$R_o = FR_w = 2.9\,\Omega\text{m and } S_w = \left(\frac{2.9}{68}\right)^{1/2} = 0.206.$$

Thus hydrocarbon saturation $= 79.4\%$.

Interval 9,811–30 *ft*: $(R_a)_{32} = 24\,\Omega$m, $(R_a)_{32}/R_m = 60$, $R_i/R_m = 99$, $R_i = 39.6\,\Omega$m, $R_{io} = 19.4\,\Omega$m, $R_{io}/R_{mf} = 77.6 \approx F$.

For $m = 2$, $\phi = 11.4\%$; $m = 2.2$, $\phi = 13.9\%$; $m = 2.5$, $\phi = 17.7\%$, $R_o = FR_w = 2.8$, $R_{lat} = 150\ \Omega$m, $R_{lat}/R_m = 375 \rightarrow R_t/R_m \approx 225$, and $R_t = 90\,\Omega$m.

$$S_w = \left(\frac{2.8}{90}\right)^{1/2} = 0.176 \text{ and thus hydrocarbon saturation} = 82.4\%.$$

Interval 9,879–91 *ft*: $(R_a)_{32} = 3\,\Omega$m, $(R_a)_{32}/R_m = 7.5$, $R_i/R_m = 6.4$, and $R_i = 2.57\,\Omega$m.

This lower zone is known to be partly wet, so a lower residual oil saturation in the invaded zone is assumed, with $S_{wi} = 0.85$, which yields $R_{io} = 1.93\,\Omega$m and $R_{io}/R_{mf} = 7.7 \approx F$.

For $m = 1.8$, $\phi = 32\%$; $m = 2$, $\phi = 36\%$; and $m = 2.2$, $\phi = 39\%$.

Because of the large resistivity contrast with the immediately overlying, very dense formation and the disturbing effect of the bottom of the hole, it is not felt that the long lateral reading can be relied upon to yield R_t. Therefore, no S_w determination is made for this zone.

The right-hand trace of Fig. 6 gives porosities for this section of the Wolfcamp as determined by analysis of cores taken from the same well. As shown in this figure, qualitatively the limestone curve gives excellent delineation of the porous zones. For quantitative agreement it is necessary to assume that in the upper interval the cementation factor, m, equals 2.2, in the central porous zone $m = 2.5$, whereas in the bottom zone $m < 1.8$. The use of $m = 2$ in all three zones would have given porosity values off by more than 5 porosity percents. It is possible that hole diameter departures would contribute to some extent to these variations. The large implied uncertainties in the value of the cementation factor within a single formation shows the need for porosity determinations, which are independent of the formation factor-porosity relationship. A number of such methods are discussed later. The long lateral readings used for the S_w calculations are probably too high because of the effects of dense adjacent beds, so that the low S_w values are overoptimistic. In modern logging techniques for carbonate formations, the long laterals are replaced by focusing devices, for which the adjacent bed effects are far less severe.

Induction-Electric Log

(References 38–46)

One of the logging combinations aimed at alleviating the handicap of severe bed thickness effects characteristic of the longer-spaced conventional resistivity curves is the induction-electric log, which consists of a focused induction device and a short normal.

The induction log was originally developed mainly for logging holes drilled with air or oil-base mud. The principle of induction logging involves the establishment of an alternating magnetic field surrounding the borehole, with the magnetic flux lines running largely parallel to the axis of the hole. Such a field induces in conductive formations eddy currents which are by and large concentric with the drill hole. The eddy currents in turn produce a secondary magnetic field, the strength of which is measured by the pickup coils of the induction sonde. The induction log response is proportional to the conductivity of the surrounding formations, but can be inverted to register on a resistivity scale. Most induction logs present both a conductivity and a resistivity scale. The simple concentric geometry of the induced currents lends itself well to limiting the thickness of the zone of investigation by adding focusing coils to the device, which cancel the primary field in the adjacent bed areas. The control of bed thickness effects and the depth of investigation are determined by the number and the arrangement of the focusing coils. Two commonly used arrangements are the five-coil device, denoted as 5FF40, and the six-coil, termed the 6FF40 system.

Interpretation of the induction-electric log combination is based on the fact that usually the short normal will be responsive primarily to the invaded zone resistivity, whereas the focused induction curve is strongly influenced by R_t. If one plots simplified departure curves for these two devices, the values obtained for R_i and R_t, corresponding to a given set of readings of $(R_a)_{16}$ and $(R_a)_{\text{ind.}}$, will be found to vary with the depth of invasion. For a large range of invasion diameters, however, the resulting ratio R_i/R_t is of fairly limited variation. Inasmuch as for given R_{mf} and R_w the ratio R_i/R_t is indicative of the water saturation, induction-electric log interpretation charts are often presented directly in terms of S_w, as shown in Fig. 8. In

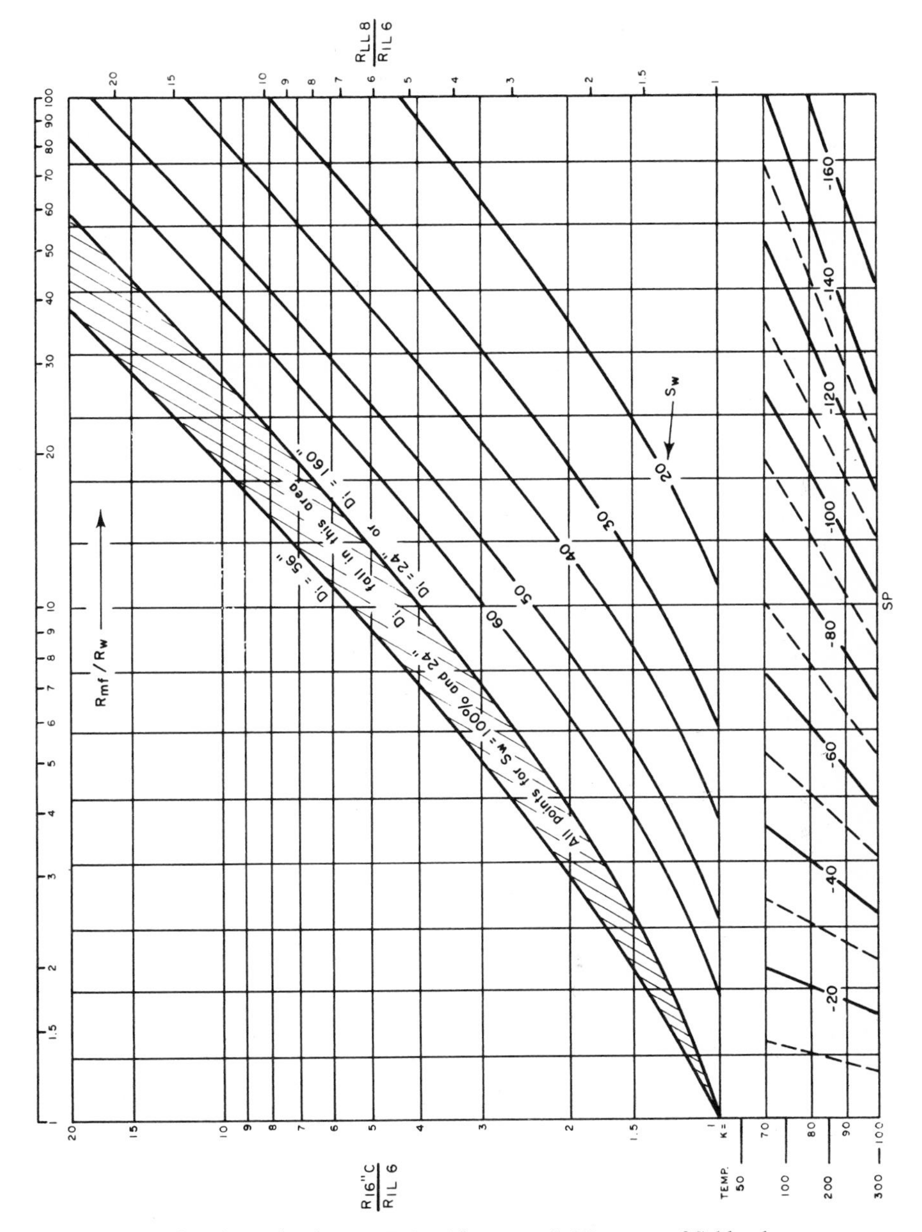

FIG. 8. Saturation determination; 6FF40, 16-in. normal. (Courtesy of Schlumberger Technology Corp.)

these charts the static SP deflection is used to determine the R_{mf}/R_w ratio.

The induction-electric log was not specifically designed for application in carbonate rocks, but rather was developed as a general successor to the standard electric log. For limestone applications it would seem preferable to combine the limestone curve with a deep-investigation induction log. To the writer's knowledge this combination is not presently available on a single sonde.

The short normal spacing for the induction electric surveys is 16 in. (Schlumberger, *Dresser Atlas*[39]) or 18 in. (Welex). Figures 9A and 9B show induction-electric log surveys for two wells through part of the Edwards Lime at the Dilworth Dome Gas Field. Figure 9A is the log of a producing field well, whereas the log in Fig. 9B represents an edge well which was wet and consequently abandoned. On the SP plot (Fig. 2) these wells represent points B and C, which have a definitely clay-free facies, and one can use $R_w = 0.017$ Ωm as determined from this plot. The interval at 11,366–82 ft in the edge well (Fig. 9B) is analyzed first.

The induction log resistivity can be obtained more accurately from the conductivity scale reading than from the small resistivity deflection; 6FF40 conductivity = 144 mmhos/m.

$$(R_a)_{ind.}\ (\equiv R_{IL}) = \frac{1{,}000}{144} = 6.94\ \Omega\text{m}$$

$$(R_a)_{16} = 37\ \Omega\text{m},\ (R_a)_{16}/R_{IL} = 5.35,\ \text{SP} = -142\ \text{mV}.$$

Formation temperature ≈ 280°F, $R_m = 0.49$ Ωm at BHT, $R_{mf} = 0.33$ Ωm at BHT.

Using these values in Fig. 8, one obtians $S_w = 0.47$.

Alternatively, $(R_a)_{16}/R_m = 75.5$ ($d_o = 7\frac{7}{8}$ in.), and from the appropriate borehole correction chart:

$$R_i/R_m \approx R_{16\,\text{corr}}/R_m = 80 \text{ and } R_i/R_{mf} = 119.$$

Assuming no residual hydrocarbons,

$R_{io}/R_{mf} = F = 119$, $R_o = FR_w = 2.02$, and $S_w = (2.02/6.94)^{1/2} = 0.54$. This indicates the presence of hydrocarbons.

For 15% residual gas in the invaded zone, $F = R_{io}/R_{mf} = 86$, $R_o = 1.46$, and $S_w = 0.46$.

This S_w value agrees quite well with that obtained from the interpretation chart. Both values are probably too low, however, because in the last analysis one should expect that $R_{IL} > R_t$ (owing to the effect of invasion) and $R_{16\,\text{corr}} < R_i$ (owing to the effect of R_t). This example is discussed again later.

The porous zone at the 10,678–700 ft interval in Fig. 9A, which has

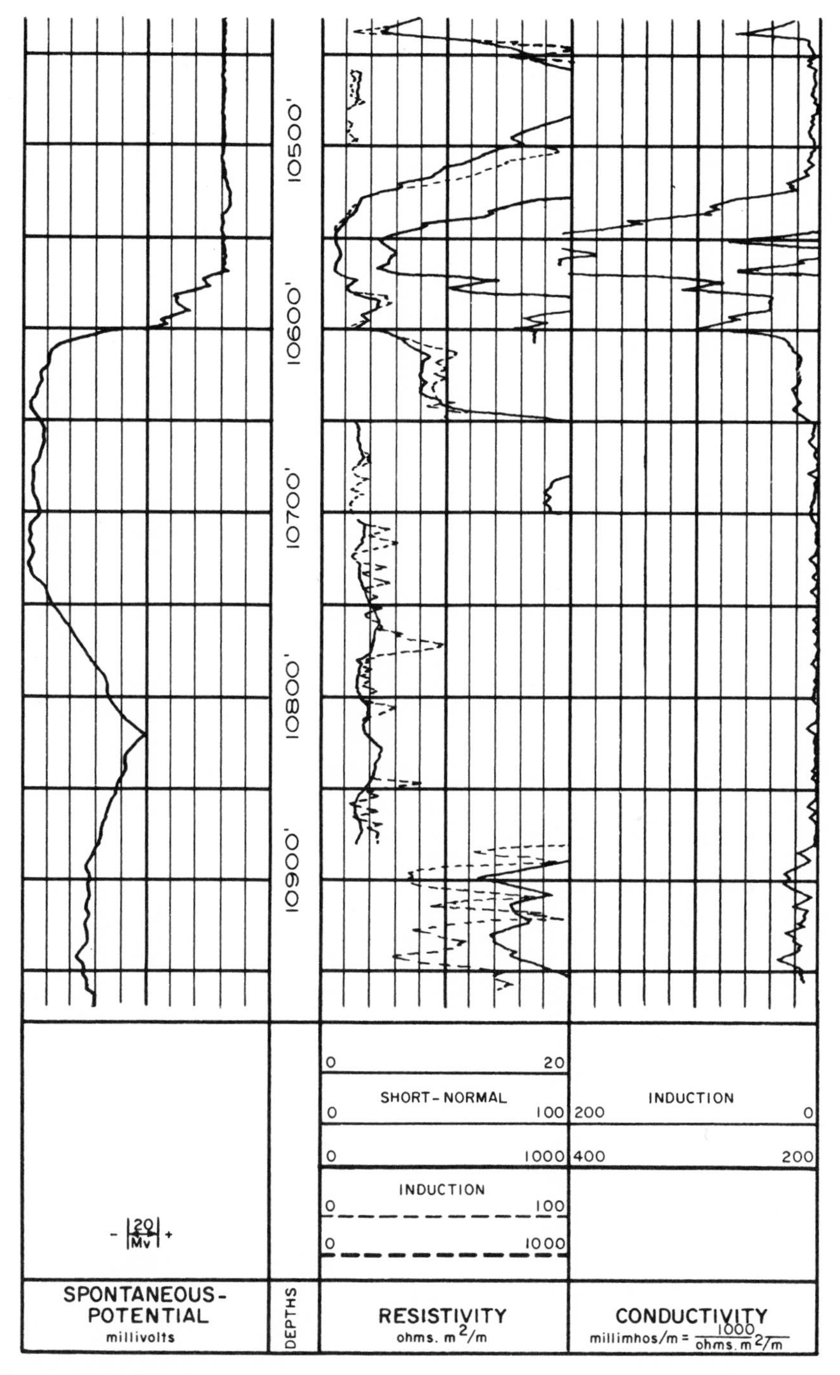

FIG. 9A. Induction-electrical surveys through part of Edwards Lime, Dilworth Dome Gas Field, McMullen County, Texas.

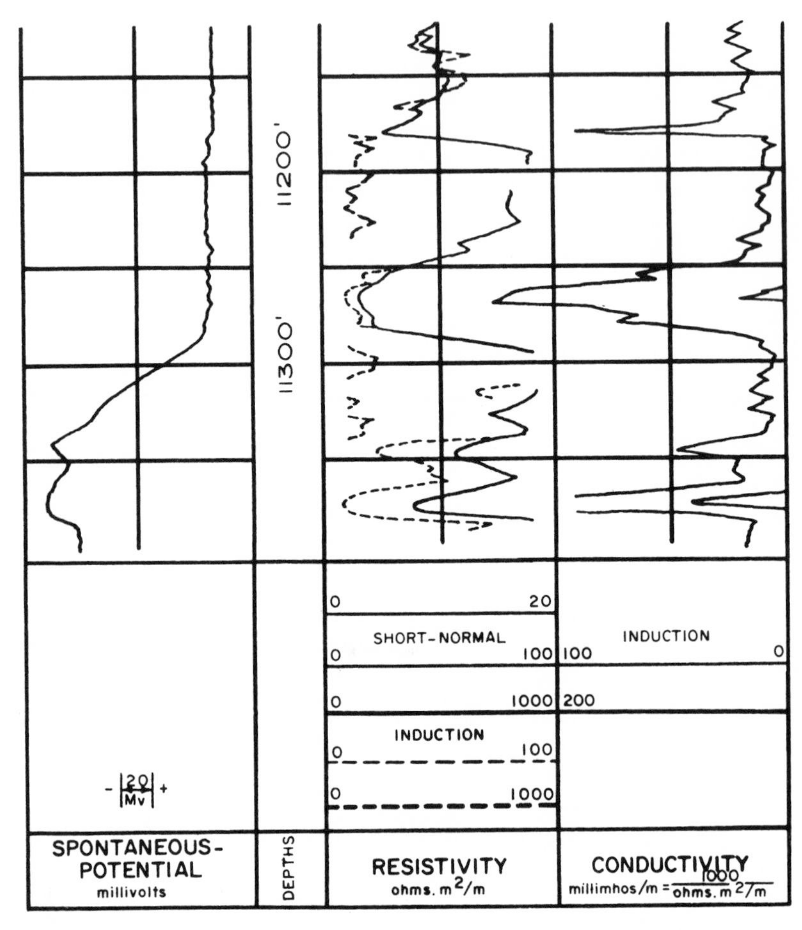

FIG. 9B.

roughly the same stratigraphic position as the zone in the previous example, is analyzed next. In this case, $(R_a)_{16} = 90\ \Omega\text{m}$, $(R_a)_{\text{ind.}} = 130\ \Omega\text{m}$, and $R_{16}/R_{IL} = 0.69$.

The cases where $R_{16}/R_{IL} < 1$ are not covered by the chart of Fig. 8; this is a common limitation of this type of charts. The alternative procedure must be used, therefore:

$$R_m = 0.64\ \Omega\text{m at BHT},\ R_{mf} = 0.4\ \Omega\text{m at BHT},\ d_o = 7\tfrac{7}{8}\text{ in.}$$
$$(R_a)_{16}/R_m = 140,\ R_i/R_m \approx R_{16\,\text{corr}}/R_m = 175,\text{ and } R_i/R_{mf} = 280.$$

Using $S_{wi} = 0.80$ gives $R_{io}/R_{mf} = F = 178$. Thus, $R_o = FR_w \approx 3\ \Omega\text{m}$, $R_t \approx (R_a)_{\text{ind.}} = 130\ \Omega\text{m}$, $S_w = (3/130)^{1/2} = 0.15$, and hydrocarbon (gas) saturation $= 85\%$.

In this case the result is probably fairly accurate, because R_i and R_t are roughly of the same magnitude so that the cross influences on $(R_a)_{\text{ind.}}$ and $(R_a)_{16}$ are not severe. Initial production of this well was 36 MMCF/D through perforations at the 10,670–765 ft interval.

Microlog; Minilog; Contact Log; Permalog
(References 47, 48, 38, 39)

The microlog and its equivalents are short-spaced devices mounted on an insulated pad, which sample resistivity immediately adjacent to the bore-face. The pad is held against the boreface by means of two springs and is molded on one of them. The other arm may hold an electrode that registers the micro-SP, and the two springs combined may be provided with instrumentation for recording a microcaliper curve. Common spacings for the microlog are $1\frac{1}{2}$ in. for microlateral and 2 in. for micronormal. The microlateral has $AM = MN = 1$ in. and is often referred to as the 1 in. × 1 in. micro inverse. The insulated pad serves to eliminate effects of the mud column to a large extent. The more compact part of the mud cake resulting from the invasion of porous formations, however, remains between the pad and the boreface.

Where no mud cake is present, as in shales and dense formations, and when the drill face is sufficiently smooth, the two curves will register the same apparent resistivities, that is, they show no separation. The presence of mud cake has a stronger influence on the micro inverse than on the 2-in. normal curves. Since invaded zone resistivities are higher than R_{mc}, the 2-in. normal registers higher than the $1\frac{1}{2}$-in. lateral, for permeable invaded strata. In certain types of swelling or heaving shales, the swollen layer, having imbibed fresh drilling fluid, may have a higher resistivity than the unaffected shale and reverse or so-called negative separation occurs.

As with all side-wall logging devices, the microsonde is lowered with the springs clamped together. Before the springs are expanded for the uphole logging run, the sonde is frequently used to register the bottom hole resistivity value of the drilling mud.

Delineation of formation boundaries and layering on the micro recordings is very sharp. Intervals with homogeneous porosity will, because of the mud cake, register low apparent resistivities and exhibit positive separation. Erratic or broken porosity will cause rapidly varying peaks and lows still having, on the average, a positive separation.

When the mud cake resistivity is known, mud cake thickness (t_{mc}) and the flushed zone resistivity can be found from microlog departure curves[38, 48] of the type shown in Figs. 10A and 10B. For the older nonhydraulic pad designs, the influence of hole diameter (boreface curvature) on the responses

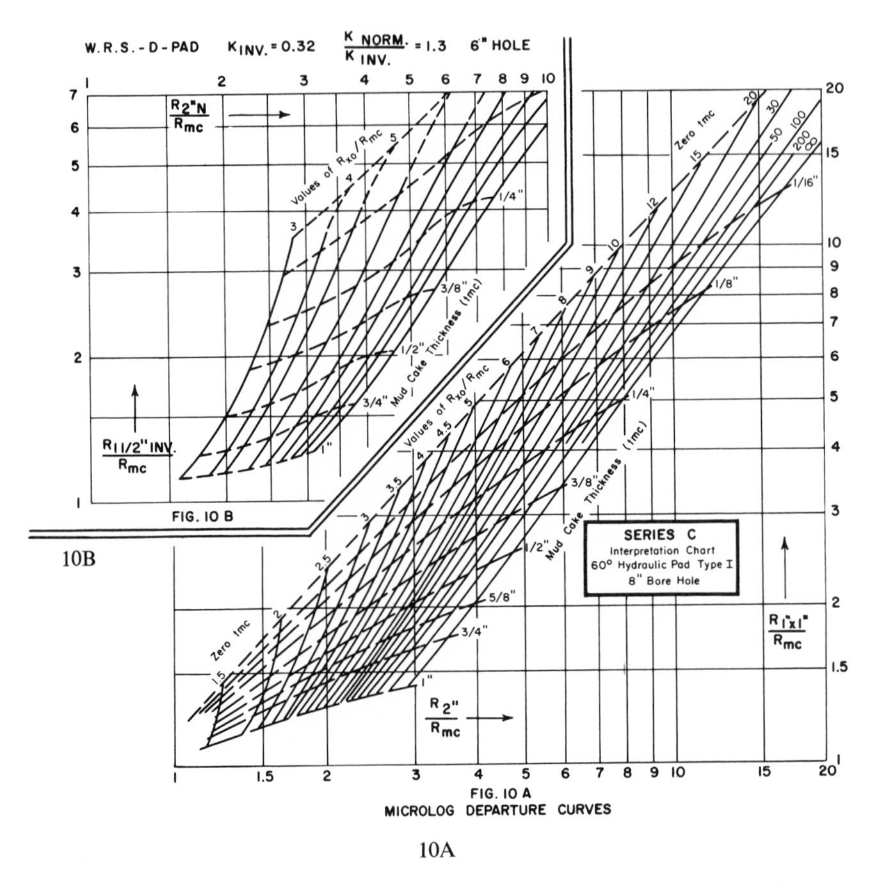

FIG. 10. Microlog departure curves. (Courtesy of Schlumberger Technology Corp.)

was sufficient to require separate departure curves for different hole diameters. For the hydraulic pads presently used the effects are small enough so that the curves made for an 8-in. diameter are reasonably applicable to most common hole sizes.

Mud cake resistivities may be obtained from empirical charts as functions of mud resistivity, temperature, and mud weight or from measurements made at the surface. The surface measurements must be corrected to the proper value at formation temperature. These corrections frequently introduce a degree of uncertainty. As can be seen from Figs. 10A and 10B, resolution of the departure curves for $R_{xo}/R_{mc} > 30$ is very poor, so that quantitative microlog interpretation of formations with less than 10% porosity is rather futile.

Errors in R_{mc} will displace the position of a point on the microlog interpretation charts along a 45° line. If mud cake thickness can be obtained from the microcaliper ($t_{mc} = \frac{1}{2}$ negative separation between bit-size line and caliper), one can move the plotted point at 45° to intersect the line for the proper t_{mc}. For the new point, $R_{xo}/R_{mc'}$ is read from the chart and $R_{2\text{ in.}}/R_{mc'}$ from the bottom scale; R_{xo} is found from:

$$R_{xo} = \left(\frac{R_{2\text{ in.}}}{R_{2\text{ in.}}/R_{mc'}}\right)\frac{R_{xo}}{R_{mc'}} \tag{17}$$

Figure 11 shows a microlog and IE (induction-electrical log) survey through a section of the Upper Hunton Lime (Bois d'Arc), in the Anadarko Basin, Caddo County, Oklahoma. The R_{mc} value derived from R_m by using a conversion chart is 0.5 Ωm. The interval of interest occurs at 19,524–62 ft. The microlog for the central portion at the 19,540–50 ft interval is analyzed here.

$$R_{1\frac{1}{2}\text{ in.}} = 1\ \Omega\text{m},\ R_{2\text{ in.}} = 1.7\ \Omega\text{m},\ R_{1\frac{1}{2}\text{ in.}}/R_{mc} = 2, \text{ and } R_{2\text{ in.}}/R_{mc} = 3.4$$

From Fig. 10A, $R_{xo}/R_{mc} = 20$ and $t_{mc} = 0.58$ in.

The microcaliper shows 0.35 in. for the mud cake thickness. This value is a lower limit. The value found from the microlog indicates that the hole diameter is 0.46 in. larger than the bit size. This seems to be in agreement with the values shown on the caliper above and below the porous interval. $R_m = 0.49$ Ωm at 298°F $= 0.52$ Ωm at 285°F; $R_{mf} = 0.36$ Ωm at 285°F; and $R_{xo}/R_{mf} = 27.8$. For $S_{wi} = 0.85$, $R_{io}/R_{mf} = 20 = F$; and for $m = 2$ this gives $\phi = 22.4\%$.

The saturation is calculated as follows:
SP $= -74$ mV; $K_t = 97$ at 285°F $\rightarrow R_{we} = 0.064$ Ωm $= R_w$; $R_{\text{ind.}} \approx 300$ Ωm $\approx R_t$; $R_o = FR_w = 1.28$; and $S_w = 0.07$. Thus, hydrocarbon saturation = 93%.

Figure 12 shows a microlog and an ES (electrical log survey) through the H zone of the Rodessa Limestone Member, Trinity Group, of the Lower Cretaceous Comanche Series, in Prairie Lake Field, Anderson County, Texas.

The following data are available for the microlog interpretation:
Zone—8,976–90 ft; $R_m = 0.31$ Ωm at 200°F; $R_{mc} = 0.66$ Ωm; $R_{mf} = 0.2$ Ωm; $d = 6.5$ in.; $R_{1\frac{1}{2}\text{ in.}} = 1.65$ Ωm; $R_{2\text{ in.}} = 3.2$ Ωm; $R_{1\frac{1}{2}\text{ in.}}/R_{mc} = 2.5$; and $R_{2\frac{1}{2}\text{ in.}}/R_{mc} = 4.85$.

From Fig. 10B (chart for $d = 6$ in.), $t_{mc} \approx \frac{3}{8}$ in.; $R_{xo}/R_{mc} = 42 \rightarrow R_{xo}/R_{mf} = 143.9$; using Eq. 8 with $S_{wi} = 0.7 \rightarrow F = 70$; and $\phi = 12\%$ (using $m = 2$).

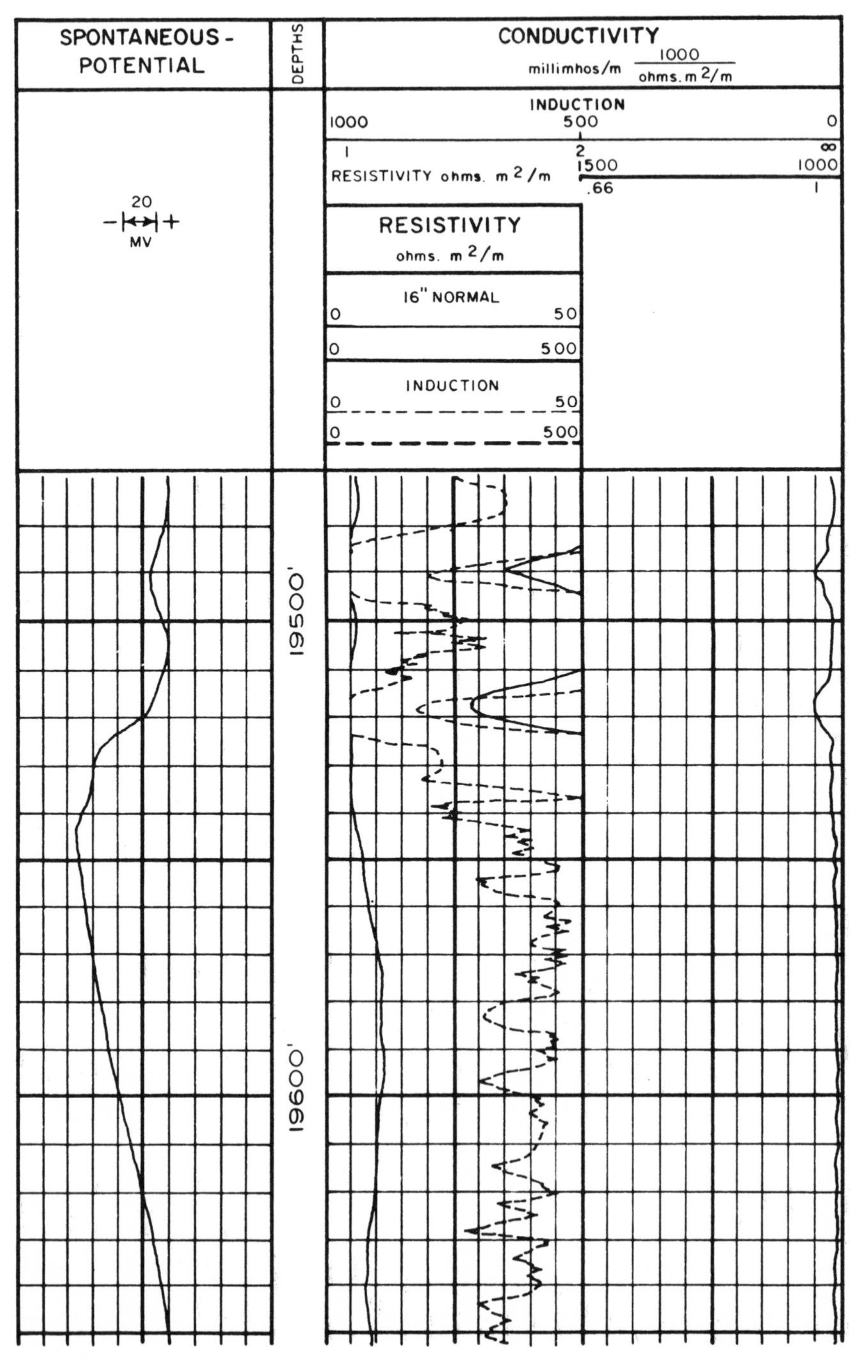

FIG. 11. Induction-electrical log and microlog through Upper Hunton Lime (Bois d'Arc), Anadarko Basin, Caddo County, Oklahoma.

FIG. 11 *continued.*

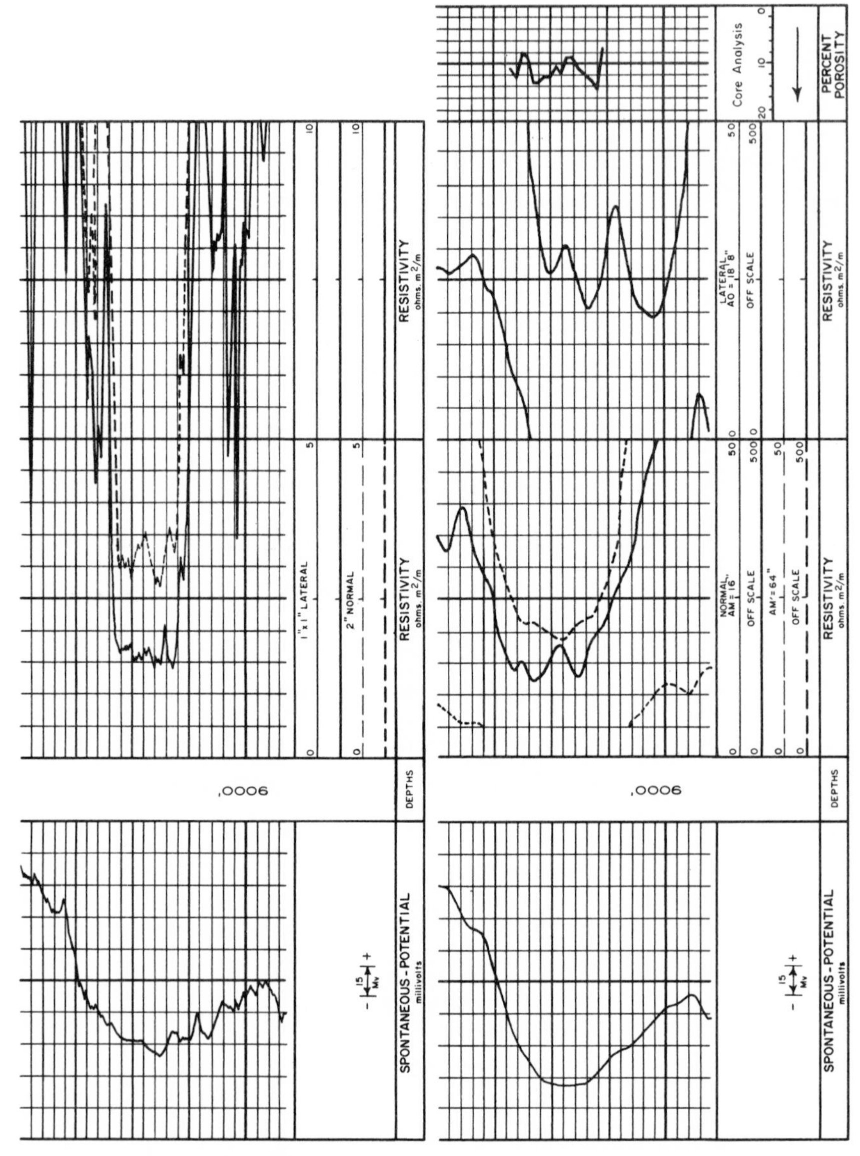

FIG. 12. Microlog and electrical log through Rodessa Limestone Member, Trinity Group, Prairie Lake Field, Anderson County, Texas. (Courtesy of Continental Oil Co.)

The average of the core analysis porosities for the zone is 11.3%. From the SP curve, SP = -99 mV; $K_t = 86$ at 200°F; $R_{we} = 0.014\ \Omega m \rightarrow R_w = 0.023\ \Omega m$; and $R_o = FR_w = 1.6\ \Omega m$.

From the long lateral, $R_{lat}/R_m \approx 65 \rightarrow R_t/R_m = 48$; $R_t = 14.9\ \Omega m$; and $S_w = (R_o/R_t)^{1/2} = 0.32$. Thus, the hydrocarbon saturation = 68%.

A drill-stem test (DST) at 6,962–7,007 ft yielded 1,800 ft gas cut water cushion, 180 ft oil cut mud (OCM), and 180 ft light green oil.

Microlaterolog (Microguard, Mini-focused log)
(References 38, 39, 49, 50)

The preceding section showed that the quantitative interpretation of the microlog is limited to formations having fairly high porosity and is handicapped by the difficulty of obtaining reliable in-place values for the mud cake resistivity. The microlaterolog partly overcomes these limitations by using a focused current beam, which strongly reduces the effect of the mud cake.

The earlier models of the microlaterolog were usually run in combination with a micro inverse, and departure curves of the type presented in Fig. 10 were used to find R_{xo}/R_{mc}. This practice has unfortunately been abandoned. The present method is to find corrections to the microlaterolog apparent resistivity as a function of mud cake thickness, the latter being obtained from the microcaliper survey. This type of correction is very simple to apply. It has the drawback, however, that the caliper gives correct values of t_{mc} only if the borehole is drilled to gauge, that is, has a diameter equal to the bit size. In cases of borehole enlargement, the mud cake thickness is underestimated, resulting in values of R_{xo} which are too low and porosity estimates which are overoptimistic.

In saline muds, mud cake thicknesses are usually very small and the microlaterolog reading frequently gives a close approximation of R_{xo}. Where mud cake is present, however, the same overoptimistic interpretation results, so that one should at least check the microcaliper for mud cake indications. Examples of microlaterologs run in saline mud are given later in the discussion of the salt mud survey and the movable oil plot.

In movable oil calculations the assumption is made that for saline muds the mud cake is so thin that the effect on the microlaterolog apparent resistivity is negligible. When mud cake effects are present, overoptimistic values for porosity and movable oil are obtained.

Salt Mud Survey (Movable Oil Plot; Laterolog—Microlaterolog Combination)
(References 51–55, 38, 39)

In saline muds (R_m at formation temperature $< 0.1\ \Omega m$) conventional unfocused electric logs lose much of their usefulness. The shunting effect of the highly conductive mud column severely increases the influence of adjacent beds so that neighboring zones are largely averaged, and no representative readings are obtained for thin porous intervals of interest. Furthermore, the high R_t/R_m ratios decrease the resolution of the logs even for thick homogeneous beds. As can be seen from the departure curves for $R_t/R_m \geqslant 100$ (Figs. 4 and 5), a small change in the apparent resistivity can mean a very significant change in the true resistivities.

A very useful survey method for salt mud wells is the laterolog-microlaterolog combination. These devices employ focused sheets of current that minimize adjacent bed effects. The thickness of the current sheet at the boreface is of the order of 3 ft for the laterolog[51] and a few inches for the microlaterolog. The interpretation of the responses of both devices is actually simpler in saline than in fresh muds. The invasion of porous formations by saline mud filtrate will make the invaded zone resistivity quite low, so that usually $R_i \leqslant R_t$. As a consequence of this, for moderate invasion the laterolog apparent resistivity becomes a good approximation of R_t. The microlaterolog response is simplified by the fact that salt muds give rise to little or no mud cake, and hence the microlaterolog reading is a reasonable approximation of R_i.

Inasmuch as in porous calcareous rocks interstitial clays usually play a negligible role, when the formation waters are saline one can use Archie's formula for representation of the hydrocarbon saturations in the invaded zone and undisturbed formation. The difference between these saturations represents the fraction of the hydrocarbons flushed out of the invaded zone. This portion, which should be equal to that recoverable by waterflooding, is referred to as movable oil. The formulas for the calculation of the movable oil saturation, S_{mo}, are developed as follows:

$$\begin{aligned} S_{mo} = S_{wi} - S_w &= \left(\frac{R_{io}}{R_i}\right)^{1/2} - \left(\frac{R_o}{R_t}\right)^{1/2} \\ &= \left(\frac{R_{io}}{R_i}\right)^{1/2} \left[1 - \left(\frac{R_w}{R_{mf}} \cdot \frac{R_i}{R_t}\right)^{1/2}\right] \\ &= S_{wi} \left[1 - \left(\frac{R_w}{R_{mf}} \cdot \frac{R_i}{R_t}\right)^{1/2}\right]. \end{aligned} \tag{18}$$

For oil-bearing porous limestones S_{wi} normally varies from 0.7 to 0.85, whereas for gas zones it is of the order of 0.8–0.9. For the case where both R_w and R_{mf} represent nearly saturated solutions, $R_w \approx R_{mf}$ and one can use the simplified relation:

$$S_{mo} = 0.75 \left[1 - \left(\frac{R_i}{R_t}\right)^{1/2}\right]. \tag{19}$$

The porosity can be obtained from the microlaterolog reading by using:

$$\phi = F^{-1/2} = \left(\frac{R_{io}}{R_{mf}}\right)^{-1/2} = \frac{(R_{mf}/R_i)^{1/2}}{S_{wi}} \tag{20}$$

An alternative way of calculating movable oil content is to determine an apparent formation factor and porosity from R_t, ignoring the possible presence of hydrocarbons, and another set of values for these quantities from R_i. The difference in the two apparent porosities is the amount of movable oil.

With modern equipment these calculations can be performed while logging by an electronic computer in the instrument truck. The continuous recording of the result is called a movable oil plot. Similar well site computations and continuous recordings are performed for water saturation, porosity, irreducible oil, etc., using various logging combinations. Schlumberger refers to these recordings as synergetic logging systems.

Figure 13 shows copies of a microlaterolog and laterolog through the Red River Formation in the Buffalo Field, South Dakota, which contains several dolomitized porous zones. The pertinent data for the 8,558–73 ft porous interval are as follows.

Microlaterolog apparent resistivity $R_{MLL} = 0.5$–$1\ \Omega$m; laterolog-7 apparent resistivity $R_{LL7} = 11.5$–$23\ \Omega$m; and $R_{mf} = 0.05\ \Omega$m at 60°F $= 0.019\ \Omega$m at 168°F.

From production tests of a well in the same section, $R_w = 0.49\ \Omega$m at 68°F $= 0.2\ \Omega$m at 168°F. Thus, $R_{MLL}/R_{LL7} \approx R_i/R_t \approx 0.04$ and $R_w/R_{mf} \approx 10$.

Assuming S_{wi} has an average value of 0.75, one obtains $S_{mo} = 0.75\ [1 - (0.4)^{1/2}] = 0.28$.

The total oil saturation is equal to that of the residual oil plus the movable oil: $S_o = 0.25+0.28 = 0.53$ or 53% of pore space. The value of porosity is found to be equal to $\phi = (R_{mf}/R_i)^{1/2}/0.75 = 0.183$–$0.258$, that is, the porosity is 25.8% in the most porous part at the 8,563.5–66 ft interval and 18.3% in the less porous parts at 8,558–63.5 and 8,571–73 ft intervals. A dense streak occurs from 8,568 to 8,571 ft. The average porosity for the producing Red River intervals in the Buffalo Field found from core analyses is 21%.

The calculated oil saturation of 53% seems in reasonable agreement with the initial production figure of 173 B/D oil and 65 B/D water.

Inasmuch as the microlaterolog is more detailed than the laterolog, point by point readings cannot be directly compared in highly varying or laminated beds. For intervals where the microlaterolog shows appreciable variations within thicknesses of less than 3 ft, a harmonic average, R_{av}, of its reading

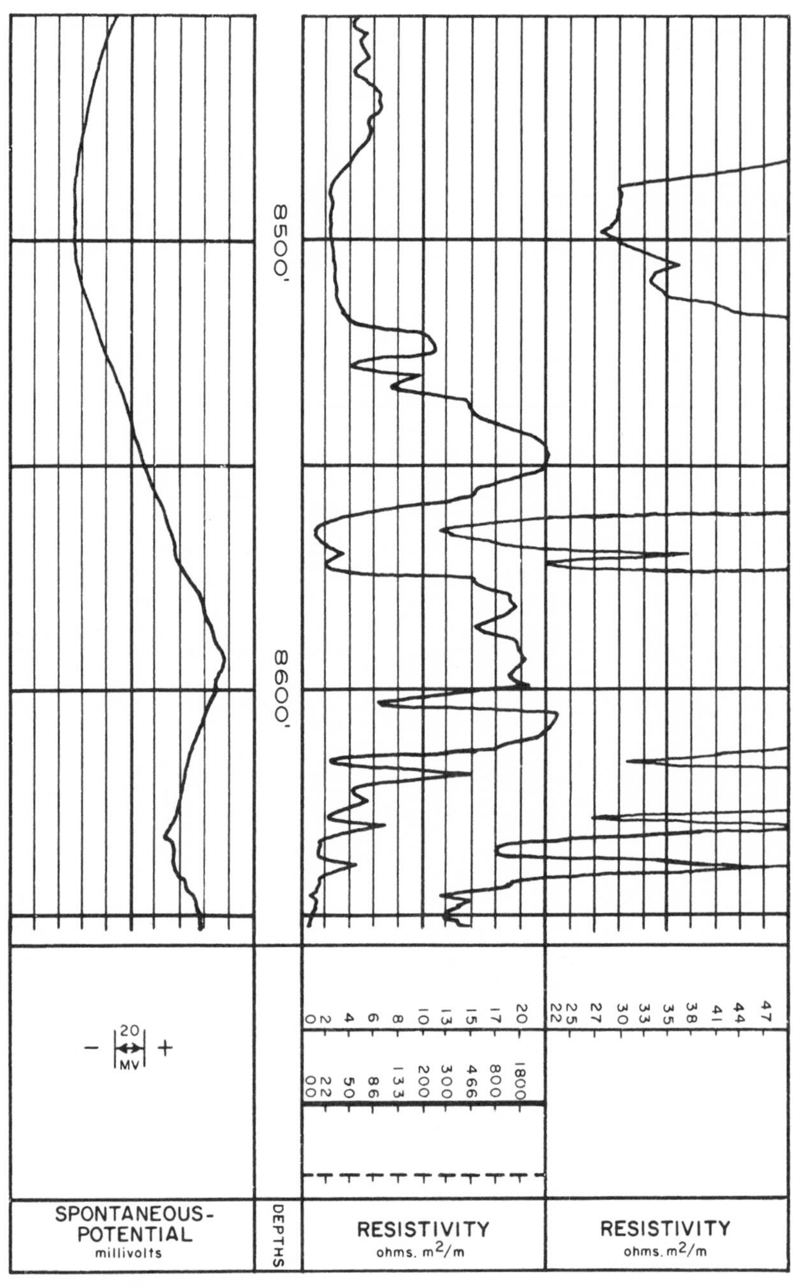

FIG. 13. Salt mud survey, Red River Formation, Buffalo Field, Harding County, South Dakota. Laterolog 7; $R_m = .1$ at 60°F; $R_{mf} = .05$ at 60°F; $R_{mc} = .19$ at 60°F; BHT = 168°F.

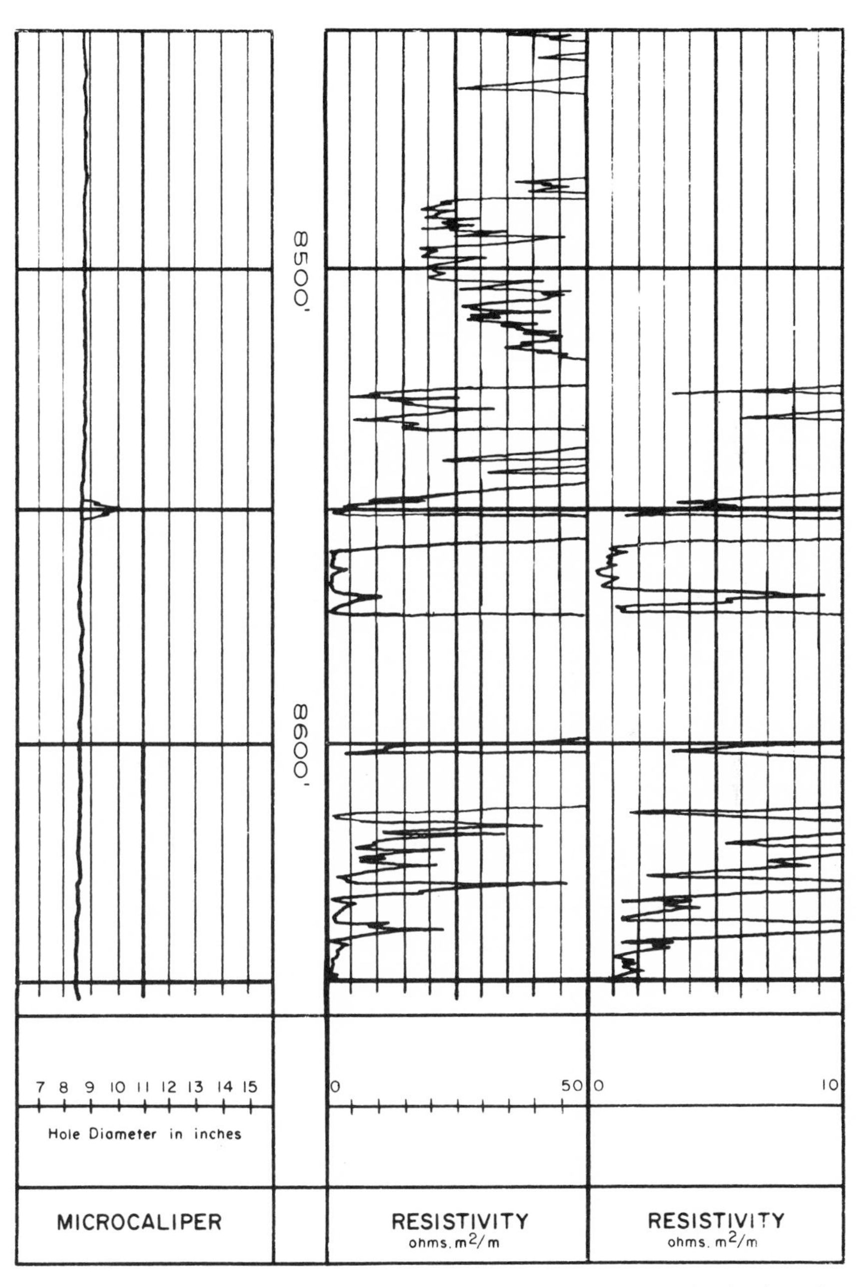

FIG. 13, *continued;* microlaterolog.

over consecutive 3-ft intervals should be taken. The harmonic average R_{av} is formed by:

$$\frac{H}{R_{av}} = \frac{h_1}{R_1} + \frac{h_2}{R_2} + \ldots \tag{21}$$

where H is the thickness over which the average is taken, and h_j and R_j ($j = 1, 2, \ldots$) are the thicknesses and the resistivities, respectively, of the successive layers within the averaged interval.

When using the movable oil plot one must make sure that the instrument setting for R_w is approximately correct. Inasmuch as saline mud columns are frequently rather inhomogeneous, the SP curve may show severe base line drifts and cannot always be relied upon. Where possible, R_w values should be obtained from wells in the area producing with a water cut.

A second example of the salt mud survey is shown in Fig. 14 for a section of Ellenburger Limestone in the Crawar Field, Ward County, Texas. The microlaterolog shows the very rapid lithological changes typical of the Ellenburger Formation. The laterolog in this case is referred to as laterolog-3, which is more commonly known as a guard electrode device. The focusing of current in this device is achieved by two long (5-ft) metal guards above and below the central current electrode. The surface of the guards forms an equipotential surface parallel to the axis of the borehole. The current flow, which is perpendicular to the equipotential surfaces, will penetrate the formations as a thin sheet emanating from the current electrode. The latter had a length of 1 ft for the run in Fig. 14. Other lengths of electrode are used, sometimes as small as 3 in. The depth of penetration of laterolog-3 is generally larger than that for laterolog-7 (used in the example of Fig. 13). It is clear from comparison of the details of the microlaterolog and laterolog-3 curves that the latter still averages the responses of several feet of formation.

The porous 8,164.5–67.5 ft interval shown on the laterolog is analyzed next: $R_{LL3} = 150\ \Omega\text{m} \approx R_t$. The microlaterolog shows laminations in this interval with thicknesses from top to bottom of $\frac{1}{2}$ ft at 80 Ωm, $1\frac{1}{2}$ ft at 35 Ωm, and 1 ft at 2 Ωm.

For porosity calculations each of these layers should be treated separately. For the movable oil computation one can form the harmonic average: $3/R_i = 0.5/80 + 1.5/35 + \frac{1}{2}$, which gives $R_i = 5.5\ \Omega\text{m}$.

$R_{mf} = 0.039\ \Omega\text{m}$ at 82°F and $0.027\ \Omega\text{m}$ at 118°F.

From the flat SP response one can assume $R_w \approx R_{mf}$ and thus:

$$S_{mo} = 0.75\left[1 - \left(\frac{5.5}{150}\right)^{\frac{1}{2}}\right] = 0.605.$$

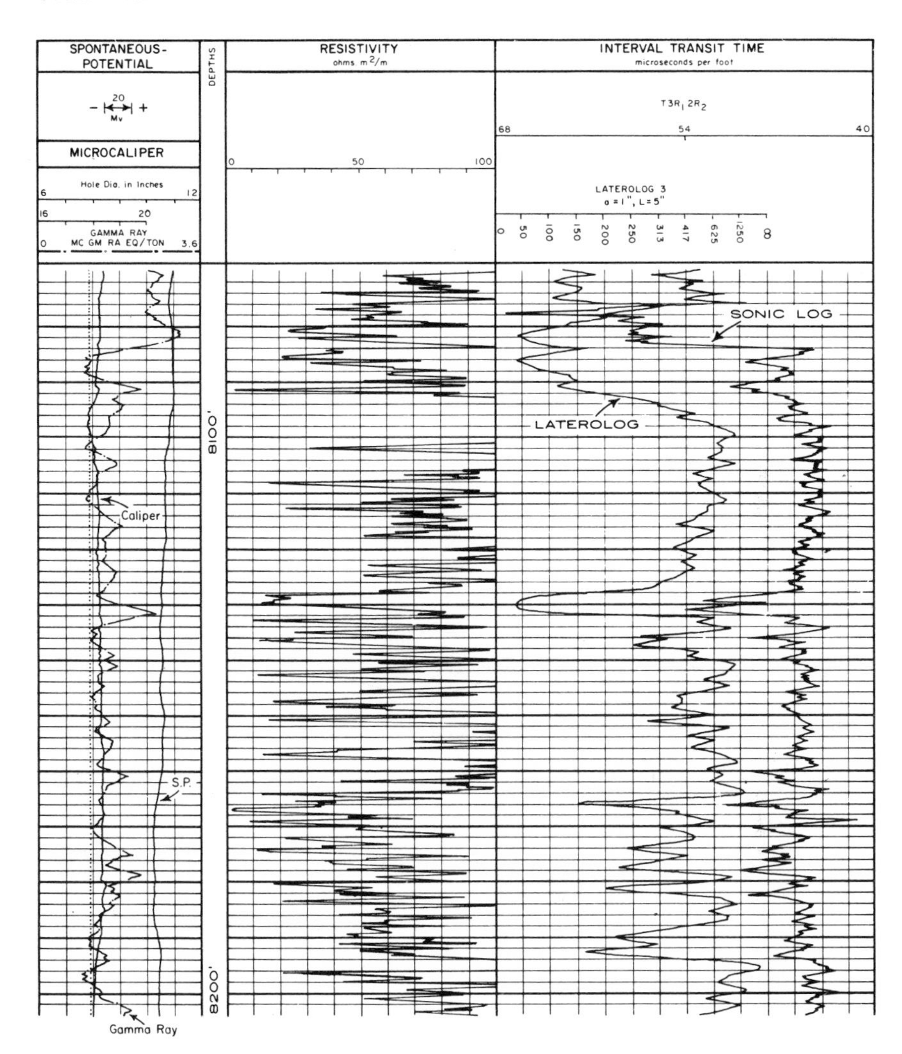

FIG. 14. Microlaterolog, laterolog-3, and sonic log through Ellenburger Formation, Crawar Field, Ward County, Texas; salt mud survey. (Courtesy of McCulloch Oil Corp. of California.)

Initial production was 386 B/D oil with no water from perforations between depths of 8,080 and 8,174 ft, including the interval analyzed. Assuming $S_{wi} = 0.75$, one has R_{io} for the three different laminations equal to 45.6, 20, and 1.14 Ωm, respectively. The corresponding values of F are 1,695, 741, and 41.5; and the porosities (for $m = 2.2$) are 3.4, 5, and 18%, respectively. The average porosity for the 3-ft interval is 9%. Later, this value is compared with that obtained from the sonic log also shown in Fig. 14.

Dual Induction-Laterolog

(References 38, 39, 45, 56, 57)

One of the most sophisticated combinations to replace conventional logging in fresh muds is the dual induction–laterolog-8 survey, which includes the following logs all recorded simultaneously: (1) a deep-investigation induction log, the 6FF40; (2) a medium-investigation induction log; (3) a shallow-investigation laterolog, with eight electrodes, the laterolog-8; and (4) an SP curve. The apparent resistivities of the first three curves are registered on a logarithmic scale and are denoted by R_{ILd}, R_{ILm}, and R_{LL8}, respectively.

With the aid of simplified departure curves as shown in Fig. 15, the ratios R_{LL8}/R_{ILd} and R_{ILm}/R_{ILd} allow determination of R_{xo}/R_t, R_t/R_{ILd}, and the depth of invasion, provided the beds are sufficiently thick and invasion is not too deep. For beds less than 6 ft in thickness, bed thickness corrections are required for the induction curves. The shallow-investigation laterolog-8 frequently requires corrections for hole effects. For deep invasion ($D_i > 50$ in.) the induction log apparent resistivities need additional corrections for residual skin effect, which is a nongeometrical decrease of field strength with distance from the source for alternating electromagnetic fields. The various correction charts and details of the techniques employed are given in the Schlumberger *Induction Log Correction Charts*.[45]

Figure 16 shows an example of a dual induction–laterolog-8 survey through the Lower Pennsylvanian Canyon-Strawn Formation in Hockley County, Texas. The interval of interest occurs at a depth of 10,062–82 ft. The upper part of this zone at 10,062–68 ft is analyzed. The following data apply: $d_o = 7\frac{7}{8}$ in.; $R_m = 0.52\ \Omega\text{m}$ at 154°F; $R_{mf} = 0.39\ \Omega\text{m}$ at 154°F; $R_{LL8} = 143$ Ωm; corrected for hole effect, $R_{LL8c} = 140\ \Omega\text{m}$; $R_{ILm} = 37\ \Omega\text{m}$; $R_{ILd} = 26$ Ωm; $R_{LL8c}/R_{1Ld} = 5.38$; and $R_{ILm}/R_{ILd} = 1.42$.

From Fig. 13, $D_i = 44$ in., $R_{xo}/R_t = 8.3$, and $R_t/R_{ILd} = 0.915$. The value of D_i indicates that residual skin effect corrections may be neglected. The determination of R_{xo}/R_t would suggest the use of Eq. 13 for the calculation of S_w. This equation, however, assumes that $R_w = R_{we}$, and the large SP deflection suggests very saline water and a correspondingly large R_{we} correction. The more explicit relationships are, therefore, used.

From the above values, $R_t = 22.9\ \Omega\text{m}$ and $R_{xo} = 190\ \Omega\text{m}$. SP $= -156$ mV, and $K_t = 81$ at 154°F; thus, $-156 = -81 \log(0.39/R_{we})$, $R_{we} = 0.00625\ \Omega\text{m}$, and $R_w = 0.017\ \Omega\text{m}$. Also, $R_{xo}/R_{mf} = 488$. For $S_{wi} = 0.85$, $F = R_{io}/R_{mf} = 353$, $R_o = 6\ \Omega\text{m}$, and $S_w = 0.513$.

In a clean limestone this value of the water saturation predicts a very significant water cut, whereas the high value of F shows a porosity of 7% or less.

A drill-stem test at the 10,053–67-ft interval yielded 5,952 ft gas, 43 ft oil, 165 ft oil and gas cut mud, and 180 ft sulfur water.

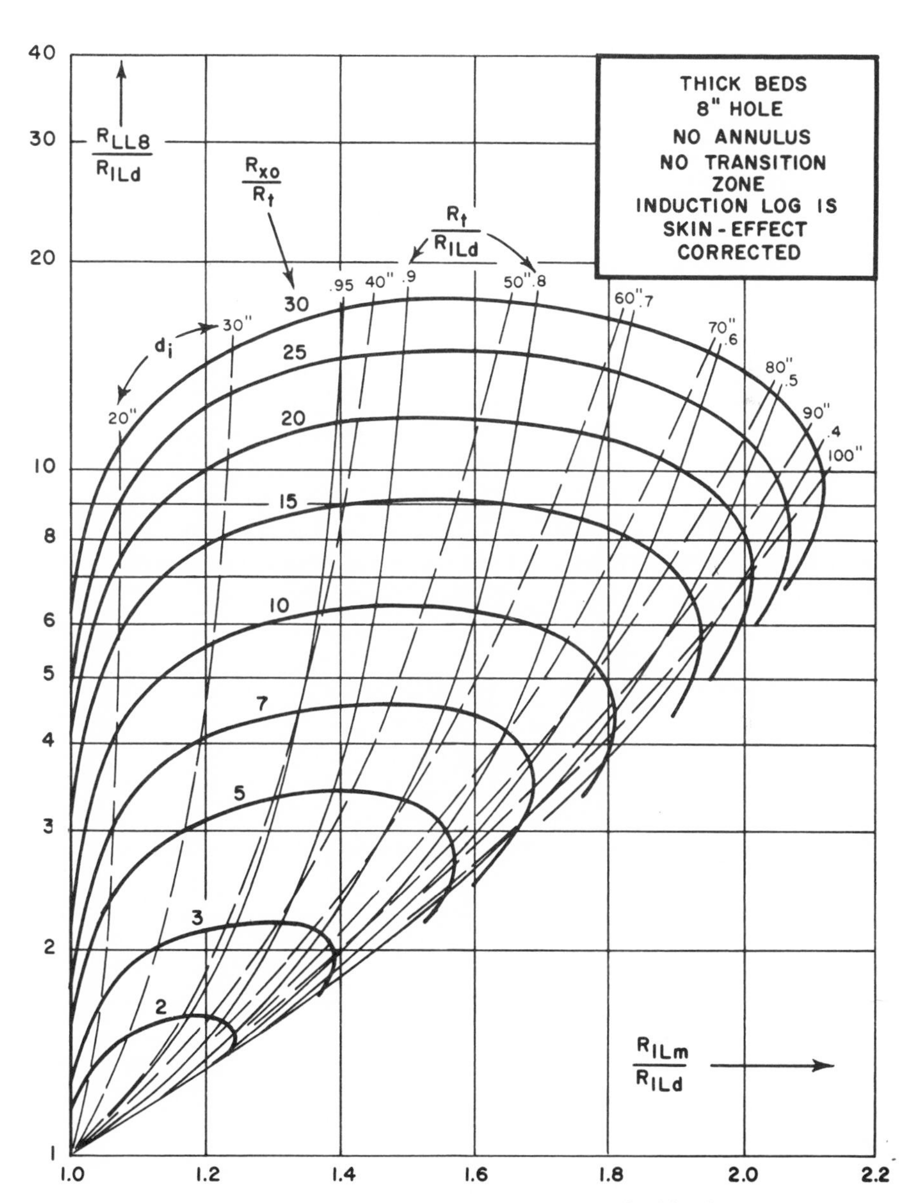

FIG. 15. Dual induction-laterolog; ILd-ILm-LL8. (Courtesy of Schlumberger Technology Corp.)

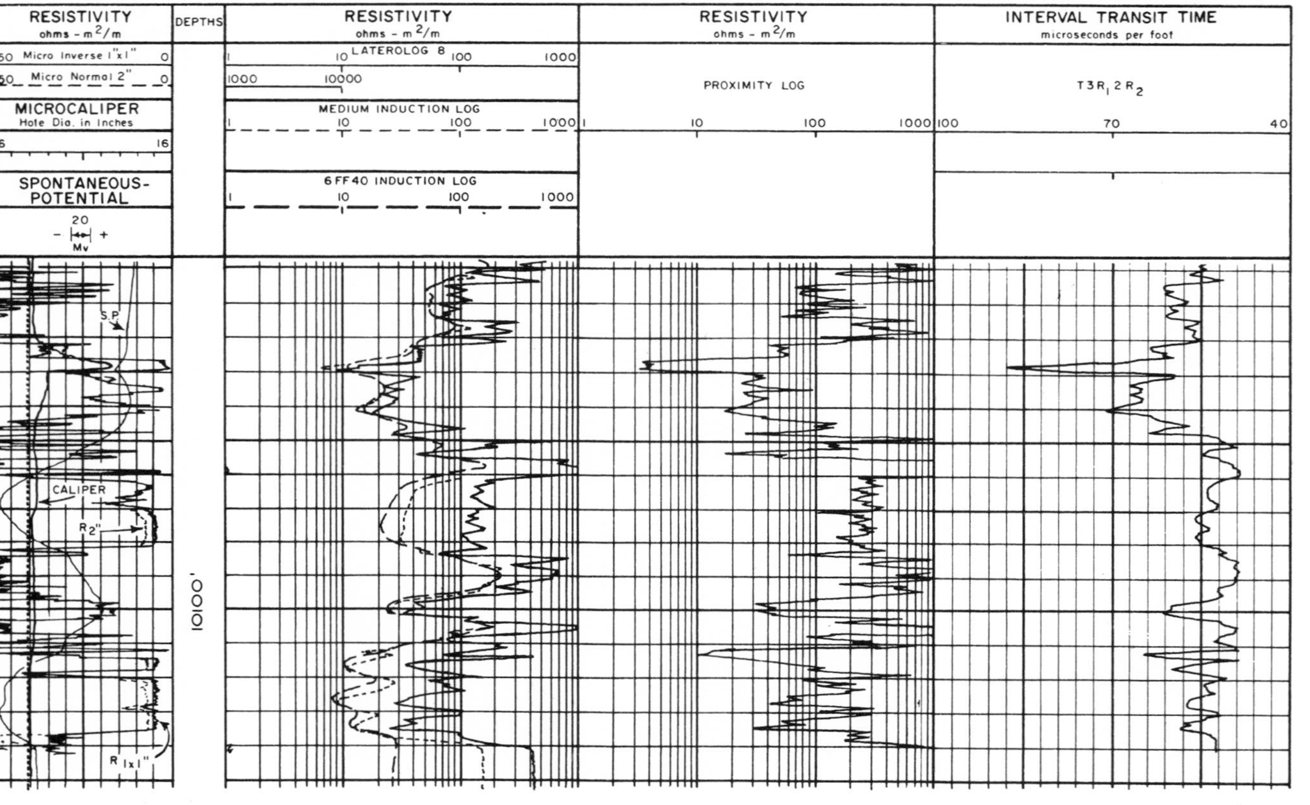

FIG. 16. Dual induction–laterolog–8, proximity log-microlog, and sonic log through Canyon-Strawn Formation, Hockley County, Texas.

Thus the R_{we} correction was appreciable. If Eq. 13 was used, ignoring this correction, then:

$$-156 = -81\left(\log 8.3 + 2\log \frac{S_{wi}}{S_w}\right)$$

which gives $S_{wi}/S_w = 4.18$; and $S_{wi} = 0.85$ yields $S_w = 0.204$. This result would normally signify water-free hydrocarbon production and, therefore, would have been quite misleading. The use of Eq. 13 should be restricted to cases where the R_{we} correction is expected to be small.

The Proximity Log
(Reference 38)

The proximity log is a contact resistivity device with deeper investigation capability than the microlaterolog. Its apparent resistivity, R_{PL}, is virtually independent of mud cake effects, but requires at least moderate depth of invasion in order not to be affected by R_t. When $R_{xo} \leqslant 0.1R_t$, an invasion diameter of 20 in. is needed to have negligible influence of R_t; whereas for $R_{xo} \leqslant R_t$, a value of $D_i \geqslant 30$ in. is required. For $R_{PL}/R_{mc} < 25$, the effect of mud cake is negligible for $t_{mc} \leqslant \frac{3}{4}$ in. For $R_{PL}/R_{mc} = 100$, mud cake effects can be neglected for $t_{mc} \leqslant \frac{1}{4}$ in. but some correction is required for larger thicknesses.

Figure 16 shows a proximity log through the Canyon-Strawn Formation, run together with a microlog and microcaliper devices. For the 10,062–82 ft interval, R_{mc}=0.9 Ωm at 154°F. From the microlog, $t_{mc} \leqslant 0.22$ in., so that mud cake effects on R_{PL} are very minor. For the upper part of the interval, $R_{PL} \approx R_{xo} \approx 260$ Ωm. The dual induction-laterolog combination yielded a lower value for R_{xo} (= 190 Ωm). Using the more accurate R_{xo} value from the proximity log and the Schlumberger *Induction Log Correction Charts*, one can reinterpret the earlier results: $R_{xo}/R_t = 11.5$, $R_t = 22.6$ Ωm, and $R_{xo}/R_{mf} = 666$.

These values yield, for $S_{wi} = 0.85$, $F = 484$, $R_o = 8.24$ Ωm, and $S_w = 0.605$; and for $S_{wi} = 0.75$, $F = 374$, $R_o = 6.35$ Ωm, and $S_w = 0.53$.

The additional information provided by the proximity log shows the porosity to be of the order of 6.5% and the water saturation around 55%. These values are somewhat more pessimistic than those obtained from the dual induction-laterolog alone. The differences are rather slight, however, and reflect favorably on the consistency of the logging systems and interpretation techniques.

Sonic Log (Acoustic Log)
(References 58–64)

In porosity determinations from the electrical measurement of R_i or R_{xo}, a number of complications and uncertainties are introduced by the effects of interstitial clays, residual oil or gas, and variations in cementation factor. Some of these factors are avoided in various types of nonelectrical porosity logs; however, each one of the porosity logs has its own complexities. Modern techniques frequently attempt to resolve the difficulties by running a combination of several types of porosity logs. One device with important applications in carbonate rocks is the sonic or acoustic velocity log.

The sonic log measures the travel time of acoustic waves through formations. The signal is created by an acoustic transducer, and the travel time, Δt, is recorded as the difference in the times of first arrivals at two receivers. The two-receiver system largely eliminates the effects of the travel or linkage of the signal through the mud column to the boreface. Typical spacings for the sonic log are 3 ft from transducer, T, to first receiver, R_1, and 1–3 ft for the receiver span R_1–R_2. The logging trace records the travel time through the formations in microseconds per foot (μsec/ft).

Acoustic velocities are higher, and hence travel times are shorter, in dense rocks than in porous formations and shales. In order to facilitate correlation with resistivity logs, the travel time scales are inverted, with the low travel times to the right and the high Δt values to the left. The curves go off-scale to the left and reappear on the next higher scale at the right-hand edge. This somewhat confusing arrangement could be avoided by recording the logs on a velocity scale (ft/μsec), which would enhance the similarity to resistivity logs because the traces would go off-scale to the right and reappear at the left. A travel time scale could be added for detail in low-velocity regions, in analogy to the induction log recordings. In common practice, however, the Δt scales are employed almost exclusively.

Porosity determinations from the sonic log are usually based on an empirical relationship between travel times and porosities established by Wyllie *et al.*[61] This relationship, referred to as the time-averaging formula, is of the form:

$$\Delta t = \frac{\phi}{v_{fluid}} + \frac{1-\phi}{v_{matrix}} \tag{22}$$

where v_{fluid} is the acoustic velocity in the interstitial fluids, and v_{matrix} is the velocity in the matrix material. The formula correctly indicates the theoretical limits for zero porosity ($\Delta t = 1/v_{matrix}$) and for 100% porosity ($\Delta t = 1/v_{fluid}$).

For carbonate rocks, v_{matrix} ($\equiv v_m$) may vary from 21,000 to 26,000 ft/sec and v_{fluid} ($\equiv v_f$) from 5,300 to 6,800 ft/sec. Whereas the electrical formation factor-porosity relation contains the variable cementation factor, the velocity-porosity formula has to contend with two unknown variable characteristics. The matrix velocity may have some relation to grain density and/or cementation factor, whereas the fluid velocity should be dependent on water or mud filtrate salinity and residual hydrocarbon content. Such relations, if quantitatively established, could appreciably enhance precision in interpreting logging combinations.

Interstitial clays or thin shale laminations reduce the sonic velocity and increase the apparent porosity. Schlumberger[38] suggested a correction of the form:

$$\phi = \frac{\phi_a}{2 - \alpha} \tag{23}$$

where ϕ_a is the apparent porosity obtained from Eq. 22, and α is the ratio of the SP deflection of the shaly bed to that of a clean formation at the same R_{mf}, R_w, and formation temperature. The ratio or so-called SP reduction factor can be obtained from comparison with adjacent horizons or from an SP plot.

The right-hand trace of Fig. 14 shows a sonic log with spacing R_1–R_2 = 1 ft, through the interval discussed in connection with the salt mud survey. The detail on the sonic log is somewhat better than that of the laterolog-3. The Δt value for the interval at 8,164.5–67.5 ft is 52 μsec/ft. The porosity values obtained from Eq. 22 for various combinations of v_m and v_f (Table 2) show very considerable dispersion (3.0–12.5%). The average interval porosity of 9% found from the microlaterolog suggests one of two combinations: (1) v_m = 23,000 ft/sec and v_f = 6,800 ft/sec or (2) v_m = 26,000 ft/sec and v_f = 5,300 ft/sec.

TABLE 2. Sonic Log Porosities (%) for Different Fluid and Matrix Velocities —Δt = 52 μsec/ft

v_f, ft/sec / v_m, ft/sec	5,300	5,800	6,300	6,800
	Porosity, %			
19,500	0.35	0.59	0.69	0.75
21,000	3.0	3.5	4.0	4.6
23,000	5.8	6.6	7.4	8.2
26,000	9.1	10.1	11.2	12.5

For larger porosity values (higher travel times) the dispersion becomes even more severe. The above combinations at Δt = 70 μsec/ft, for instance, yield ϕ = 21% and ϕ = 25.5%, respectively. It is clear that for critical quantitative interpretations the sonic log is not very reliable unless local experience has indicated the appropriate values of v_m and v_f.

Figure 17 shows two runs of a sonic log with a spacing R_1–R_2 = 2 ft through the Edwards Lime section in the same well for which the IE appears

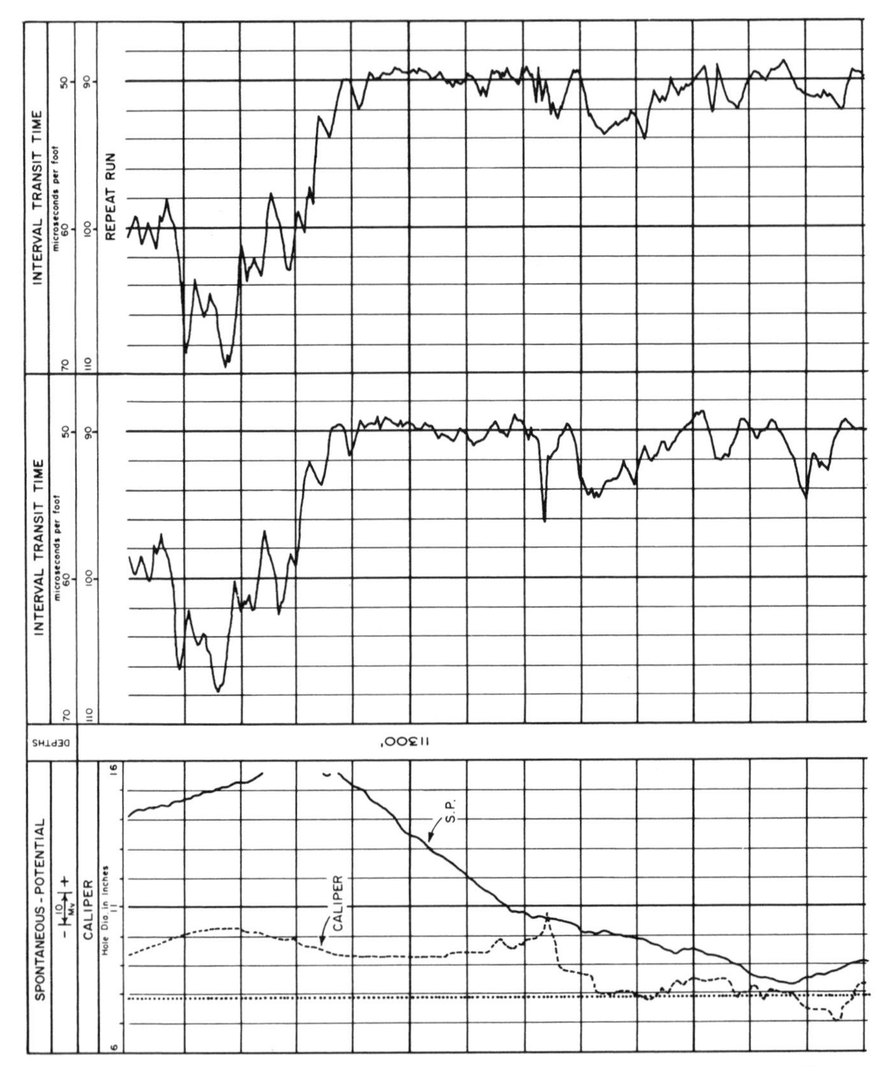

FIG. 17. Sonic log with repeat run through Edwards Lime, Dilworth Dome, McMullen County, Texas.

in Fig. 9B. The sonic log shows the porosity of the interval discussed for Fig. 9B to be limited to the 11,368–77 ft interval. The average travel time for the two runs is again 52 μsec/ft. From the short normal on the induction electric log, $F = R_{io}/R_{mf} = 86$, which gives $\phi = 13.1\%$ for $m = 2.2$ and $\phi = 10.8\%$ for $m = 2$.

As shown in Table 2, this porosity range requires a high v_f and v_m = 26,000 ft/sec. This matrix velocity is usually encountered only in dolomites. For the more typical limestone matrix velocity of 23,000 ft/sec, the obtained porosity is 7–8%. Even this value may be too high because of effects of unflushed gas. In all probability, therefore, the porosity and hence the hydrocarbon saturation obtained from the induction electric log were too high, as indicated in the discussion of the IE example.

If the formation containing the porous intervals under investigation also includes some thick very dense zones, one can reduce the uncertainty in the value of v_m by the following procedure. Find R_t, R_w, and hence F for the densest zone from resistivity curves and the SP. Calculate ϕ by using Eq. 3. For this very small ϕ and an average v_{fluid} value (v_f = 5,800 ft/sec), find v_m from Eq. 22. The uncertainty in v_f will have only a minor effect on the v_m determination, because of the weighting with the small ϕ value of the first term in the right-hand side of Eq. 22.

Radioactivity Logging

Radioactivity (RA) logs measure either the natural radioactivity of formations or the radiation induced in the formations. The interpretation of RA surveys relates these measurements to formation characteristics such as clay content, porosity, density, or other lithological properties and in some cases to gas saturation.

The outstanding advantages of RA logs are that the surveys can be made in cased holes and that the conductivity of the drilling fluid has no appreciable influence on the readings. These features are useful in reworking old holes and in wells drilled with air or oil-base mud. Also, the feasibility of running a casing collar locator simultaneously with the radiation logs has made them of importance for selective perforating.

The most common types of RA logs in use are (1) the gamma-ray log or natural radioactivity log, (2) the neutron log, (3) the density or gamma-gamma log, and (4) the thermal decay time log. The neutron logs may be subdivided into several classes, depending on the type of radiation detected in conjunction with the neutron bombardment of the formations. Current varieties include combination of a fast neutron source with detection of either induced gamma rays or of thermal, epithermal, or medium-fast neutrons.

Gamma-ray Logs
(References 65–70)

Gamma-ray surveys measure the spontaneous radioactivity of the rocks traversed in the drill hole by detecting the gamma rays emitted by the radioactive minerals present in these rocks. Gamma rays are extremely high-frequency electromagnetic waves with wavelengths between 0.07×10^{-8} and 0.4×10^{-8} cm. The most common radioactive element in sedimentary rocks is potassium, which is particularly abundant in argillaceous minerals. Other minerals which are compounds of the rare radioactive elements uranium and thorium are often in colloidal state when deposited and tend to occur in the same facies as the argillaceous materials. Clays and shales are, therefore, highly radioactive in comparison with sandstones and limestones. The radioactivity of the latter types of rocks depends largely on their clay content.

Normally, the gamma-ray log depicts roughly the amount of clay present as a fraction of bulk volume. There are, however, a number of exceptions to this rule.

1. Potash salts (polyhalite, sylvite) have, because of their potassium content, a high gamma-ray intensity even when completely free of clays. They occur frequently in evaporite sequences and can be distinguished by their relatively high resistivities.

2. Formations (often sandstones) containing carnotite or other uranium or thorium salts in quantity show anomalously high levels of radioactivity.

3. Igneous rocks usually have higher radioactivity than sedimentary rocks. Beds derived directly from igneous sources without a long transportation process by water are usually more radioactive than normal sediments. Among these are conglomerates and breccias derived directly from igneous plugs (e.g., the Bend Conglomerate Series of north Texas).

4. At erosional surfaces, clay minerals of exposed shales may undergo a secondary enrichment of potassium, thus providing a marker for erosional unconformities (e.g., the unconformity in the Mississippian Joana Limestone of Nevada).

5. In old fields, circulating or produced ground waters may deposit radioactive scale at the perforations in the liners or casing. This gives rise to extremely high intensities on the gamma-ray logs.

The scales used for the registration of gamma-ray curves may be based on comparison of deflections with those for a standard source (inches of deflection, API units). Other scales employed are absolute radiation units (microroentgens), equivalent radium content (micrograms of Ra/ton), or simply counts per second.

For carbonate rocks the gamma-ray curve is seldom used quantitatively. Relative clay content or shaliness is judged by comparison with shale deflections and the deflections for the cleanest parts of the formations. On older logs the gamma-ray curve is often displaced on the depth scale from the neutron curve, because the gamma-ray detector is usually placed at a distance of 6–9 ft from the neutron source. Modern logging devices automatically correct for the detector offset.

Neutron Logs
(References 38, 39, 71–92)

Neutron logs measure the radioactivity induced in formations by bombardment with high-energy (fast) neutrons. Either the neutron density or the intensity of gamma rays induced by neutron capture is detected at some distance ($\approx$ 1–2 ft) from the neutron source.

Commonly employed sources for the fast neutrons are mixtures of either radium, polonium, plutonium, or americium with beryllium. The alpha particles from the radioactive element are captured by the beryllium nuclei under emission of a neutron according to the nuclear equation:

$$_4B_e{}^9 + {}_2H_e{}^4 \rightarrow {}_6C^{12} + {}_0n^1.$$

Neutrons are uncharged particles of approximately the same weight as a hydrogen nucleus and have a large power of penetration compared to electrically charged particles.

The energy of neutrons is usually expressed in electron volts (1 eV = 1 electron volt = 1.602×10^{-12} erg). The fast neutrons leaving the source have energies ranging from 100 keV to more than 10 MeV, depending on the type of source (k = 1,000; M = 1,000,000).

After emission, the neutrons are slowed down by collisions with nuclei in the surrounding material (drilling fluid, mud cake, formations). Most nuclei are very much larger than the neutrons, which retain most of their energy upon collision. The hydrogen nucleus, however, has approximately the same mass as the neutron, and collisions with hydrogen will slow neutrons down very rapidly. After slowing down, the neutrons may be captured by the nuclei with which they collide. The probability of capture depends on the type and energy level of the colliding nucleus and the energy of the neutron. For most of the elements encountered in rock, the peaks of high capture probability (resonance peaks) fall in the neutron energy range from 50 eV down to thermal energies (0.025 eV). The capture of a neutron is always accompanied by emission of a gamma ray. The neutron-gamma logs detect these induced gamma rays, whereas the neutron-neutron logs measure the intensity of neutrons in some energy range prevailing in the vicinity of the detector. Theoretically, the higher the energy level of neutrons to which the detector is sensitive, the smaller the effects of the capture processes.

The neutron intensity near the detector depends on how fast the neutrons are slowed down with distance from the source, and this in turn is a function of the hydrogen content of the formations. The neutron logs, therefore, reflect primarily the presence of liquid-filled pore space and of bound water associated with rock minerals. The most common rock constituents with appreciable chemically bound water content are the clay minerals, the presence of which can usually be detected on the gamma-ray curve. Gypsum ($CaSO_4 \cdot 2H_2O$) has very low natural radioctivity and appears on the gamma-ray log as "clean." Its bound water causes the neutron curve to indicate a very high apparent porosity (equivalent to limestone with 49% porosity), whereas its actual porosity is virtually zero, as indicated by extremely high resistivity values.

In most clean formations, however, the neutron curve is essentially a porosity log. The effect of lithological composition (other than clay content) on the porosity determination is smaller for the neutron-neutron devices than for the neutron-gamma logs, and smaller for the fast and epithermal detection systems than for the thermal neutron logs. Although the lithological effects have been reduced in the fast and epithermal neutron logs, they are by no means negligible, and to use the neutron log independently for porosity determination a knowledge of the formation lithology is required. Modern interpretation techniques tend to combine the neutron log with either a sonic log or a formation density log. Such a combination allows construction of a so-called lithology plot and makes possible quantitative porosity determination in formations of diverse compositions.[115]

Drill hole factors, such as hole diameter, mud composition, presence and size of casing, mud cake thickness, and temperature, all have a distinct effect on the recorded neutron intensities. The hole diameter and mud composition effects have been largely eliminated in the side-wall type of neutron-porosity logs. For older neutron surveys, the uncertainties due to hole effects can frequently be reduced by using resistivity logs to "calibrate" the neutron-porosity curve.

The hydrogen concentration is very much lower in low-pressure hydrocarbon gas than in oil or water. In formations of homogeneous porosity, the neutron curve can be used to detect oil-gas contacts. In carbonate rocks, however, thick sections of homogeneous porosities are not encountered very frequently.

A large number of empirical studies and determinations in calibration test wells have shown that in clean carbonate rocks the neutron deflection (D_n) is roughly proportional to the logarithm of porosity. Neutron-porosity correlations are usually plotted on two-cycle semilog paper with the porosity on the logarithmic scale and the neutron deflection on the linear scale. The correlation will then be approximated by a straight line. For some devices, especially those with smaller spacings (e.g., $15\frac{1}{2}$-in. GNT log) the slope of the line increases at porosities below 5%. The intercept and the slope of the correlation lines depend on the hole diameter, number and size of casings, type of fluid in the hole, temperature, and the lithology of the formation.

Figure 18 shows a gamma-ray–neutron log and a plot of core analysis porosities for a section of Lansing-Kansas City Lime in a well of the N.W.

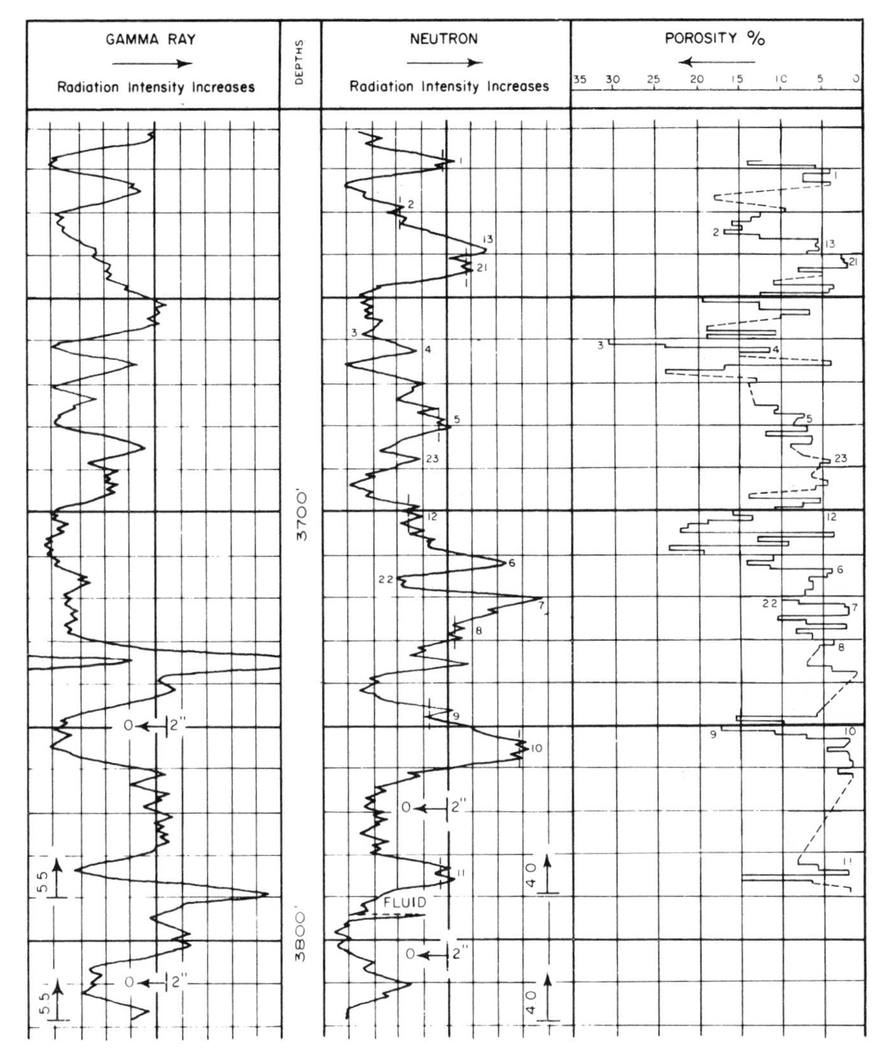

FIG. 18. Gamma-ray–neutron survey and core porosities, well A, Lansing-Kansas City Lime, N.W. Adell Field, Kansas. (Courtesy of Continental Oil Co.)

Adell Pool in Kansas. Figure 19A shows the corresponding neutron-porosity plot. The circled points 20–23 represent shale intervals. The amount by which the porosity for these points falls below that predicted by the neutron plot correlates roughly with the magnitude of the gamma-ray deflection. Figure 19B shows this correlation. Similar empirical correlations can be used to correct the neutron-derived porosities for clay effects. The neutron-porosity correlation of Fig. 19A applies only to the formation, hole and casing size,

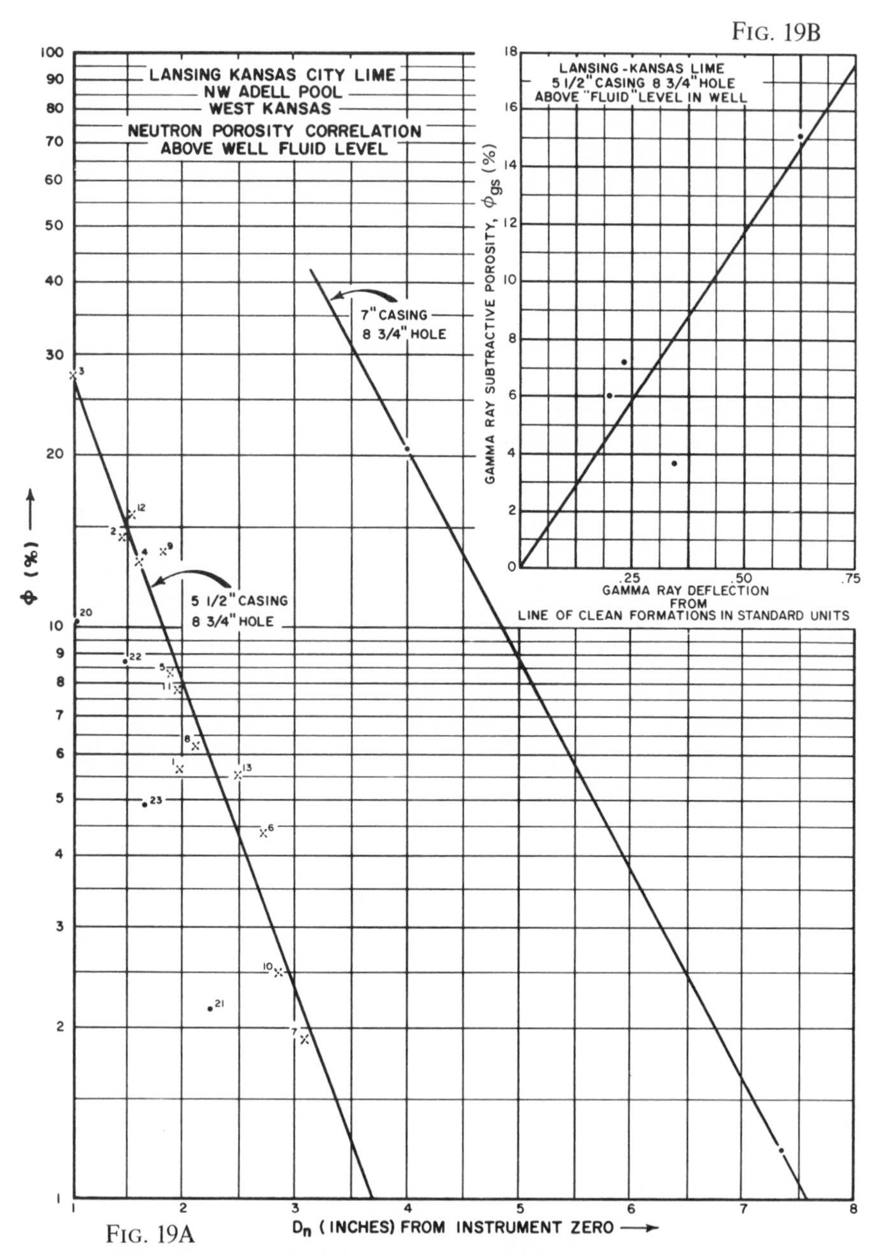

Fig. 19. Neutron-porosity correlation above well fluid level, Lansing-Kansas City Lime, N.W. Adell Pool, west Kansas. (Courtesy of Continental Oil Co.)

and type of neutron survey for which it was established. Such plots are, therefore, rather limited in their application.

A useful parameter that can be obtained from these local studies is the apparent porosity of certain marker shale horizons. The apparent shale porosities are frequently quite constant over a large area for a given horizon, and can be used as pivot points for neutron-porosity calibrations for other wells in the area with different hole conditions. As an example, the calcareous shale section at the 3,766–74 ft interval in Fig. 18 has a D_n value of 1.23 in. (measured from instrument zero), which from Fig. 19A represents an apparent porosity of 20.5%. Figure 20 shows a gamma-ray–neutron log and a laterolog-7 for another well in the area in which a different casing size was used. The same shale marker occurs here at the 3,799–806 ft interval and has D_n = 4.0 in. Combined with the apparent shale porosity value of 20.5%, this gives one point for the neutron-porosity graph. The second point is obtained by resistivity interpretation. The densest interval in the section at 3,786–94 ft is selected. The laterolog apparent resistivity = 650 Ωm, and R_m = 1.7 Ωm at 102°F; R_{mf} = 1.43 Ωm. From noninvasion departure curves for the laterolog-7, $(R_{LL7})_{corr.}$ = 595 Ωm ≈ R_t. For the largest SP deflection in the section (−94 mV), $-94 = -73 \log (1.43/R_{we})$; whence R_{we} = 0.073 Ωm and R_w = 0.087 Ωm. The formation factor $F = 595/0.087 = 6{,}850$, and ϕ = 1.21% (for $m = 2$). Combined with D_n = 7.35 in. for the same zone, this gives the second calibration point. The neutron-porosity correlation graph established by the two points is shown as the dashed curve in Fig. 19A.

An interesting variation on the combined usage of neutron and resistivity curves is the interpretation technique developed by Halliburton for its radiation-guard survey.[92] The Halliburton guard-electrode device has a 3-in. central electrode and 5-ft guards. Its apparent resistivity is a function of R_i, D_i, R_t, and some slight hole effects. An approximate relation for this dependence is given by:

$$R_a = d'[0.633 (\log D_i/d) R_i + (1 - 0.633 \log D_i/d) R_t] \tag{24}$$

where d' is a function of the hole diameter, with a value of 1 for $d_o = 3\frac{5}{8}$ in. and 0.8 for d_o = 9 in.

The neutron curve yields porosity after correction for hole effects, which are made with the aid of a nomograph. From the porosity, the formation factor is derived and then one finds $R_o = FR_w$ and $R_i = FR_{mf} \times S_{wi}^{-2}$. This approximate value of R_i allows calculation of R_t from Eq. 24 for average values of D_i/d. The combination of R_t and R_o, in turn, yields the water saturation, S_w. These latter operations can be incorporated with Eq. 24 in the expression:

$$\frac{R_a}{d'FR_w} 0{\cdot}633 \left(\log \frac{D_i}{d}\right) \frac{R_{mf}}{R_w} \cdot S_{wi}^{-2} + \left(1 - 0{\cdot}633 \log \frac{D_i}{d}\right) S_w^{-2}. \tag{25}$$

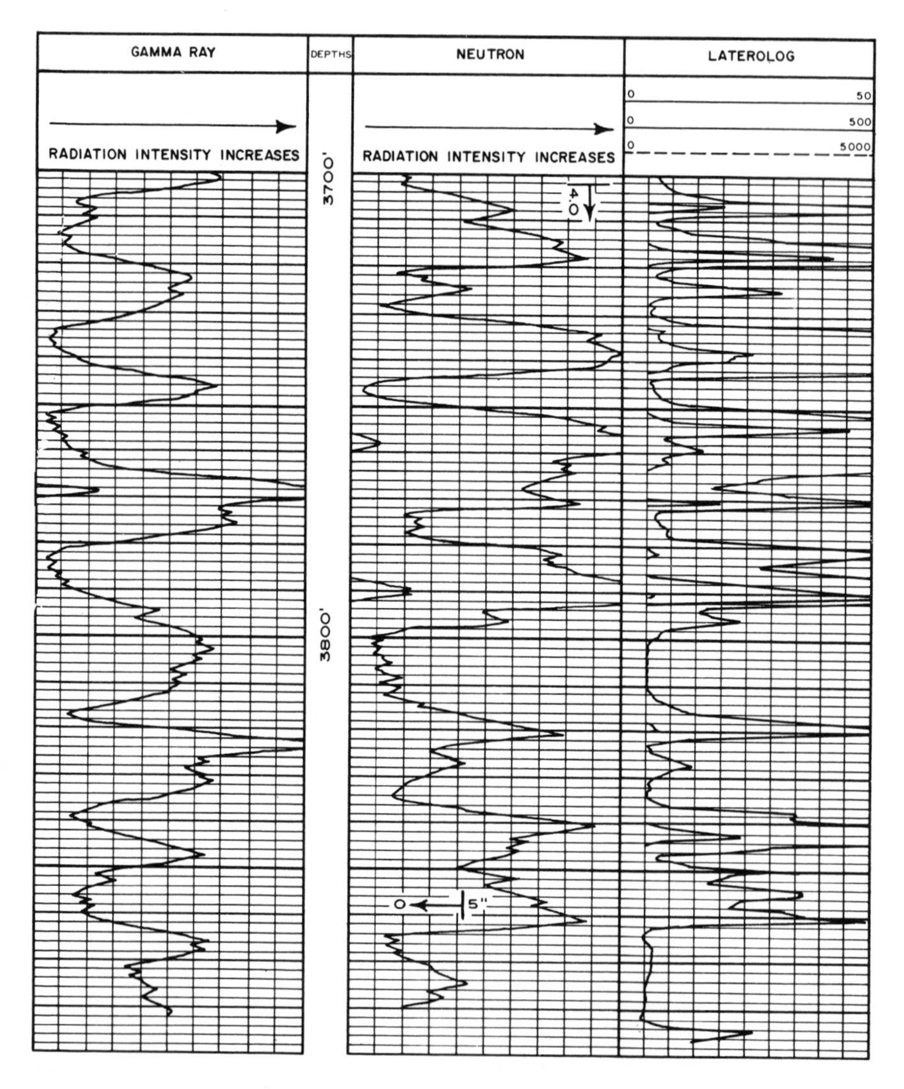

FIG. 20. Gamma-ray, neutron, and laterolog-7 surveys, well B, Lansing-Kansas City Lime, N.W. Adell Field, Kansas. (Courtesy of Continental Oil Co.)

Because of the large number of assumptions and approximations involved in Eq. 25, it is advisable to first calculate R_i and R_t from Eq. 24 with the aid of the neutron curve and to attempt to verify the results at each step by other independent means.

For open-hole neutron logging, Schlumberger has introduced a series of interpretation charts which make allowance for borehole effects, based on hole diameter, mud weight, mud cake thickness, and temperature. The charts furnish a porosity index value as a function of the neutron deflection in API units. This porosity index is the percentage porosity that would prevail if the formation were limestone. For other known lithologies a set of neutron-porosity equivalence curves then permits conversion of the limestone porosity index to the true porosity. If the lithology is not known, one can use the resistivity-type calibration of the neutron-porosity relationship or combine the neutron deflection with the response of either a sonic or a density log to make a so-called lithology-porosity plot. An example of this method is given later.

For cased holes or unknown hole conditions, the resistivity calibration is recommended, provided there is a sufficiently dense bed within the section, for which the assumption of 100% water saturation is valid ($\phi \leqslant 2.5\%$). If the equivalent porosity of adjacent shales is unknown, a good approximation can usually be obtained by taking the equivalent porosity of the most purely argillaceous shale (shale with lowest resistivity and SP) as 35%.

A significant improvement in open-hole neutron-porosity logging was achieved by the introduction of the side-wall epithermal neutron-porosity (SNP) log.[90] This device, which is run as a contact log against the boreface and is shielded from the rest of the hole, is calibrated for a $7\frac{7}{8}$-in. hole. The small effects of hole diameters different from the calibration standard are automatically corrected for in the logging panel. The use of epithermal neutron detection somewhat reduces the effect of lithology on the porosity determination but does not eliminate it. The SNP log presents computed porosity directly on a linear scale for any one of three rock matrices: limestone, sandstone, or dolomite. When the lithology is known, the corresponding scale is selected for direct porosity readings. For unknown lithological sequences, the log yields the limestone porosity index, which can subsequently be used in a lithology-porosity plot.

The SNP curve is affected by mud cake. The device is, however, equipped with a simultaneous hole caliper curve. Corrections for mud cake thicknesses up to $\frac{1}{2}$ in. can be made with the aid of a simple chart. The lower limit for mud cake thickness is obtained from the caliper survey by taking $t_{mc} \geqslant \frac{1}{2}(d_{\text{caliper}}\text{-bit size})$. Corrections for temperature and pressure variations from the standard calibration conditions are made automatically. Another automatic correction is made for drilling fluid salinity by an appropriate panel setting. A separate panel selection is available for logging gas-filled holes.

Formation Density Log; Gamma-Gamma Log; Densilog (References 93–102)

Density logs are based on the fact that the absorption of gamma rays traversing a medium by Compton scattering is roughly proportional to the density of the medium. The logging device is a contact type, pressed against the boreface and shielded at the side which faces the borehole. Density logging was introduced by Stanolind Oil and Gas Company (Pan American Petroleum Corporation) some fifteen years ago as a porosity determination tool.

The porosity is related to the measured bulk density, ρ_B, of the rocks by:

$$\phi = \frac{\rho_G - \rho_B}{\rho_G - \rho_F} \tag{26}$$

where ρ_G is the matrix or grain density, and ρ_F is the density of interstitial fluids.

Early density logs were fairly sensitive to factors such as hole diameter, mud density, mud cake thickness and density, and boreface rugosity (borehole effects). Modern logs use two detectors, one of which is very close to the source and quite sensitive to the hole effects. From the combination of the signals a correction to the long-spacing detector recording is computed. The corrected signal is registered directly in terms of bulk density in grams per cubic centimeter on a linear scale. In addition, a second trace records the amount of correction (or compensation) made, and sometimes either the uncompensated bulk density or the short-spacing curve is also recorded. Frequently, when both compensated and uncompensated curves are given, it is difficult to distinguish between the two. One can identify the correct (compensated) curve by the relation:

$$\rho_B \text{ (compensated)} = \rho_B \text{ (uncompensated)} + \Delta\rho \text{ (correction)} \tag{27}$$

where $\Delta\rho$ is added algebraically with whatever sign it shows on the log. For hole diameters larger than 9 in., an additional manual correction must be made to the compensated bulk density.

In many cases one or more porosity scales are also shown on the log. The sandstone porosity scale is based on an average grain density of 2.65 for sandstone. The limestone porosity scale is based on $\rho_G(\equiv\rho_{ma}) = 2.71$. In the absence of such scales, porosity is found from bulk density, using the graphs shown in Fig. 21. The grain density for dolomites is 2.87 g/cc. The value for anhydrites is still higher ($\approx$ 3.0 g/cc). In areas where rapid compositional changes occur, accurate porosity determination from density alone becomes difficult, and the density log should be combined with an SNP log to make a lithology-porosity plot, as described below.

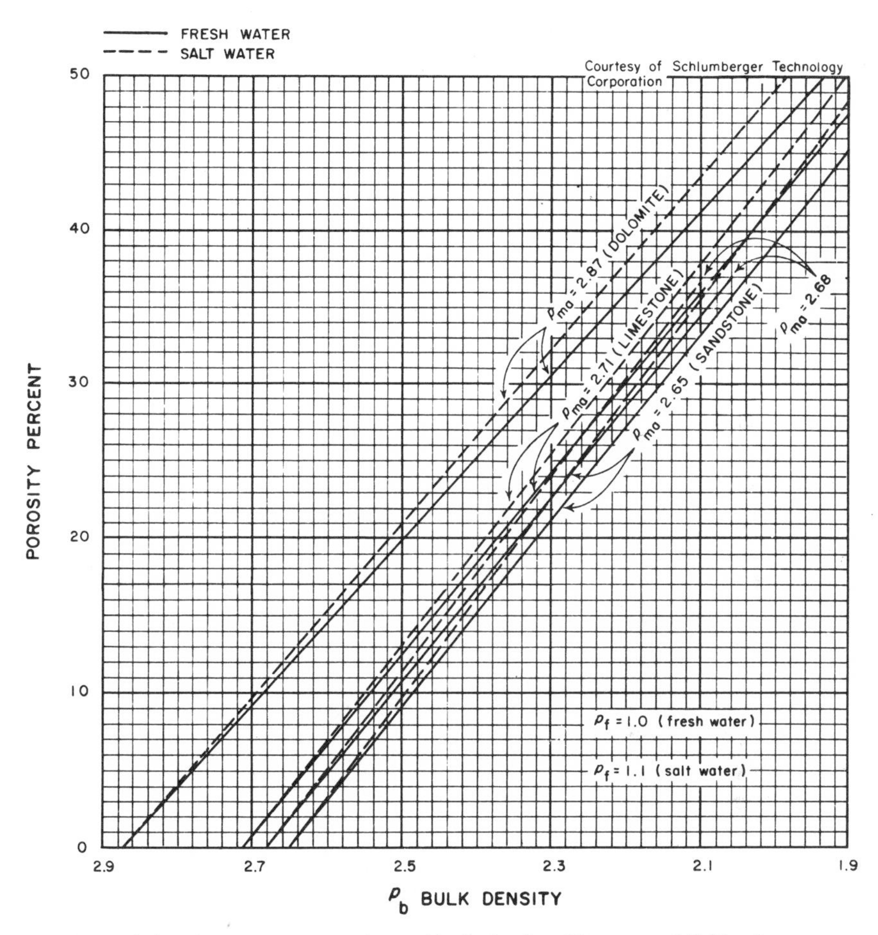

FIG. 21. Relationship between porosity and bulk density. (Courtesy of Schlumberger Technology Corp.)

Figure 22 shows a gamma-ray–neutron (GNTF-19.5-in.) and density log through the Paleozoic section of the S.W. Williston Basin, from the Permian Minnekanta Formation to the Cambrian Deadwood sandstones. Down to 2,850 ft the hole diameter was $8\frac{3}{4}$ in., and below that depth it was $6\frac{1}{4}$ in. For the limestone neutron-porosity calibration in the $6\frac{1}{4}$-in. hole, the dense zone at 3,610 ft, with an electric log porosity of 1.77%, and the Winnipeg Shale at 3,665–735 ft, with an apparent shale porosity of 35%, were used. For the

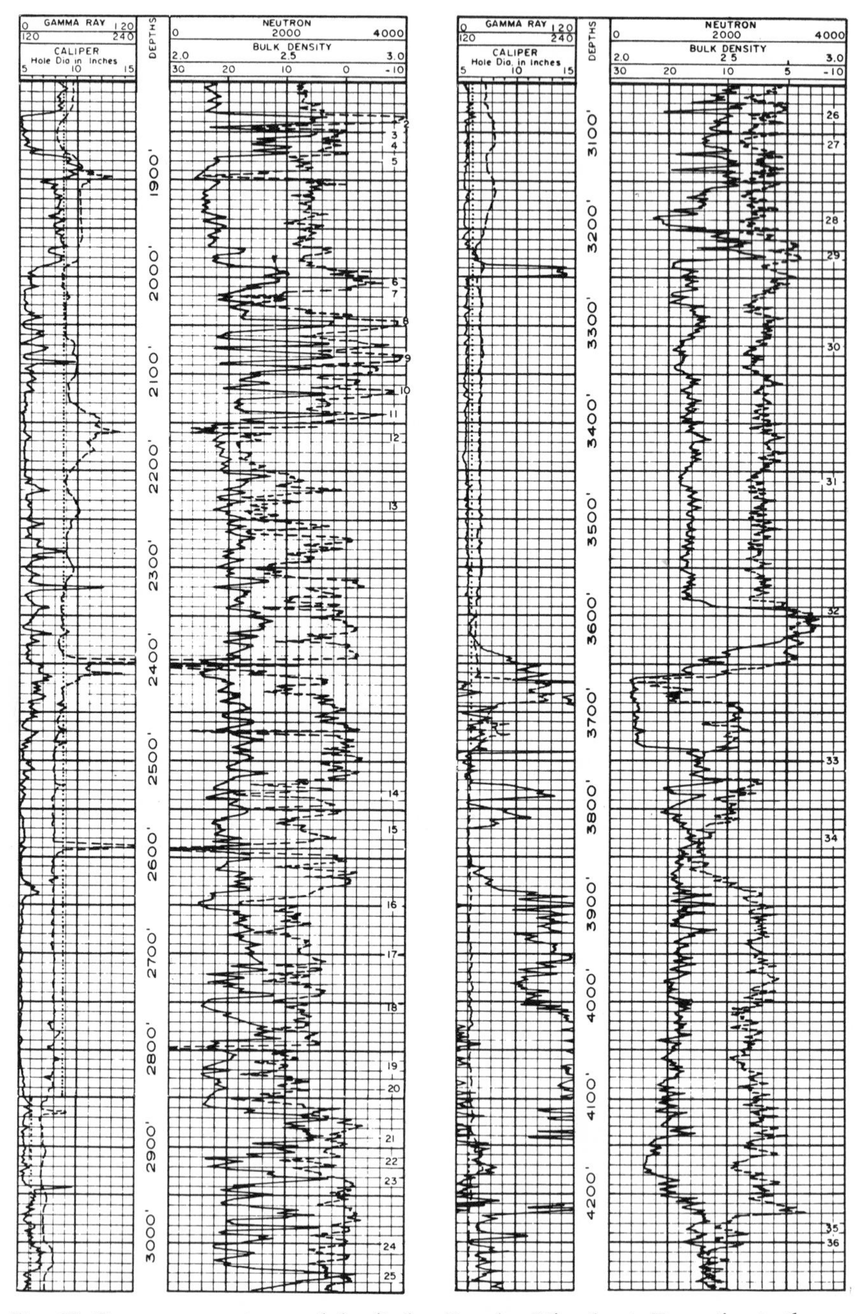

FIG. 22. Gamma-ray–neutron and density log, Permian Minnekanta Formation to the Cambrian Deadwood Formation, S. W. Williston Basin.

upper part of the hole, the 1% porosity point was obtained from the 19.5-in. GNT neutron departure curves and the shale point from the Opeche Shale at 1,895 ft. The two correlation graphs are indicated in Fig. 23. The density-log bulk densities of the numbered points of Fig. 22 for different formation intervals are plotted against the equivalent limestone porosities, obtained

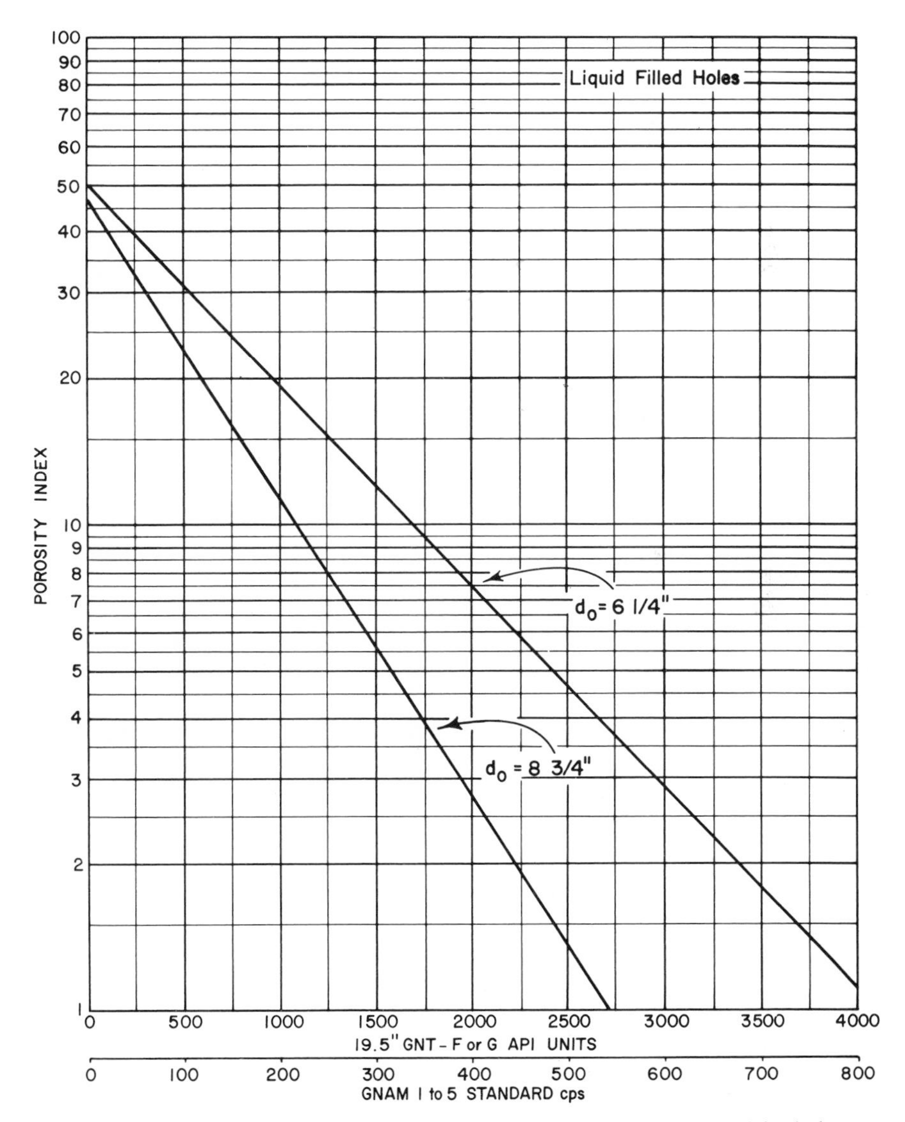

FIG. 23. Limestone porosity index versus neutron deflection for $6\frac{1}{4}$-in. and $8\frac{3}{4}$-in. holes.

with the aid of the correlation graphs, in Fig. 24. Superimposed on the joint plot are the characteristic empirical lines for different matrix compositions.

Points 1, 2, and 3 are in the Permian Minnekanta Formation and represent, respectively, dolomitic lime, anhydrite, and limestone beds. Points 4 and 5 are dolomitic Opeche limestones. Points 6–13 are in the Pennsylvanian Minnelusa Formation, with points 6 and 10 representing dolomitic limestone; 8, 9, and 11, anhydrite beds; and 7, 12, and 13, sandstone intervals. The horizon at point 14 is in the Charles Formation and probably contains

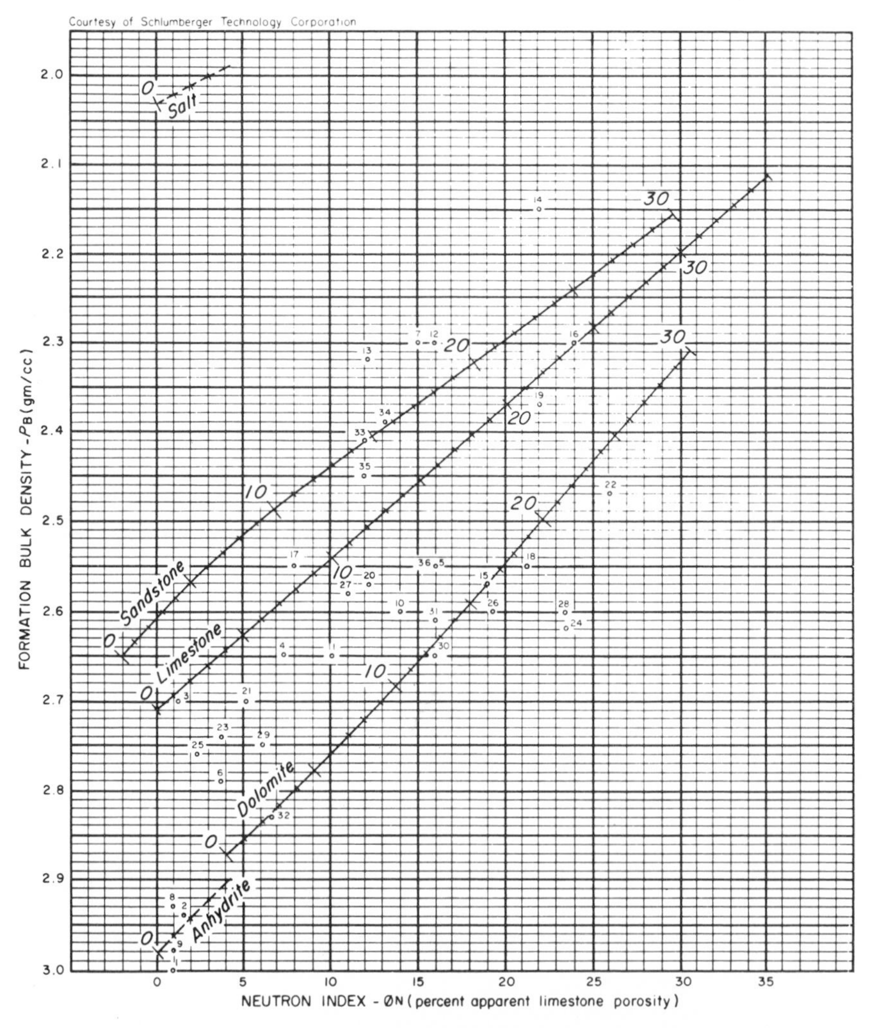

FIG. 24. Lithology cross-plot: formation bulk density (g/cc) versus neutron index (% apparent limestone porosity). (Courtesy of Schlumberger Technology Corp.)

an appreciable amount of rock salt, 15 is a Charles dolomite, and 16 is typical Charles limestone. Points 17, 19, and 20 are limestones of the Lower Mississippian Lodgepole Formation, whereas 18 and 22 are dolomites; 21 and 23 are Lodgepole dolomitic lime. Point 24 probably represents dolomite with vugular porosity. Points 25, 27, and 29 represent dolomitic limestone of the Upper Red River Formation (Ordovician), whereas 26, 30, 31, and 32 represent the pure Red River dolomites. Points 33–35 are in the Cambrian Deadwood sandstones, whereas point 36 is a dolomitic limestone in the Lower Deadwood Formation.

The porosities indicated in Fig. 24 range from practically zero in the anhydrites to more than 20% in some of the limestone and dolomite horizons.

Detection and Evaluation of Fractures

(References 103–114)

In carbonate rocks part of the void spaces frequently consists of irregular holes or vugs and of fracture systems. When the distribution of fractures is sufficiently dense and homogeneous, the log characteristics are fairly similar to those for rocks with purely intergranular porosity (e.g., the fractured cherts of the Monterey Formation, Santa Maria Basin, California). For less dense and irregular distributions of fractures and vugs, the common logging devices measure, in general, the mixed formation characteristics, incorporating both the effects of the matrix interstitial pore space and the void space due to irregular features.

Vugs can contribute significantly to total void volume, and their contribution is automatically included in density, neutron-porosity, velocity, and resistivity measurements. Inasmuch as their distribution is irregular, their presence causes rapid, erratic local changes in the formation parameters. Most sondes measure averaged characteristics. The microlaterolog and proximity log, however, reflect at least part of the erratic variations due to vugs, and when their readings are combined with those of other logs in a particular interpretation technique, appropriate interval averages must be used.

With the exception of cases with very dense fissure distributions, fractures normally make only a minor contribution to reservoir void space. They may add very significantly, however, to formation permeability and recovery factors. Rocks having low interstitial matrix porosities, and correspondingly low permeabilities in the absence of fractures, may constitute commercial reservoirs when fractures provide local short-range drainage for the matrix and effective transportation channels to the wellbore. For this reason it is often desirable to obtain qualitative and quantitative indications of the presence, density, and orientation of fracture systems.

One approach to estimating the reservoir volume contributed by fractures is to compare neutron-derived porosities with plug core porosities. The former indicate total void space, whereas the latter represent only the interstitial porosity of the matrix. The difference is representative of fracture volume and volume of large vugs.

Studies of transient pressure behavior during injection and withdrawal of fluids also may provide clues to the presence of significant fracture systems; fractured reservoirs follow linear models, which differ from the models for homogeneous reservoirs. There is evidence that fracture systems reflect a larger *kh* (permeability-thickness) value during injection than during withdrawal and show an increase in calculated *kh* with increase in injection rate.

Anomalously large calculated k/ϕ (permeability/porosity) ratios are also indicative of fracturing. Among the continuous surveying techniques, two recent developments have contributed significantly to improved evaluation of fractured reservoirs: the acoustic amplitude-microseismogram surveys and the borehole televiewer. A brief discussion of fracture-detecting techniques is presented here.

Acoustic Amplitude-Microseismogram; Variable-Intensity Display; Sonic Seismogram; Acoustic Character Log

The sonic devices generate acoustic wave trains that traverse borehole and formation and consist of mixed sequences of compressional and shear waves. The detected signals correspond to waves that have traveled parallel to the borehole axis for some distance. For homogeneous reservoir rocks, the wave trains passing the detector show a regular sequence of oscillations representing the successive arrivals of the different wave types, each having its characteristic transit time for the transducer-detector path.

Ordinary sonic logs record the transit time between two vertically spaced detectors for a distinguishable first arrival of one of the waves. The acoustic amplitude log registers the amplitude of a preselected wave peak, whereas the microseismogram records the amplitudes of the entire wave train as a variable-intensity (light to dark) display. The amplitude of any peak in the wave train will be considerably diminished by the presence of fracture planes between source and detector, because part of the wave energy will be reflected at the solid-fluid interfaces bounding the fractures. Furthermore, the multiple reflections and refractions due to the fracture interfaces will distort and dampen the regular pattern of the intensity display. The combination of the amplitude trace and the variable-intensity display gives a clear qualitative record of the presence of fracture systems.

Welex Corporation[109] has introduced a semi-quantitative system for its fracture-finder microseismogram log in the form of a fracture number scale used with the amplitude trace. The scale is based on the observation that a

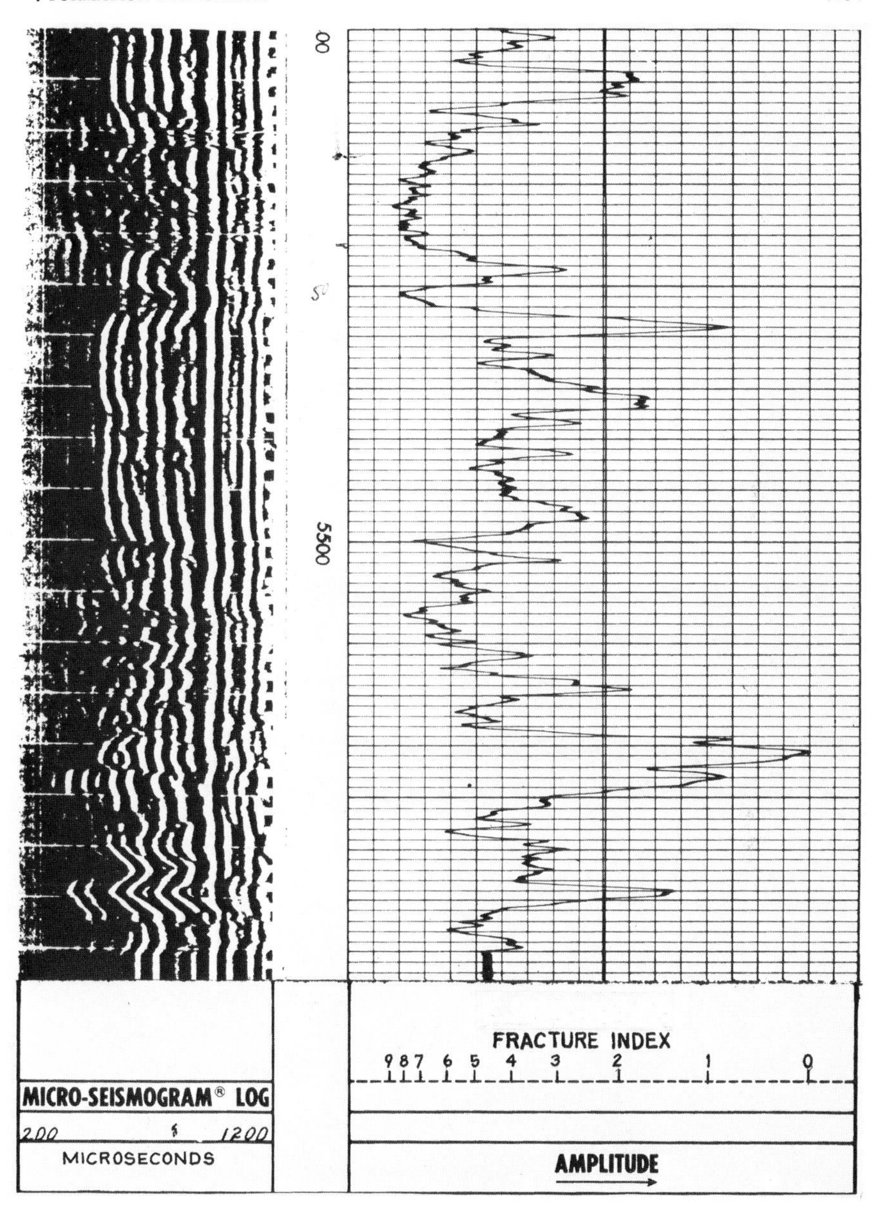

FIG. 25. Microseismogram log and fracture index, St. Louis-Mississippi Chat, Baca County, Colorado.

single pair of fracture interfaces perpendicular to the wave path decreases the signal amplitude by a factor of about 0.77. A given loss of amplitude can, therefore, be thought to correspond to a certain number of fractures within the source-detector interval (4 ft). Strict quantitative application is hindered by several complexities: (1) fractures are seldom perfectly planar and have many points of solid-to-solid contact; (2) the majority of fractures are not perpendicular to the borehole axis; (3) vertical or nearly vertical fractures do not produce amplitude reduction; and (4) lithology variations such as shaliness and porosity contrasts cause changes in amplitude, as do variations in sonde centralization. The effects of tool centralization can be overcome by the use of centralizers or identified by making several repeat runs. In spite of these limitations, the fracture number index presents an interesting and useful technique.

Figure 25 shows an amplitude-microseismogram fracture-finder log through the St. Louis-Mississippi Chat for a well in Baca County, Colorado. One can observe the regular intensity pattern for the unfractured sections, where the amplitude log readings are high, and the damped, distorted patterns obtained where the amplitude is low, as in the 5,425–43 ft interval. The fracture index scale shows an equivalent of eight horizontal fractures per 4-ft section for this interval.

Borehole Televiewer Log

The borehole televiewer developed by Mobil Research and Development Corporation[114] presents a continuous acoustic picture of the boreface, produced by a rotating ultrasonic scanner. The tool is oriented with respect to magnetic north, and a picture of the wellbore is reproduced at the surface, as if the boreface were cut by a southward-pointing vertical plane and subsequently rolled out on a flat surface. This representation makes boreface traces of inclined fractures appear as sinusoids, from which the dip (θ) of the fracture planes can be readily determined by using the relation:

$$\theta = \tan^{-1} h/d_o \qquad (28)$$

where h is the peak-to-peak distance of the sinusoid, and d_o is the hole diameter.

Figure 26 shows a borehole televiewer picture of vuggy porosity and natural fractures in west Texas Permian dolomite. From such pictures quantitative estimates of vuggy porosity and fracture density can be made.

Figure 27 shows the lithological resolution of the borehole televiewer in a sequence of thin beds. Resolution of better than 1 in. can be easily obtained when a sufficiently large vertical scale is selected. Sonic, gamma-ray, and caliper logs through the same section are shown for comparison.

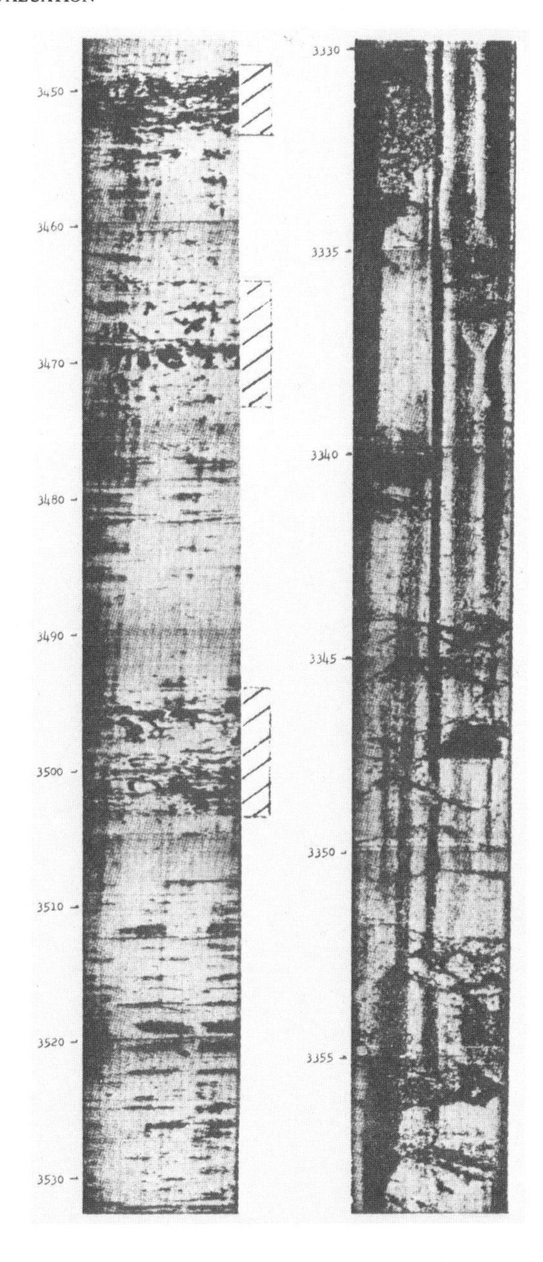

FIG. 26. Borehole televiewer in west Texas Permian dolomite, showing vuggy porosity and natural fractures. After Zemanek et al.[114]. (Courtesy of Mobil Field Research Laboratories, reprinted with permission of the SPE.)

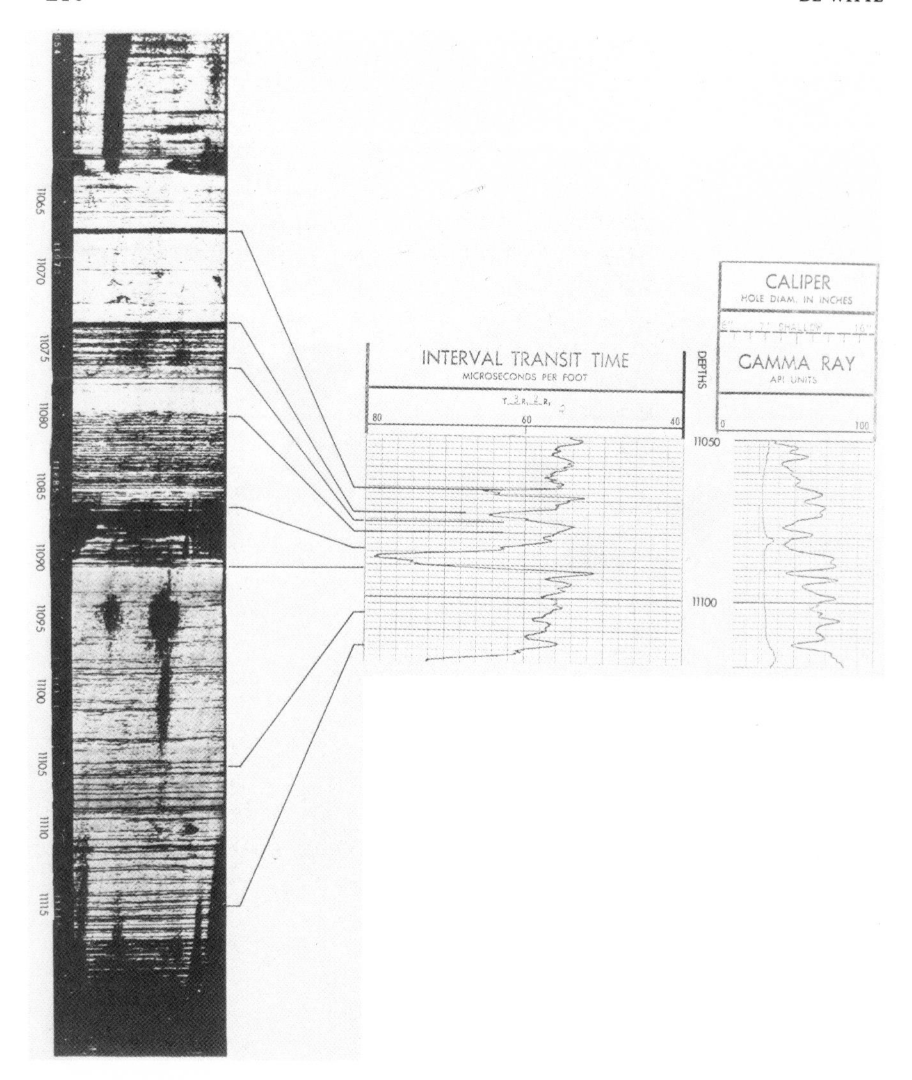

FIG. 27. Borehole televiewer log illustrating resolution for lithology identification. After Zemanek et al.[114] (Courtesy of Mobil Field Research Laboratories, reprinted with permission of the SPE.)

References

1. Archie, G. E.: "Classification of Carbonate Reservoir Rocks and Petrophysical Considerations", *Bull.*, AAPG (1952) Vol. 36, No. 2, 278–298.
2. Archie, G. E.: "Electrical Resistivity Logs as an Aid in Determining Some Reservoir Characteristics", *Pet. Trans.*, AIME (1942–1943) Vol. 146, 54.

3. Archie, G. E. "Introduction to Petrophysics of Reservoir Rocks", *Bull.*, AAPG (1950) Vol. 34, 943.
4. Doll, H. G. and Martin, M.: "Electrical Logging in Limestone Fields", *Proc.*, World Pet. Cong., 3d Session, The Hague, Sec. 2. (1951) 395–417.
5. de Witte, L.: "Resistivity and Saturation Distribution in Infiltrated Zones of Porous Formations", *Oil and Gas J.* (July 27, 1950) Vol. 49, No. 12, 246.
6. de Witte, L.: "A Study of Electric Log Interpretation Methods in Shaly Formations", *Pet. Trans.*, AIME (1955) Vol. 204, 103–110.
7. Frickie, H.: "A Mathematical Treatment of the Electric Conductivity and Capacity of Disperse Systems, I", *Phys. Rev.* (1924) 575.
8. Guyod, H.: "Electrical Well Logging, Part XIII, Electrical Properties of Oil-Bearing Reservoirs", *Oil Weekly* (Nov. 13, 1944) Vol. 115, No. 11, 40; "Electrical Well Logging, Parts XIV and XV, Electrical Potentials in Bore Holes", *Oil Weekly* (Nov. 20, 1944) Vol. 115, No. 12, 44 and (Nov. 27, 1944) No. 13, 2.
9. Guyod, H. "Interpretation of Electric Logs in Limestone", *World Oil* (Mar., 1948) Vol. 127, 90.
10. Milne, W. P.: "Relation of Electric Log Resistivities in Limestone to Oil Production", *Proc.*, Okla. Acad. Sci. (1950) Vol. 29, 50–56.
11. Mounce, W. D. and Rust, W. M., Jr.: "Natural Potentials in Well Logging", *Trans.*, AIME (1946) Vol. 164, 288–295.
12. McCardell, W. M. and Winsauer, W. O.: "Ionic Double Layer Conductivity in Reservoir Rock", *J. Pet. Tech.* (May, 1953) Vol. 5, No. 5, 129.
13. McCardell, W. M., and Winsauer, W. O.: "Origin of the Electric Potential Observed in Wells", *J. Pet. Tech.* (Feb., 1953) Vol. 5, No. 2, 41.
14. Patnode, H. W.: "Relationship of Drilling Mud Resistivity to Mud Filtrate Resistivity", *J. Pet. Tech.* (Jan., 1949) Vol. 1, 14.
15. Patnode, H. W. and Wyllie, M. R. J.: "The Presence of Conductive Solids in Reservoir Rocks as a Factor in Electric Log Interpretation", *J. Pet. Tech.* (1950) Vol. 2, No. 2, 47.
16. Poupon, A., Loy, M. E. and Tixier, M. P.: "A Contribution to Electric Log Interpretation in Shaly Sands", *Trans.*, AIME (1954) Vol. 201, 138.
17. Scotty, C. B.: "Quantitative Log Interpretation of the San Andres Dolomite", *World Oil* (1951) Vol. 133, No. 1, 166–176.
18. Tixier, M. P.: "Porosity Index in Limestone from Electric Logs", Part I, *Oil and Gas J.* (Nov. 15, 1951) Vol. 50, No. 28; Part II, *Oil and Gas J.* (Nov. 22, 1951) Vol. 50, No. 29, 63.
19. Wright, T. R. and Pirson, S. J.: "Porosity Determination from Electric Logs", *Bull.*, AAPG (1952) Vol. 36, No. 2.
20. Wyllie, M. R. J. "An Investigation of the Electrokinetic Component of the Self Potential Curve", *J. Pet. Tech.* (1951) Vol. 3, No. 1, 1–18.
21. Wyllie, M. R. J.: "A Quantitative Analysis of the Electrochemical Component of the S.P. Curve", *J. Pet. Tech.* (Jan., 1949) Vol. 1, No. 1, 17.
22. Wyllie, M. R. J. and Rose, W. D.: "Some Theoretical Considerations Related to the Quantitative Evaluation of the Physical Characteristics of Reservoir Rocks from Electric Log Data", *Trans.*, AIME (1950) Vol. 189, 105–118.
23. Gondouin, M., Tixier, M. P. and Simard, G. L.: "An Experimental Study on the Influence of the Chemical Composition of Electrolytes on the SP Curve", *J. Pet. Tech.* (Feb., 1957) Vol. 9, No. 2, 58–70.
24. Overton, H. L. and Lipson, L. B.: "A Correlation of the Electrical Properties of Drilling Fluids with Solids Content", *Pet. Trans.*, AIME (1958) Vol. 213, 333–336.
25. Moore, E. J., Szasz, S. E. and Whitney, B. F.: "Determining Formation Water Resistivity from Chemical Analysis", *J. Pet. Tech.* (Mar., 1966) Vol. 18, No. 3, 373–376.
26. Dunlay, H. F. and Hawthorne, R. R.: "The Calculation of Water Resistivity from Chemical Analyses", *J. Pet. Tech.* (Mar., 1951) Vol. 3, No. 3, 17.

27. Wyllie, M. R. J.: "A Statistical Study of the Accuracy of Some S.P. Log Data", *Bull.*, AAPG (1949) Vol. 33, No. 11, 1892–1900.
28. Doll, H. G.: "The SP Log: Theoretical Analysis and Principles of Interpretation", *Pet. Trans.*, AIME (1949) Vol. 179, 146–186.
29. de Witte, L.: "Simplified Departure Curves in Resistivity Logging", *Oil and Gas J.* (Feb. 11, 1952) Vol. 50, No. 40, 117–118, 120–122.
30. Guyod, H.: "Electrical Well Logging, Parts I through XII", *Oil Weekly* (Aug. 7, 1944) Vol. 114, No. 10; through (Oct. 30, 1944) Vol. 115, No. 9.
31. Martin, R. I.: "Conventional Resistivity Logging", *Oil and Gas J.* (Aug. 10, 1953) Vol. 52, No. 14.
32. Martin, R. I., "The Three-Electrode Device and Its Response Opposite Thick Beds", *Oil and Gas J.* (Nov. 16, 1953) Vol. 52, No. 28.
33. Martin, R. I., "Response of Three-Electrode Device Opposite Beds of Varying Thickness", *Oil and Gas J.* (Nov. 30, 1953) Vol. 52, No. 30.
34. Schlumberger, C. and M. and Leonardon, E. G.: "Electrical Coring: a Method of Determining Bottom-Hole Data by Electrical Measurements", *Geophys. Prospecting*, AIME (1934), 237.
35. Schlumberger, C. and M. and Leonardon, E. G.: "A New Contribution to Subsurface Studies by Means of Electrical Measurements in Drill Holes", *Geophys. Prospecing*, AIME (1934) 273–288.
36. Schlumberger Well Surveying Corp.: *Resistivity Departure Curves*, Schlumberger Doc. 3 (1949).
37. Schlumberger Well Surveying Corp.: *Interpretation Handbook for Resistivity Logs*, Schlumberger Doc. 4 (July, 1950).
38. Schlumberger Technology Corp.: *Schlumberger Log Interpretation Charts*, Ridgefield, Conn. (1968).
39. Dresser Atlas Division of Dresser Industries: *Dresser Atlas Log Interpretation Chart Manual*, Houston, Tex. (1968).
40. Doll, H. G.: "Introduction to Induction Logging and Application to Logging of Wells Drilled with Oil-Base Mud", *J. Pet. Tech.* (June, 1949) Vol. 1, No. 6, 148.
41. Dumanoir, J. L., Tixier, M. P. and Martin, M.: "Interpretation of the Induction-Electrical Log in Fresh Mud", *J. Pet. Tech.* (July, 1957) Vol. 9, No. 7, 202–215.
42. Moran, J. H. and Kunz, K. S.: "Basic Theory of Induction Logging and Application to Study of Two-Coil Sondes", *Geophysics* (Dec., 1962) Vol. 27, No. 6, 829–858.
43. Tixier, M. P.: "Porosity Balance Verifies Water Saturation Determined from Logs", *J. Pet. Tech.* (July, 1958) Vol. 10, No. 7, 161–169.
44. Doll, H. G., Dumanoir, J. L. and Martin, M.: "Suggestions for Better Electric Log Combinations and Improved Interpretations", *Geophysics* (Aug., 1960) Vol. 25, No. 4, 854–882.
45. Schlumberger Well Surveying Corp.: *Induction Log Correction Charts* (1962).
46. Tixier, M. P., Alger, R. P. and Tanguy, D. R., "New Developments in Induction and Sonic Logging", *J. Pet. Tech.* (May, 1960) Vol. 12, No. 5, 79–87.
47. Doll, H. G.: "The Microlog—a New Electrical Logging Method for Detailed Determination of Permeable Beds", *Trans.*, AIME (1950) Vol. 195, 155–164.
48. Schlumberger Well Surveying Corp.: *Interpretation Charts for Microlog*, Schlumberger Doc. 5 (1952).
49. Doll, H. G.: "The Microlaterolog", *J. Pet. Tech.* (Jan., 1953) Vol. 5, No. 1, 17.
50. Smith, H. D. and Blum, H. A.: "Microlaterolog versus Microlog for Formation Factor Calculation", *Geophysics* (Apr., 1954) Vol. 19, No. 2, 310–32.
51. Doll, H. G.: "The Laterolog: A New Resistivity Logging Method with Electrodes Using an Automatic Focusing System", *J. Pet. Tech.* (Jan., 1953) Vol. 5, No. 1, 17.
52. Owen, J. E. and Greer, W. J.: "The Guard Electrode Logging System", *J. Pet. Tech.* (Dec., 1951) Vol. 3, No. 12, 347.
53. de Witte, L.: "Resistivity Logging in Thin Beds", *Trans.*, AIME (1954) Vol. 201, 155.

54. Schlumberger Well Surveying Corp.: *Introduction to Schlumberger Well Logging*, Schlumberger Doc. 8 (1958).
55. Decker, C. J. and Martin, J.: "The Laterolog and Salt-Mud Well Logging in Kansas", *Oil and Gas J.* (Feb. 16, 1952) Vol. 50, No. 41, 119.
56. Tixier, M. P., Alger, R. P., Biggs, W. P. and Carpenter, B. N.: "Dual Induction-Laterolog: A New Tool for Resistivity Analysis", paper SPE 173 presented at SPE Meeting, New Orleans, La. (1963).
57. de Witte, A. J.: "Influence of Differential Displacement in Invaded Oil and Gas Sands on Induction Log", *J. Pet. Tech.* (June, 1957) Vol. 9, No. 6, 48–50.
58. Summers, G. C. and Broding, R. A.: "Continuous Velocity Logging", *Geophysics* (Oct., 1952) Vol. 17, No. 3, 598.
59. Vogel, C. B.: "A Seismic Velocity Logging Method", *Geophysics* (July, 1952) Vol. 17, No. 2, 586.
60. Tixier, M. P., Alger, R. P. and Doh, C. A.: "Sonic Logging", *J. Pet. Tech.* (May, 1959) Vol. 11, 106–114.
61. Wyllie, M. R. J., Gregory, A. R. and Gardner, L. W.: "Elastic Wave Velocities in Heterogeneous and Porous Media", *Geophysics* (Jan., 1956) Vol. 21, No. 1, 41–70.
62. Breck, H. R., Schoellhorn, S. W. and Baum, R. B.: "Velocity Logging and Its Geophysical and Geological Applications", *Bull.*, AAPG (Aug., 1957) Vol. 41, No. 8, 1667–1682.
63. Hicks, W. G. and Berry, J. E.: "Application of Continuous Velocity Logs to Determination of Fluid Saturation of Reservoir Rocks", *Geophysics* (July, 1956) Vol. 21, No. 3, 739–754.
64. Kokesh, F. P. and Blizard, R. B.: "Geometrical Factors in Sonic Logging", *Geophysics* (Feb., 1959) Vol. 24, No. 1, 64–76.
65. Fearon, R. E.: "Gamma Ray Well Logging", *Nucleonics* (Apr., 1949) Vol. 4, No. 4, 67–75.
66. Jackson, W. J. and Campbell, J. L. P.: "Some Practical Aspects of Radioactivity Well Logging", *Pet. Trans.*, AIME (1946) Vol. 165, 241–267.
67. Russell, W. L.: "Relation of Radioactivity, Organic Content and Sedimentation", *Bull.*, AAPG (Oct., 1945) Vol. 29, No. 10, 1470–1493.
68. Russell, W. L.: "The Total Gamma Ray Activity of Sedimentary Rocks as Indicated by Geiger Counter Determination", *Geophysics* (Apr., 1944) Vol. 9, No. 2, 180–216.
69. Bell, K. E., Goodman, C. and Whitehead, W. L.: "Radioactivity of Sedimentary Rocks and Associated Petroleum", *Bull.*, AAPG (Sept., 1940) Vol. 24, 1529–1547.
70. Howell, L. G. and Frosch, H.: "Gamma Ray Well Logging", *Geophysics* (Apr., 1939) Vol. 4, No. 2, 106–114.
71. Bush, R. E. and Murdock, E. S.: "Some Preliminary Investigations of Quantitative Interpretations of Radioactivity Logs", *J. Pet. Tech.* (Jan., 1950) Vol. 2, No. 1, 19–34.
72. Bush, R. E. and Murdock, E. S.: "The Quantitative Application of Radioactivity Logs", *Trans.*, AIME (1951) Vol. 192, 191–198.
73. Fearon, R. E.: "Neutron Well Logging", *Nucleonics* (June, 1949) Vol. 4, No. 6, 30–42.
74. Mercier, V. J.: "Radioactivity Logging of Dolomites", *Oil Weekly* (Apr. 15, 1946) Vol. 121, No. 7, 36–41.
75. Pontecorvo, Bruno: "Neutron Well Logging", *Oil and Gas J.* (Sept., 1941) Vol. 40, No. 18, 32–33.
76. Russell, W. L.: "Interpretation of Neutron Well Logs", *Bull.*, AAPG (1952) Vol. 36, No. 2, 312–341.
77. Swift, G.: "Simultaneous Gamma Ray and Neutron Logging", *Geophysics* (1952) Vol. 17, No. 2, 387–394.
78. Wyllie, M. R. J.: "Procedure for the Direct Employment of Neutron Log Data in Electric Interpretation", *Geophysics* (Oct., 1952) Vol. 17, No. 4, 790.

79. Atkins, E. R., Jr.: "Fundamental Theory and Instrumentations of Radioactivity Logging", paper presented at AIME Symp. on Formation Evaluation, Houston, Tex. (Oct., 1955).
80. Brannon, H. R., and Osoba, J. S.: "Spectral Gamma Ray Logging", *J. Pet. Tech.* (Feb., 1956) Vol. 8, No. 2, 30–35.
81. Dewan, J. T.: "Neutron Log Correction Charts for Borehole Conditions and Bed Thickness", *J. Pet. Tech.* (Feb., 1956) Vol. 8, No. 2, 50–58.
82. Dewan, J. T. and Alland, L. A.: "Experimental Basis for Neutron Log Interpretation", *Pet. Eng.* (Sept., 1953) Vol. 25, No. 10, B49–50, B52–54.
83. Faul, H. and Tittle, C. W.: "Logging of Drill Holes by the Neutron-Gamma Method and Gamma Ray Scattering", *Geophysics* (Feb., 1954) Vol. 16, No. 2, 260.
84. Grosmangin, M. and Walker, E. B.: "Gas Detection by Dual-Spacing Neutron Logs in the Greater Oficina Area, Venezuela", *J. Pet. Tech.* (May, 1957) Vol. 9, No. 5, 140–147.
85. McVicar, B. M., Heath, J. L. and Alger, R. P.: New Logging Approaches for Evaluation of Carbonate Reservoirs, paper presented at Williston Basin Symposium (Oct. 9–12, 1956).
86. Muench, N. L. and Osoba, J. S.: "Identification of Earth Materials by Induced Gamma Ray Spectral Analysis", *J. Pet. Tech.* (Mar., 1957) Vol. 9, No. 3, 89–92,
87. Scotty, C. B. and Egan, E. F.: "Neutron Derived Porosity—Influence of Bore Hole Diameter", *J. Pet. Tech.* (Aug., 1952) Vol. 4, No. 8, 203.
88. Tittle, C. W., Faul, H. and Goodman, C.: "Neutron Logging of Drill Holes: the Neutron-Neutron Method", *Geophysics* (Oct., 1951) Vol. 16, No. 4, 626–658.
89. Tittman, J.: "Moderation of Neutrons in SiO_2 and $CaCO_3$", *J. Appl. Phys.* (Apr., 1955) Vol. 26, No. 4, 394–398.
90. Tittman, J., Sherman, H., Nagel, W. A. and Alger, R. P.: "The Sidewall Epithermal Neutron-Porosity Log", *Trans.*, AIME (1966) Vol. 237, 1351–1362.
91. Edwards, J. M. and Simpson, A. L.: "A Method for Neutron Derived Porosity Determination for Thin Beds", *J. Pet. Tech.* (Aug., 1955) Vol. 7, No. 8, 132–136.
92. Halliburton Oil Well Cementing Co.: *The Application of Radiation-Guard Surveys to the West Texas, New Mexico Province*, Houston, Tex. (Sept., 1955).
93. Baker, P. E., "Density Logging with Gamma Rays", *Trans.*, AIME (1957) Vol. 210, 289–294.
94. Danes, Z. F.: "A Chemical Correction Factor in Gamma-Gamma Density Logging", *J. Geophys. Res.* (1960) Vol. 65, 2149–2153.
95. Davis, D. H.: "Estimating Porosity of Sedimentary Rocks from Bulk Density", *J. Geol.* (1954) Vol. 62, 102–107.
96. Diadkin, I. G.: "On the Theory of Gamma-Gamma Logging of Bore Holes", *Izv. Akad. Nauk SSSR, Ser. Geofiz.* (1955), 323–331.
97. Rodermund, C. G., Alger, R. P. and Tittman, J.: "Logging Empty Holes", *Oil and Gas J.* (June 12, 1961) Vol. 59, No. 24, 119–124.
98. Wahl, J. S., Tittman, J., Johnstone, C. W. and Alger, R. P.: "The Dual Spacing Formation Density Log", *J. Pet. Tech.* (1964) Vol. 16, 1411-1416.
99. Tittman, J. and Wahl, J. S.: "The Physical Foundation of Formation Density Logging (Gamma-Gamma)", *Geophysics* (1965) Vol. 30, No. 2, 284–294.
100. Alger, R. P., Raymer, L. L., Hoyle, W. R. and Tixier, M. P.: "Formation Density Log Applications in Liquid Filled Holes", *J. Pet. Tech.* (1963) Vol. 15, No. 3, 321–332.
101. Raymer, L. L., Jr. and Biggs, W. P., "Matrix Characteristics Defined by Porosity Computations", paper presented at SPWLA Meeting, Oklahoma City (1963).
102. Savre, W. C.: "Determination of More Accurate Porosity and Mineral Composition in Complex Lithologies with Use of Sonic, Neutron and Density Surveys", *J. Pet. Tech.* (1963) Vol. 15, No. 9, 945–959.
103. Pickett, G. R., "Acoustic Character Logs and Their Applications in Formation Evaluation", *J. Pet. Tech.* (June, 1963) Vol. 15, No. 6, 659–667.

104. Walker, T.: "Progress Report on Acoustic Amplitude Logging for Formation Evaluation", paper SPE 451 presented at AIME 37th Annual Meeting, Los Angeles, Calif. (Oct., 1962).
105. Walker, T.: *The Interpretation of the Fracture Finder, Micro-Seismogram Log*, Welex Publ. L-20 (May, 1964).
106. Lawrence, H. W.: "Reflection, Refraction and Energy Mode Conversion as Seen on 3-D Velocity Logs", paper presented at SEG 35th Annual Meeting, Dallas, Tex. (1965).
107. Biot, M. A.: "Theory of Propagation of Elastic Waves in a Fluid-Saturated Porous Solid", *J. Acoust. Soc. Am.* (1956) Vol. 28, 168.
108. Anderson, W. L. and Walker, T.: "Application of Open Hole Acoustic Amplitude Measurements", paper SPE 122 presented at AIME Meeting, Dallas, Tex. (Oct., 1961).
109. Welex Corp.: *Location of Formation Fractures and Shallow Gas Sands with Acoustic Amplitude Logs*, Welex Tech. Bull. L-5 (June, 1961).
110. Walker, T.: "Fracture Zones Vary Acoustic Signal Amplitudes", *World Oil* (1962) Vol. 154, No. 6, 135, 137, 142.
111. Morris, R. L., Grine, D. R. and Arkfield, T. E.: "Using Compressional and Shear Acoustic Amplitudes for the Location of Fractures", *J. Pet. Tech.* (1964) Vol. 16, No. 6, 623–632.
112. Muir, D. M. and Fons, L. H.: "The New Acoustic Parameter Log", *Oil* (1964) Vol. 24, No. 5, 10–11.
113. Pickett, G. R. and Reynolds, E. B.: "Evaluation of Fractured Reservoirs", *Soc. Pet. Eng. J.* (Mar., 1969) Vol. 9, No. 1, 28–38.
114. Zemanek, J., Caldwell, R. L., Glenn, E. E., Holcomb, S. V., Norton, L. J. and Straus, A. J. D.: "The Borehole-Televiewer—a New Logging Concept for Fracture Location and Other Types of Borehole Inspection", *J. Pet. Tech.* (June, 1969) Vol. 21, No. 6, 762–774.
115. Burke, J. A., Schmidt, A. W. and Campbell, R. L.: "The Litho- Porosity Cross Plot", *The Log Analyst* (1969) Vol. 10, No. 6, 25–43.

CHAPTER 5

Estimation of Oil and Gas Reserves and the Production Forecast in Carbonate Reservoirs

ROBERT W. MANNON

Introduction

In Chapter 1 of this book, the methods of estimating oil and gas reserves and of forecasting future production are discussed in general by Dr. N. van Wingen. The purpose of this chapter is to elucidate certain portions of this discussion with specific application to carbonate reservoirs. Such an assignment is not without its problems. A theme recurring through this book is the uniqueness of oil and gas production from carbonate rocks. This uniqueness requires that certain aspects of oil and gas reserve estimation, which normally are given only cursory review, be examined in detail. The question arises as how to spend a seemingly disproportionate amount of time on certain facets of the overall technology in order to meet this end, and still be able to do justice to the general topic of reserve estimation. One solution would be to concentrate entirely on the "unique" aspects that pertain mainly to carbonate reservoirs at the expense of other, more general principles. This approach would accomplish certain objectives but would leave all but the most serious students of the subject without proper orientation. Another solution would be to strive to emphasize the phases of reserve estimation particularly applicable to carbonates in the context of a general treatment. The writer has chosen the latter course.

The term "reserves" is at times used very loosely. Lahee,[1] for example, has classified reserves as (1) drilled and proved, (2) undrilled and proved, (3) possible—likely untested structures in oil provinces or untested prospects above or below producing zones, (4) discovered possible—recoverable by secondary methods of operation, and (5) hypothetical—untested structures in a basin similar to another basin known to contain oil.

In this chapter "oil and gas reserves" refers to proved reserves in Lahee's categories 1, 2, and 4. They are the types of reserves normally included in an analytical engineering appraisal of a property or properties prepared by petroleum engineers. To include Lahee's types 3 and 5 can lead to gross misconceptions. Eggleston[2] and Arps[3] have addressed themselves specifically

References p. 253.

to this point. They feel that terms such as "probable" and "possible" reserves should generally be avoided or at least confined to the descriptive portion of an appraisal report. Arps believes, however, that there may be a need for terminology less severe than the "proved" definition. He has proposed a classification including "probable" and "possible" reserves, based on the degree of proof. These terms could be used in evaluating the results of an exploration program, or when considering the geologic potential of a given basin or area. This nomenclature is similar to Lahee's classification.

Proved reserves, as defined by Sheldon,[4] refer to oil and gas which is producible by operating techniques known today, and the production of which will provide a reasonable profit after payment of all costs and return of invested capital. Sheldon[4] has defined primary developed reserves as hydrocarbons which will be produced through boreholes capable of production at the time the appraisal is prepared. He has defined undeveloped reserves as hydrocarbons which are proved to exist to a high degree of probability, but the production of which will require the investment of additional capital. Naturally, estimates of future recoverable oil and gas for proved undeveloped lands are less reliable than estimates prepared for properties that are developed and currently producing.

Proved developed secondary reserves are those recoverable through secondary recovery techniques presently in operation on the subject property or properties. The secondary recovery methods utilized must have been proved effective on the subject properties or on properties with similar characteristics. Proved undeveloped secondary reserves are recoverable through proved secondary techniques not presently in operation on the subject property.

In the introductory chapter, N. van Wingen discusses two basic approaches to the estimation of reserves, namely, empirical and analytical. Empirical methods are based on some form of analogy with other pools and are the subject of subsequent discussion in this chapter. Analytical methods consist of oil and gas reserve estimates based on (1) the analysis of production decline curves, (2) a knowledge of the initial oil- or gas-in-place, based on volumetric or material balance calculations, and (3) production forecasts obtained by numerical procedures of relating pressure decline to oil recovery and gas/oil ratio. From a practical standpoint, these "analytical" methods are empirical to varying degrees, as pointed out later.

Reserve Estimation from Decline Curve Analysis

The analysis of past trends in production performance in order to estimate oil and associated gas reserves is frequently a very reliable method. In addition, a usually equally reliable yearly forecast of the anticipated oil or gas production is provided.

The Importance of the Production Forecast

It appears that early in the history of the oil industry operators realized the importance of having some idea of the amount of oil to be forthcoming from their wells and the length of time required to produce this amount. These early-day producers soon became aware that the net profit generated from a well is disproportionately sensitive to changes in production rate. The business of producing oil is somewhat unique in this respect. The gross revenue from a well is directly proportional to the well's production rate because the oil is sold with little difficulty at the prevailing posted price. On the other hand, the cost of producing this oil is very insensitive to the relative production rate—that is, the rate at which a well produces oil has, under normal circumstances, no significant effect on the operating expense to the producer. As a result, the net income from oil-producing operations is affected by changes in production rate to a greater extent than are profits from other businesses. It soon became apparent, also, that new wells characteristically demonstrate initial rates over a wide range, with subsequent differing types of production-decline performance. The need for reliable production forecasts became evident. Factors like fair market value, loan value, and payout of properties are very sensitive to production performance irrespective of the factor of reserves. Expert help was accordingly sought for the task of forecasting. This may have been one of the first duties of the petroleum engineer.

The Use of Decline Curve Analysis as a Tool

Engineers, in their search for techniques to provide reliable forecasts of future performance, have long recognized the possibilities of using some form of graphic representation of the production data from a well as a guide. References to this technique first appeared in the literature around the turn of the century. Since that time considerable work has been done in attempting to formulate interpretation procedures that will produce reliable forecasts.

Beginning as early as 1908 with the work of Arnold and Anderson,[5] numerous investigations through the years have attempted to define the nature of the decline in the production rate of oil wells to permit a more intelligent extrapolation of future rates. During the period from 1908 to 1917 important investigations were made in the field of production forecasting, which were summarized by Lewis and Beal[6] in 1917. It is interesting to compare the methods available at that time for the estimation of oil reserves and future rates of production with the techniques presently available. In making this comparison, one should bear in mind that the discipline of petroleum engineering as known today did not exist at that time. This period also predates all the classical work on fluid flow in porous media and reservoir and well behavior. Lewis and Beal cited in their paper three principal ways to prepare estimates of reserves and future oil production rates. They mentioned "(1) the saturation method, which is based on calculating the porosity of the sand; (2) the production curve method, which consists of determining, from the decline in production in the past, the amount of oil that will be recovered in the future; and (3) the production per acre method, by which the future production is estimated by comparing actual recoveries per acre from similar oil properties in the same or in a comparable district."

Except for the methods of numerical analysis, which are used to a relatively limited extent, present-day techniques are little more than refinements of the three methods mentioned by Lewis and Beal. The "saturation method" corresponds to estimating reserves by volumetric calculations, which now consider fluid saturations and shrinkage. In the category of the "production curve method," now called decline curve analysis, engineers have prepared graphs of virtually every imaginable type of data relating to the production of oil and gas in an endeavor to arrive at more useful relationships. Plots of "water cut," pressure, depth of oil-water contact, or cumulative gas versus cumulative oil are quite common. Various types of graph paper are used, such as Cartesian coordinate, semilog, and log-log. Most of these graphic relationships represent attempts to depict linear trends, for ease in extrapolation. A majority of the trends are nonlinear, however, although there are exceptions. Noble attempts have been made to straighten out nonlinear plots of oil rate versus time and oil rate versus cumulative oil by devising special semilog graph paper or by suitable shifting on log-log paper. Unfortunately, investigators have met with only limited success in their endeavors. In attempting to devise more reliable ways to treat performance data, it is important to determine, if possible, why more progress has not been made in analyzing decline curves. The third category of "production per acre method" relates, of course, to the concept of analogy with producing zones having similar reservoir rock and fluid properties, pool geometry, and general producing characteristics.

Some Aspects of Well Performance

In analyzing oil well production decline curves in detail it is helpful to examine the more salient features that play a role in the production performance of a well.

Muskat[7] has shown that the factors affecting the production rates of individual oil wells from the standpoint of reservoir behavior are essentially the effective permeability to oil (k_o), the net oil-zone thickness (h), the oil formation volume factor (B_o), the oil viscosity (μ_o), the wellbore radius (r_w), and the effective drainage radius of the well (r_e). These factors define the productivity index (J) of a well under steady-state conditions:

$$J = f\left[\frac{k_o h}{B_o \mu_o \ln (r_e/r_w)}\right]. \tag{1}$$

Furthermore, the production rate from a well (q_o) is equal to:

$$q_o = J(p_e - p_{wf}) \tag{2}$$

where p_e = pressure in reservoir at distance r_e from wellbore, and p_{wf} = bottom-hole flowing pressure.

Expressed in this form, the decline in the production from a well appears to be a function of the depletion of the area being drained by the well. Moreover, the relationship is not a simple one. As a reservoir is depleted, the pressure drawdown ($p_e - p_w$) normally decreases. The productivity index of the well also decreases owing to the sensitivity of the oil permeability to fluid saturation and the effect of reservoir depletion on the product of oil viscosity and oil formation volume factor. The effect of pressure decline on B_o is opposite to the effect on μ_o; the factor B_o decreases and μ_o increases, but the rate of change in μ_o is greater. Only in instances in which the oil saturations and values for oil viscosity and oil formation volume factor are relatively constant will the productivity index approach a constant value, in which case the decline will usually be exponential, as explained later. Notable examples are reservoirs producing above the bubble point and solution gas-drive pools subjected to pressure maintenance. Gas wells in the later stages of depletion, producing at low reservoir pressure and with a relatively constant producing pressure, will usually exhibit exponential decline, as will water-drive pools producing at close to bubble point. Solution gas-drive pools at pressures considerably below bubble point will demonstrate approximate exponential decline for a time, but at a lower decline rate than when reservoir pressure was above bubble point. The relationship between the productivity index and the degree of depletion is complicated further by transient behavior and wellbore damage effects incurred after completion. It appears, therefore, that the trends of decline curves depend on a complex set of variables. This fact tends to complicate any method of classification of decline curves.

Classification of Decline Curves

In the introductory chapter, van Wingen has summarized the classification of oil production decline curves, as generally accepted in the industry, as exponential, hyperbolic, and harmonic. The decline of some oil wells can be described as:

$$\frac{q_o}{dq_o/dt} = -a \tag{3}$$

where the ratio of the oil production rate (q_o) to the loss in production rate per unit time (dq_o/dt) is constant. When this ratio, called the loss ratio by Arps,[8] is constant, the decline is termed exponential.

If, on the other hand, the loss ratio is not constant, but the first difference is, the decline is classified as hyperbolic and

$$d\left(\frac{q_o}{dq_o/dt}\right)\Big/dt = -b. \tag{4}$$

In other words, in hyperbolic decline the first derivative of the loss ratio is a constant. Arps[8] showed that the decline rate (D) is equal to:

$$D = Kq_o^{\,b} \tag{5}$$

where K is a constant determined from initial conditions. Equation 5 is sometimes used to estimate the future decline performance of a well approximating hyperbolic decline, provided exponent b can be evaluated. If, for example, b is set at 0.5, the decline percentage (D) is proportional to the square root of the production rate. A well that produces 10,000 bbl/month with a 50% per year decline will be declining at 5% per year when yielding 100 bbl/month. Most individual well decline curves exhibit values for b in the range from 0 to 0.4 when fitted to Eq. 5. When $b = 0$, there is no change in decline rate with time (exponential decline). The case of $b = 1$ has been termed harmonic decline and apparently occurs only rarely.

Current practice is to attempt to identify the decline behavior of any oil well as either exponential or hyperbolic. The classification appears excessively restrictive in light of the complexity of the decline process. It is questionable whether a large segment of the decline curves is described adequately by these relatively simple relationships. The classification is used extensively, nevertheless. It does indicate the general form that virtually all production rate-time curves take when plotted on semilog graph paper, namely, either linear for exponential decline or concave upward for hyperbolic decline. To apply the relatively simple hyperbolic equation rigorously to practical situations, however, requires high-speed computers to accomplish the necessary curve-fitting.[9, 10] Locke *et al.*[11] have found, moreover, that conventional least-squares curve-fitting techniques may lead to gross errors in reserve estimates. They have employed a least-squares technique based on Taylor's theorem with greater apparent success. In any case, in light of the complexities of the problem perhaps a more satisfactory approach is not to insist that the future decline of a well comply with Eq. 5 but to employ a power series of the form:

$$y = a_o + a_1x + a_2x^2 + a_3x^3 + \ldots a_nx^n \qquad (6)$$

where y is production rate and x is time.

Important Considerations in the Analysis of Decline Curves

There are other important considerations in the analysis of decline curves that frequently invalidate any rigorous mathematical analysis of past performance. To elucidate these considerations, it is helpful to consider the status of the art of decline curve analysis as commonly practiced today. In order to plot oil rate versus time, for example, monthly production figures of the well may be obtained from company records or from the files of the government oil and gas agency involved. At this point three questions arise pertaining to the validity of the production data in the interpretation of decline curves: (1) Do the production figures represent capacity production? (2) Are the figures accurate? and (3) To what extent has the well been off production during the period represented?

The first question is vital because in decline curve analysis the production decline characteristics of a well must reflect reservoir depletion. It is apparent, from Eq. 2, that the production rate (q_o) is a function of drainage-area depletion alone only when the bottom-hole producing pressure of the well is constant. From a practical standpoint, therefore, the well must be producing at its maximum rate, that is, with the bottom-hole producing pressure approaching a minimum value.

Accuracy of production data is also of prime importance. Frequently wells on the same lease producing into a common tank battery are not subjected to individual tests as often as necessary. The reported individual well production rates are then only approximated and may be grossly in error.

The Problem of Down Time

To interpret properly the decline curve of a well, the time during which the well was off production (down time) must be known. Many investigators make allowances for down time by calculating the oil rate in barrels per producing day instead of barrels per calendar day or per month. The theory is, apparently, that plotting oil rate on a calendar day basis does not give a fair representation of the ability of a well to produce. A common procedure is to enter the data on the graph as barrels per producing day versus time, to extend the decline curve in the future, to convert the projection to a 365-day yearly forecast, and either to ignore the down time of the well or wells or to hope that the cause for the down time can be eliminated.

The practice of plotting oil rate as barrels per calendar day or barrels per month seems to be a far better method. In many cases, down time on a well or a lease is a fact that cannot be ignored. A decline curve of barrels per producing day versus time usually yields a forecast optimistic to an extent which is dependent on the amount of down time of the well or lease and on the reservoir fluid and rock characteristics. In any event, a plot of barrels per calendar day versus time provides a means for any significant down time to be represented graphically and interpreted in the light of circumstances. Usually, only in cases of excessive down time is the decline trend of the well obscured. The problem of interpreting flush production of a well after being off production can also be handled more intelligently in this manner.

The Nature of the Decline Curve under Examination

After the production rate data are represented graphically in a meaningful way, the engineer is ready to proceed with the analysis. He may be working with one or more separate graphic representations on the same well. In examining the decline data as represented, he must first decide what role he thinks the past production history should play in forecasting future production, that is, a dominant or a passive role. Can the past production data be taken at face value, or is the available production history not entirely indicative of probable future performance? If it appears that future performance will follow the trend to date, the mathematical treatments previously discussed may be applicable. In the alternative case, the past production history does not adequately reflect probable future performance.

The Experience Factor

In decline curve analysis, the engineer is frequently faced with a situation in which he sees no possibility for a formal type of analysis, apart from using experience with wells having similar characteristics and exercising intuition and judgment. Here the practical experience of the engineer comes to the forefront, and certain implications may be drawn.

Analogy to Monte Carlo Methods. Any procedure which involves the use of statistical sampling techniques to approximate the solution of a mathematical or physical problem can be classified as a Monte Carlo method.

Such methods have an ancient and honorable history and are distinguished by their experimental nature. In recent years Monte Carlo methods have been best known in connection with problems that do not yield conveniently to classical analysis, as in areas of statistical mechanics and particle diffusion. In these cases the numerical computation of approximate solutions can sometimes be approached through random sampling.

An illustration of the Monte Carlo method can be found in the pastime of playing poker. After examining his cards, the experienced poker player knows, at least subconsciously, what his chances are of filling an inside straight. He has gained this knowledge the hard way. The actual probability for a given hand could be calculated, but most poker players learn from experience rather than by direct calculation. In this case the player has profited by a Monte Carlo approach.

The Problem of Uncertainties. It is apparent that in many situations the petroleum engineer must resort to measures comparable to the Monte Carlo method in predicting the future production rates of oil wells. Unlike the poker player, however, he may face a problem that does not lend itself to rigorous analysis. He is confronted with a situation where the decline history to date does not adequately describe the probable future production trend. He is aware of the possibility of his forecast deviating substantially from the actual ultimate performance. What course should he follow in this situation? He will ultimately have to think in terms of probabilities and will choose the forecast which he considers most likely to be correct. His choice will be a judicious one, of course. It will be based on all the engineering and geological information available on the well in question and similar wells. Frequently, however, the technical data at hand are inadequate. In making the final judgment, then, the estimator must lean heavily on his professional experience with wells of similar characteristics. Because like the poker player he has also learned the hard way from predictions that may have turned out badly, he is drawing on his previous experience as an estimator to solve the present problem.

To the uninitiated it may seem that the petroleum engineer is adopting a haphazard procedure in his evaluation work. Actually, because important variables often cannot be evaluated rigorously, he is simply dealing with the situation by drawing on every resource at his disposal.

Some investigators have proposed that, where significant uncertainties exist, the engineer may resolve the problem of forecasting error by simply defining a range within which he believes the actual performance will fall. The project is thus evaluated within certain limits of reasonableness. Although this approach has merit and can be valuable at times, conditions usually dictate that a single estimate be given.

Depicting Trends in the Curves

In some instances it may be feasible to develop a mathematical equation to describe the trend of data, using the method of least squares as previously mentioned. Most decline curves, however, can be interpolated satisfactorily by using a smooth freehand curve to represent a trend through the data points. The curve, which may be linear or curvilinear, is drawn to obtain the best fit. The tendency is to draw the curve in such a way that the areas above and below the line are equal. Normally, this is a satisfactory method. When using semi-log or log-log paper, however, the fact that the scale is nonlinear should be compensated for as much as possible.

The Causes for Decline in Production Rate

In decline curve analysis, the estimator is, of course, anxious to detect "trends" in production attrition that will continue in the future. The causes of decline in the production rate of a well fall into three main categories: (1) reservoir depletion, (2) fluctuation in bottom-hole producing pressure, and (3) changes in conditions in or immediately adjacent to the wellbore.

Reservoir Depletion. In regard to the first category, Eqs. 1 and 2 show that the decrease in productivity index (J) reflects depletion in the reservoir or reservoirs from which the well is producing. Under primary production operations, depletion is directly related to reservoir withdrawals. Hence data on decline in production performance, which is mainly the result of reservoir depletion, are ordinarily clearly defined and usually can be interpreted reliably.

Two factors to consider regarding reservoir depletion are gravity drainage and well interference. These may complicate the extrapolation of the curve, as illustrated in the following example.

Example of Gravity Drainage. The example concerns the performance of four wells in the Townlot extension of the Huntington Beach Field, Orange County, California. The reservoir is the Jones Sand, but the type of behavior applies to highly permeable carbonate reservoirs as well. In this case the permeability of the zone is upwards of 8,000 md and the oil gravity is 20° API. Figure 1, showing a portion of the field, indicates very close spacing and substantial formation dip. Figures 2 and 3 are the decline curves of wells 1 and 2 together and wells 3 and 4 together, respectively. An abrupt change in the performance of both well groups clearly took place during the fifth year. If an investigator extrapolating the curves prior to this time did not anticipate, at least in part, this gross change in performance, his forecasts would be

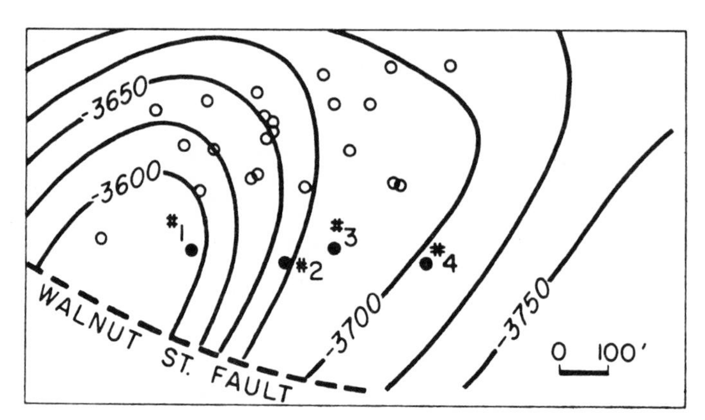

FIG. 1. Map of portion of Huntington Beach Field, Orange County, California. Subsurface contours on "AD" Marker. (After Hunter et al.,[18] courtesy of Calif. Div. of Oil and Gas).

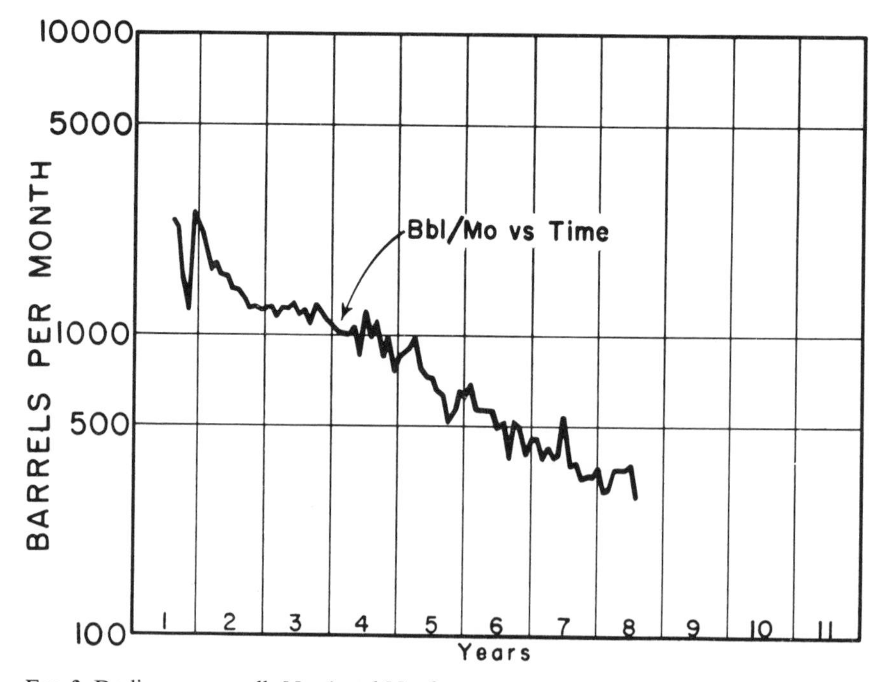

FIG. 2. Decline curve, wells No. 1 and No. 2.

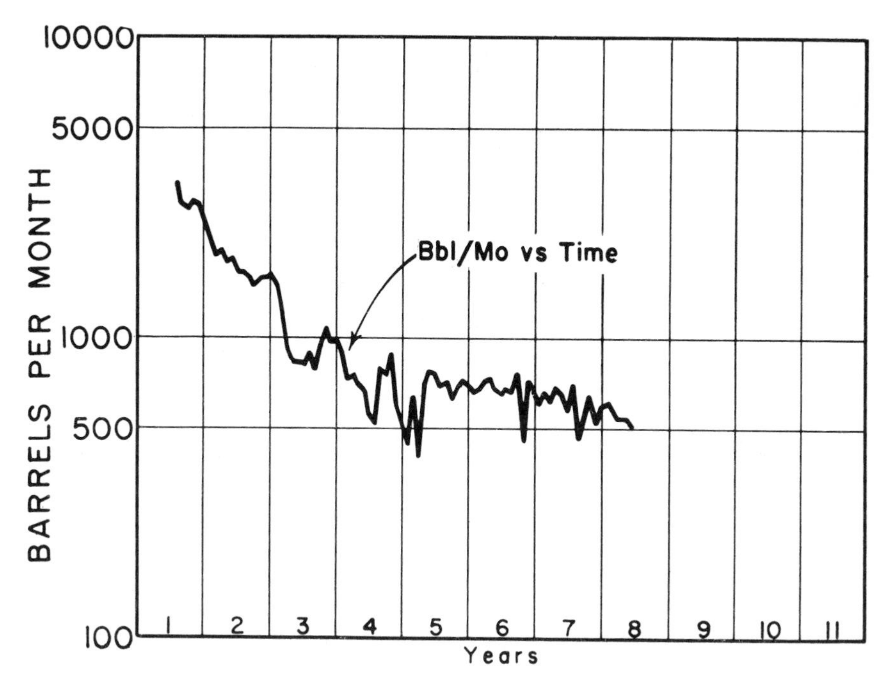

FIG. 3. Decline curve, wells No. 3 and No. 4.

greatly in error. Although the example is extreme, it illustrates the fact that a knowledge of reservoir parameters may be vital. In this case a combination of factors—closeness of wells, high zone permeability, and formation dip—provided for a drastic downdip movement of oil in the latter years of depletion.

Fluctuation in Bottom-Hole Producing Pressure. Fluctuation in bottom-hole producing pressure will obscure the decline characteristics of a well. Wells, both on natural flow and on artificial lift, that are not producing at capacity may demonstrate this fact. Paying particular attention to the efficiency of artificial lift equipment may help to alleviate this problem.

Changes in Wellbore Conditions. In the third category—changes in conditions in or immediately adjacent to the wellbore—a number of factors that can lower well productivity should be considered. Most of these factors serve to decrease the effective permeability to oil. Some of the more common of these factors involve plugging of the wellbore by formation materials (clay and silt) and asphaltene and wax deposition and accumulation in and near the wellbore. Reduction in oil permeability because of increased free gas saturation, water production, or clay swelling can also occur. Some of these changes in wellbore conditions can be remedied by well treatment, whereas others will not ordinarily respond to such measures. Transient effects, such as flush production after shut-in, may also be included in this category.

Summary of Decline Curve Analysis and Production Forecasting

Various graphic representations of production performance are employed from time to time to meet specific situations. Ideally, a method should be available for estimating the remaining reserves apart from the analysis of production decline curves. It may be possible to make a trustworthy estimate of reserves by performing reservoir material balance calculations, by employing the volumetric method, or by interpreting such meaningful trends as produced water cut versus cumulative oil production. When the data for an independent estimate are lacking, the engineer must rely solely on the production decline data.

Obviously, there is no infallible method for predicting, in all situations, the future rates of production from a well. It would be unwise to set forth a specific procedure to follow; the problem of decline curve analysis is much too complex to be solved in this manner. The investigator must exercise ingenuity and imagination. He must have knowledge of the various facets in interpreting production data and be aware of the possible effect of related geological and engineering factors on future production performance. In analyzing a given situation, he must also be sensitive to conditions that dictate a modification in method in order to prepare a trustworthy prediction.

For each well a decision must be made concerning the extent to which the past production history of the well will represent the future performance. Every effort should be made to develop a separate estimate for remaining reserves and for future rates of production, particularly on wells with decline histories that are difficult to interpret. The key to analyzing decline curves with reliability is a careful, thoughtful approach, a thorough understanding of the physical and chemical factors involved, adequate experience in decline curve analysis and in other aspects of oil production technology, and good judgment.

Tank Oil-in-Place by Volumetric Methods

The estimation of initial oil-in-place by volumetric methods requires the determination of representative values for porosity (ϕ), interstitial water saturation (S_w), and net pay thickness (h) in the reservoir. This holds true for sandstone as well as for carbonate reservoirs. The values of these parameters, of course, will vary throughout the reservoir both horizontally and vertically. If one had the opportunity to sample the reservoir at random he would find a multitude of sets of values for ϕ and S_w. A rigorous calculation of the hydrocarbon pore volume would require (1) assigning representative values of ϕ and S_w to individual segments of the reservoir, (2) calculating the hydrocarbon pore volumes of the individual segments, and (3) summing the volumes of the individual segments to determine the overall hydrocarbon pore volume. A detailed procedure of this type would be justified only in cases where rapid changes in the character of the reservoir rock occur from place to place and where sufficient data are available. Normally, the initial oil- or gas-in-place in sandstone reservoirs can be estimated satisfactorily through the use of average values for ϕ and S_w, although there are exceptions. In carbonate reservoirs, on the other hand, more elaborate techniques are very often required to achieve reliable oil- or gas-in-place values because of the heterogeneous character of the rock.

As discussed by Jodry in Chapter 2, carbonates are typified by abrupt changes in lithology and petrophysical properties, as well as a wide variation in porosity and permeability owing to the nonuniform distribution and orientation of vugs and fractures. These conditions create, for the engineer and the geologist estimating in-place oil or gas in carbonates, problems that are not normally encountered in sandstones, except in very low-permeability zones. Large amounts of oil or gas, for example, may be bound up in rocks having low porosity and permeability and, possibly, low productivity. The question arises as to whether the low-permeability intervals contribute a substantial portion of the total flow rate from the zone. Oil-productive measures that do not account for a significant portion of the well's total production should not be included in the reserve calculations. Carbonate rock heterogeneity also complicates the estimation of oil or gas in the transition zone. These difficulties and others encountered in reserve estimation are discussed later in the chapter.

Relationship of Porosity, Water Saturation, and Net Productive Thickness

The parameters of ϕ, S_w, and net productive thickness (h), so critical in reserve estimation, are not unrelated in the reservoir; meaningful relationships exist among them.[12–16] Inasmuch as reservoirs cannot be sampled at random to determine ϕ, S_w, and h, quantitative knowledge of these interrelations is of utmost importance. In carbonate reservoirs, the relationships depend largely on the degree of heterogeneity of the zone and can be quite complex. To date, the phenomenon of capillary pressure seems to be the best common denominator to correlate zonal properties.

Capillary pressure *per se*, also discussed in Chapters 2 and 3, refers, of course, to the capillary forces that result when two or more immiscible fluids are present in a porous media. The principles of capillary pressure in the volumetric estimation of oil-in-place are utilized in determining (1) the distribution of the fluids in the reservoir, and (2) the mobility of the fluids in the reservoir.

Distribution of Fluids in the Reservoir

To gain a better understanding of the nature of the distribution of oil and water and, sometimes, free gas under conditions of equilibrium in the reservoir, it is helpful to consider in a concise, simplified manner the conditions prevailing during the migration and accumulation of the oil and gas in its final resting place in the pore space. Most investigators agree that the reservoirs were 100% water saturated before the influx of oil and gas. As the hydrocarbons encroached into areas that provided favorable conditions for the trapping of oil and gas, gravity forces were initiated because of the differing densities of oil, gas, and water.

As demonstrated by the J-tube illustration in Chapter 2 (p. 47), the gravity forces give rise to pressure differential, inasmuch as the pressure in the oil phase exceeds that in the water phase. The oil, therefore, tends to displace the water from individual pore capillaries throughout the section. The vertical extent throughout which this takes place is dependent on the supply of oil, the vertical continuity of porous and permeable zones, and the amount of closure. If this condition of displacement of water by oil were allowed to go to completion, oil and gas reservoirs would contain no interstitial water. Gravity segregation of the fluids would be complete with 100% free gas, when present, at the top, an underlying 100% oil-saturated zone, and water at the bottom. There would be a sharp line of demarcation between the gas, oil, and water zones, with no transition zones.

Capillary forces, of course, do not permit the water to be displaced entirely from the reservoir void space. As the displacement process progresses, capillary effects serve to retard the oil movement. Finally an equilibrium reflecting a balance of gravity, capillary, and, to some extent, geothermal effects is achieved. The manner in which equilibrium is reached is somewhat complex and is beyond the scope of this book; Muskat[7] has discussed the mechanics in some detail. It appears that imbibition is operative in the final stages before equilibrium, at least in the lower portion of the zone.

In the simplified case of a zone of uniform rock characteristics, the equilibrium condition that exists in the reservoir before discovery can be illustrated by the capillary pressure curve in Fig. 4 for a San Andres Dolomite (k = 10 md, ϕ = 18.8%). The difference in pressure between the two phases, which equals the capillary pressure, increases with distance above the free water table. The displacement of water by oil, therefore, is more nearly complete in the upper portions of the zone. The corresponding distribution of oil and water for this rock type (Fig. 4) in relation to height above the water table is shown in Fig. 5 (solid curve to the left). The water saturation is the "irreducible water saturation" below which natural forces of gravity are unable to drain the water at a given level above the free water table. In the above example a uniform porosity and permeability of 18.8% and 10 md, respectively, are assumed to exist throughout the zone. Figure 5 shows that in a zone of this type the water saturation would vary from 80% at the base of the transition zone to approximately 15% at a distance of 80 ft above the

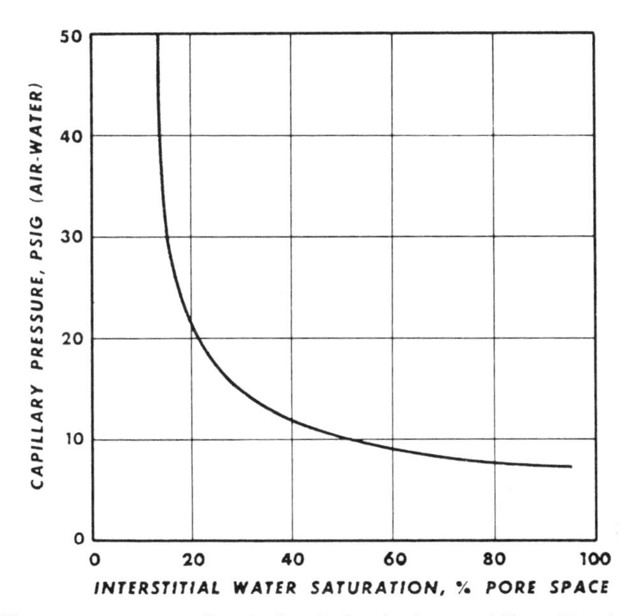

FIG. 4. Capillary pressure curve, San Andres Dolomite (permeability = 10 md, porosity = 18.8%). (After Aufricht and Koepf,[13] Fig. 1, courtesy of AIME.)

bottom of the transition zone. The saturation discontinuity across the free water level from 80% to 100% water saturation suggests the existence of imbibition before attainment of equilibrium, as previously mentioned.

The laboratory values of capillary pressure (P_c) may be translated into values of height above the base of the transition zone by first converting P_c from laboratory $(P_c)_L$ to reservoir conditions $(P_c)_R$:

$$(P_c)_R = \frac{\sigma_{wo} \cos \theta_{wo} (P_c)_L}{\sigma_{wg} \cos \theta_{wg}} \tag{7}$$

where σ_{wo} = interfacial tension (dynes/cm) between water and oil in the reservoir; σ_{wg} = interfacial tension (dynes/cm) between water and air in the

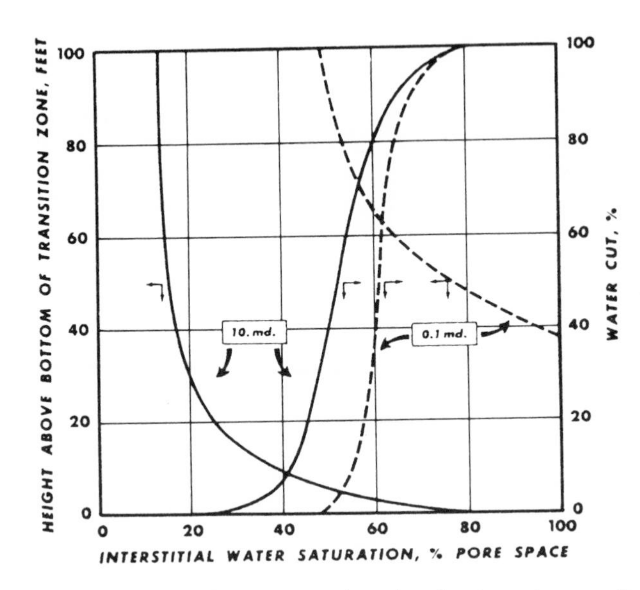

FIG. 5. Water saturation distribution curves and produced water cut curves, San Andres Dolomite, 10-md and 0.1-md zones. (After Aufricht and Koepf,[13] Fig. 3, courtesy of AIME.)

laboratory; and θ_{wo} and θ_{wg} = water-oil and water-air contact angles, respectively. To simplify the calculation, the values for θ_{wo} and θ_{wg} are commonly set equal to each other.

The height (h) in feet above the bottom of the transition zone is then found by the following equation:

$$h = \frac{144P_c}{\rho_w - \rho_o} \tag{8}$$

where ρ_w = density of formation water (lb/cu ft), and ρ_o = density of reservoir oil (lb/cu ft). (See Appendix B.)

Mobility of the Fluid

If the solid curves in Fig. 5 were representative of the entire reservoir, the transition zone would be about 25 ft thick. Above and below the transition zone, water-free oil and 100% water are produced, respectively. It can be seen that within the transition zone both oil and water are produced. Figure

5 (solid curve to the right) shows the relative amounts of water and oil production versus height. For example, in going from a height of 5 ft to a height of 10 ft above the base, the water cut decreases from about 60% to approximately 7%. This curve is constructed through use of the relative permeability curves illustrated in Fig. 6 for the 10-md carbonate zone under investigation.

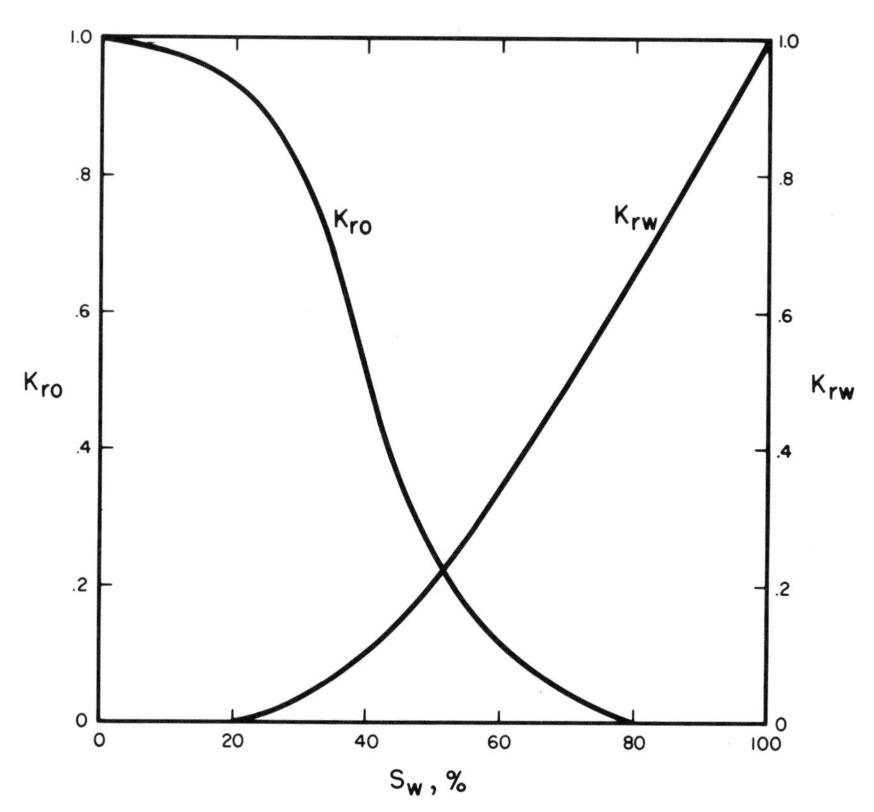

FIG. 6. Relative permeability curves representative of San Andres Dolomite. (After Aufricht and Koepf,[13] courtesy of AIME.)

Darcy's law for the flow of oil (q_o) and water (q_w) in barrels per day may be expressed as:

$$q_o = \frac{7.08 k_o h (p_e - p_{wf})}{B_o \mu_o \ln (r_e/r_w)} \tag{9}$$

$$q_w = \frac{7.08 k_w h (p_e - p_{wf})}{B_w \mu_w \ln (r_e/r_w)}. \tag{10}$$

The produced water cut is taken as $q_w/(q_w + q_o)$, and, therefore:

$$\text{Water cut percent} = \frac{100}{1 + (k_o/k_w)(B_w \mu_w / B_o \mu_o)}. \tag{11}$$

In the calculation, the anticipated initial reservoir pressure can be employed to determine the $B_w\mu_w/B_o\mu_o$ ratio.

It would be fortuitous to assume that a single set of reservoir parameters, as represented by the 10-md zone of Fig. 5, would suffice in the reserve calculations. Productive measures of lower permeability, for example, would demonstrate vastly different curves. Plots of height above the bottom of the

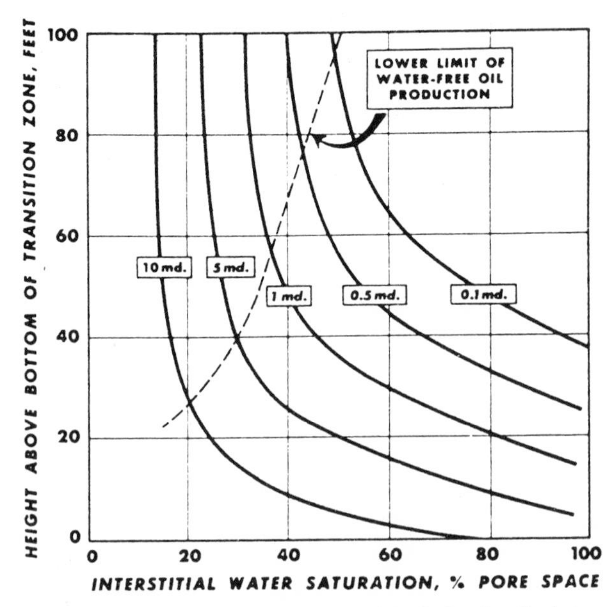

FIG. 7. Water saturation distribution curves, San Andres Dolomite, for intermediate permeabilities. (After Aufricht and Koepf,[13] Fig. 5, courtesy of AIME.)

transition zone and produced water cut versus S_w for a 0.1-md zone in the San Andres Dolomite are also depicted in Fig. 5. As in the case of the 10-md permeability, the 0.1-md curves are derived from (1) basic laboratory and relative permeability data for the 0.1-md zone, and (2) Eqs. 8 and 11 to plot the capillary height and produced water cut curves. The extreme variation in formation properties over the range of a 100-fold change in permeability is evident. For example, at a height of 50 ft above the base of the transition zone, S_w is 15% and 77% for the 10-md and 0.1-md zones, respectively. The corresponding water cuts are 0% and 98%.

As an aid to a better definition of the overall producibility of a zone, Fig. 7 shows the vertical distribution of S_w and water cut for intermediate permeabilities in the 10–0.1 md range. These curves were determined in the same manner as those for the 10-md and 0.1-md permeability zones. The existence of low-permeability measures in a producing section results in some initial water production from intervals that are as much as 100 ft above the base of the zone; this is true in the case of the 0.1-md rock. The presence of high water cut intervals may, however, have only minor effect on the overall water cut because of the low productivity of these sections.

Determination of Tank Oil-in-Place

A Coregraph Using a Permeability Correlation

Application of the data from Figs. 4–7 is illustrated by the coregraph in Fig. 8, which was presented by Aufricht and Koepf.[13] Values for k, ϕ, S_w, and produced water cut are shown for each foot in a 50-ft interval of San Andres Dolomite. The k and ϕ values represent laboratory data, and the S_w and water cuts were determined from a complete suite of curves comparable to Figs. 4–7, using permeability as the correlating physical parameter. From the data on the coregraph the detailed calculations of initial oil- and gas-in-place previously mentioned can be made.

Tank oil-in-place (TOP) for each foot of oil section is

$$\text{TOP} = \frac{7758\phi(1-S_w)}{B_{oi}} \quad \text{(bbl/acre-ft)} \tag{12}$$

where B_{oi} = oil formation volume factor at initial reservoir pressure, and ϕ and S_w are unique values for each foot from the coregraph.

The calculated value for tank oil barrels in place for the well from this zone would, of course, be equal to the average oil-in-place figure (bbl/acre-ft) for the entire producing section multiplied by the estimated areal extent (acres) drained by the well.

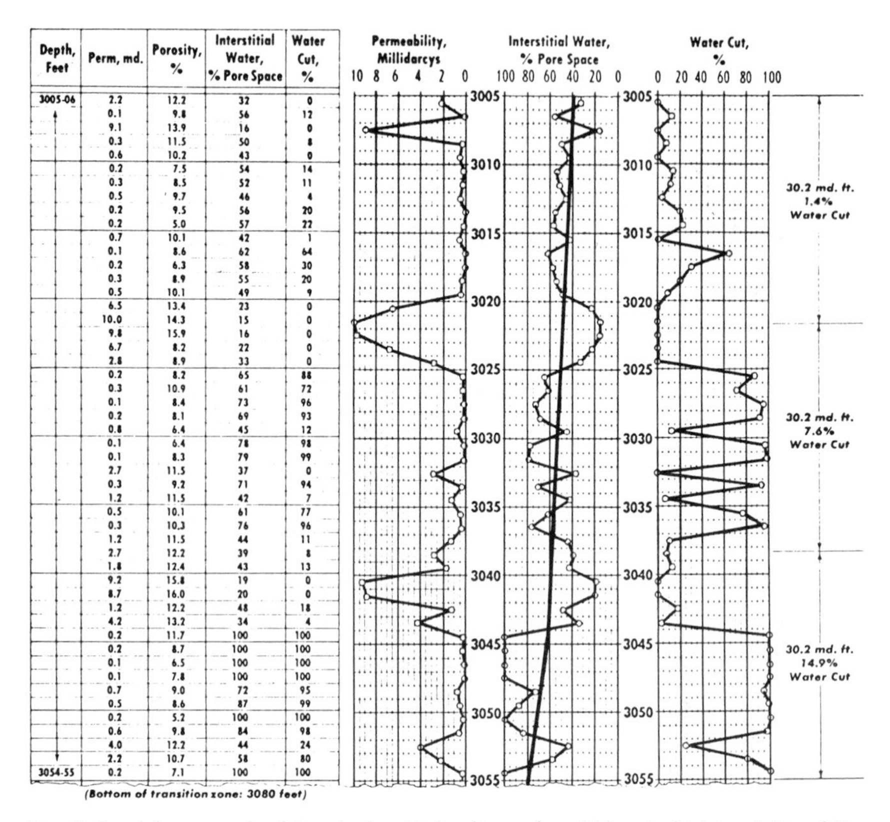

Depth, Feet	Perm, md.	Porosity, %	Interstitial Water, % Pore Space	Water Cut, %
3005-06	2.2	12.2	32	0
	0.1	9.8	56	12
	9.1	13.9	16	0
	0.3	11.5	50	8
	0.6	10.2	43	0
	0.2	7.5	54	14
	0.3	8.5	52	11
	0.5	9.7	46	4
	0.2	9.5	56	20
	0.2	5.0	57	22
	0.7	10.1	42	1
	0.1	8.6	62	64
	0.2	6.3	58	30
	0.3	8.9	55	20
	0.5	10.1	49	9
	6.5	13.4	23	0
	10.0	14.3	15	0
	9.8	15.9	16	0
	6.7	8.2	22	0
	2.8	8.9	33	0
	0.2	8.2	65	88
	0.3	10.9	61	72
	0.1	8.4	73	96
	0.2	8.1	69	93
	0.8	6.4	45	12
	0.1	6.4	78	98
	0.1	8.3	79	99
	2.7	11.5	37	0
	0.3	9.2	71	94
	1.2	11.5	42	7
	0.5	10.1	61	77
	0.3	10.3	76	96
	1.2	11.5	44	11
	2.7	12.2	39	8
	1.8	12.4	43	13
	9.2	15.8	19	0
	8.7	16.0	20	0
	1.2	12.2	48	18
	4.2	13.2	34	4
	0.2	11.7	100	100
	0.2	8.7	100	100
	0.1	6.5	100	100
	0.1	7.8	100	100
	0.7	9.0	72	95
	0.5	8.6	87	99
	0.2	5.2	100	100
	0.6	9.8	84	98
	4.0	12.2	44	24
	2.2	10.7	58	80
3054-55	0.2	7.1	100	100

FIG. 8. Special coregraph of San Andres Dolomite section. (After Aufricht and Koepf,[13] FIG. 6, courtesy of AIME.)

Meaningful interpretation of oil-in-place values in order to arrive at estimates of reserves requires careful examination of the relative mobilities of the oil and water. A detailed examination of the section on the coregraph reveals that about three fifths of the zone is composed of rocks having permeabilities of less than 1 md. These tight sections, moreover, produce some water even at the very top of the zone, and they are predominately water productive below 3,025 ft, which is still some 55 ft above the base of the transition zone. In spite of these high water cuts, the calculations indicate that the overall water cut in the 3,038–55 ft interval is only 14.9%, with the entire 17 ft open to production. It is evident that the permeable stringers in the upper part of the interval account for most of the oil production, along with minor amounts of formation water. The lower portion of the interval is largely water productive, but the rate of fluid entry is less than that above because of the lower permeabilities.

Porosity Correlation

The above example employs permeability as a correlating parameter; porosity has also been used effectively. Some investigators[13] have obtained better results in carbonates of relatively uniform porosity by using permeability for correlation. In fractured or vuggy sections, correlating porosity appears to be more satisfactory.

Another method presented by Rockwood *et al.*[14] for determining tank oil-in-place in carbonate reservoirs uses porosity and rock type as parameters for correlation. The technique does not make use of relative permeability curves and is somewhat more qualitative, but can be useful and serves to emphasize some important concepts. In an example given by the authors, capillary pressure curves were obtained by mercury injection methods for 28 cores from a carbonate section in a well drilled in west Texas. Figure 9 shows plots of the averaged P_c versus S_w data at various porosities. In the diagram, the pressure levels of 5, 11, 21, and 62 atm have been arbitrarily selected. The 62-atm pressure level represents a position in the section a great distance above the free water table, normally well above the top of the transition zone. The 5-atm pressure level, on the other hand, relates to

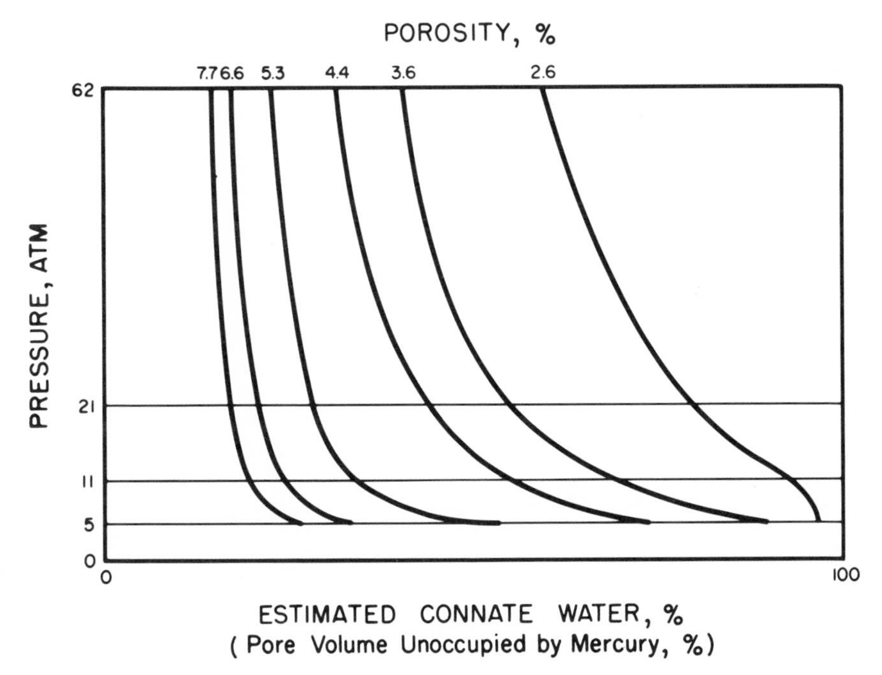

FIG. 9. Average mercury capillary pressure curves, Devonian Formation, Andrews County, Texas. (After Rockwood et al.,[14] Fig. 2, courtesy of AIME.)

producing intervals close to the water table, and the 11- and 21-atm levels are between the two extremes. The curves in Fig. 9 show the relationship between porosity and water saturation at these various positions in the producing section. The shapes and relative positions of the curves representing the range of porosities from 2.6 to 7.7% in this figure reveal information on petrophysical rock properties that is valuable in estimating oil and gas reserves.

Pore-Size Distribution

The productive measures close to the free water table at the 5-atm level in Fig. 9 contain hydrocarbons only in the larger capillary spaces. The smaller pore spaces in this region are 100% water saturated owing to the weak capillary environment. In moving up the oil section in the direction of the 62-atm level, increasingly smaller void spaces are found to contain oil. From the standpoint of the estimation of reserves and productibility, it is important, of course, to have information on the proportion of large and small pore spaces in an oil-producing interval. A zone with the pore volume composed of predominantly small capillaries holding the oil would be less interesting than one consisting mainly of large capillaries. Moreover, many carbonate reservoirs contain thick sections of such low permeability and porosity that it is questionable whether these sections should be included in the calculation of reserves.

Figure 9 indicates that the pore-size distribution is related to porosity in this formation. On the basis of the total pores that contain oil at a pressure of 62 atm, it is seen that the 7% porosity sections consist of approximately 83% large pores. The lower-porosity intervals included down to 2.6% porosity show successively lower portions of large capillaries. As previously noted in this figure, the pores invaded by hydrocarbons at the 5-atm level are considered to consist of the large pores in the rock.

Porosity and Permeability Cutoff

The problem of excluding oil and gas measures of low permeability and porosity which do not contribute significantly to the production of the well has been mentioned. Parameters of porosity and permeability have both been used in delineating intervals that should not be included in estimating reserves. Some qualitative measure of the productibility of these tight intervals must be made so that a judgment is possible as to whether or not to include them. Permeability is a direct measure of the ability of an interval to produce fluids. Porosity, on the other hand, is indirectly related to productivity because of its relationship to permeability.

Rockwood *et al.*[14] have found that for carbonate reservoirs porosity provides a better correlation to determine the cutoff point for reserves. It is

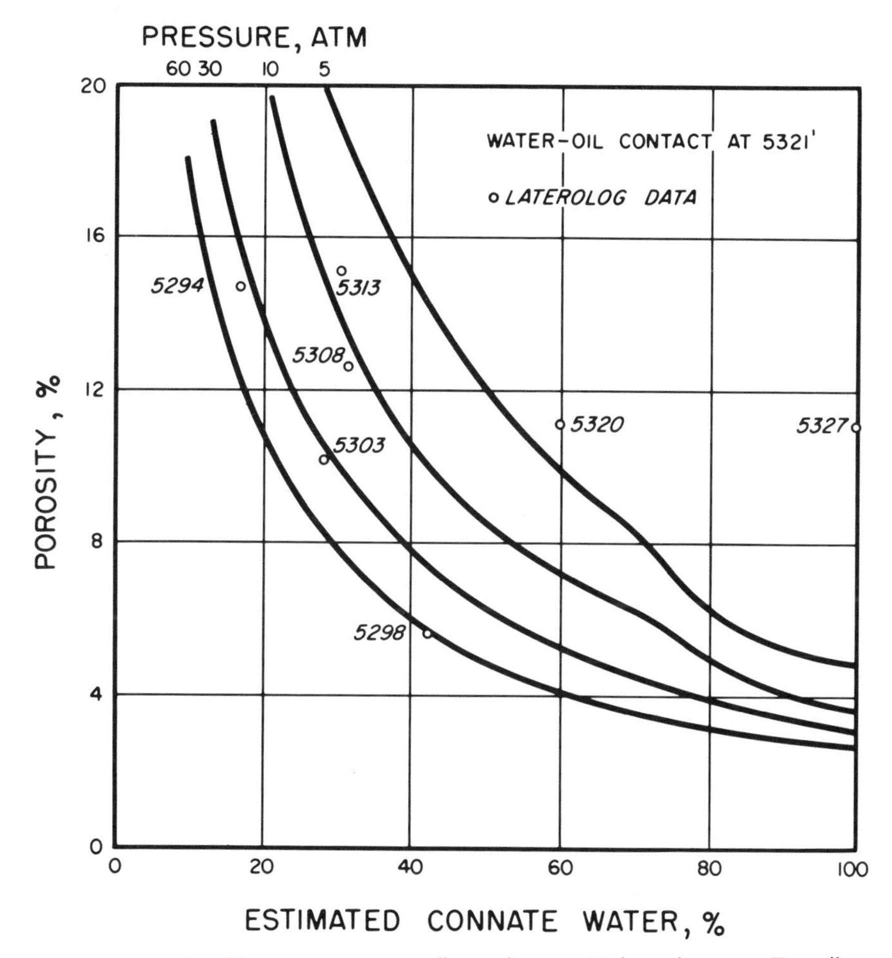

FIG. 10. Effect of capillary pressure on cutoff porosity, Des Moines Limestone, Toenail Strawn Field, Schleicher County, Texas. (After Rockwood et al.,[14] Fig. 14, courtesy of AIME.)

also more convenient because ϕ readings can be determined *in situ*. Figure 10 is their plot of S_w versus ϕ for the Des Moines Limestone, Toenail Field, Schleicher County, Texas. The data for the curve were obtained from the laterolog rather than capillary tests as in Fig. 9. Either type of data can be used along with oil-base mud core data. The indicated depths of the laterolog data of Fig. 10 have been expressed in terms of capillary pressure (atm). The typical behavior is observed of water saturations increasing sharply with decreased porosity in the lower porosity ranges. This is particularly significant in this application, inasmuch as low porosity readings not only indicate reduced in-place values for oil but normally correspond to lower production rates. A cutoff porosity can, therefore, be chosen in the region of this sharp increase. The cutoff porosity will, of course, vary with distance from the water table; high in the zone (62-atm curve) the cutoff porosity might be on the order of 5–6%, compared to 10–12% near the water table (5-atm curve). The same general comments hold for excluding measures of extremely low productivity, using permeability as a parameter by means of Fig. 7.

Leverett's *J* Function

In cases of extreme heterogeneity, which is typical of carbonates, neither permeability nor porosity may be satisfactory as a correlating parameter. The *J* function developed by Leverett[12] describes the rock characteristics more adequately by combining porosity and permeability in a parameter for correlation. The *J* function has received very little attention in the literature in recent years, and consequently certain advantages gained by its use have not been fully exploited.

The capillary pressure function developed by Leverett,[12] which is dimensionless, is:

$$J(S_w) = \left(\frac{P_c}{\sigma \cos \theta}\right) \left(\frac{k}{\phi}\right)^{0.5} \tag{13}$$

where σ = interfacial tension between the two fluids (dynes/cm), and θ = the contact angle between the interface separating the two fluids and the rock surface ($\theta = 0°$ for completely water-wet cores).

Ideally, a function of this type attempts to reduce the capillary pressure-saturation relationship to a common curve irrespective of the rock type. In order to accomplish this, the equation must express the basic properties of the system, which govern the fluid behavior in the reservoir. According to Rose and Bruce,[15] these properties are related to (1) the geometrical configuration of the interstitial spaces, (2) the physical and chemical nature of the interstitial surfaces, and (3) the physical and chemical behavior of the fluid phases in contact with the interstitial surfaces. To approach a satisfactory general correlation of these basic properties for all rock types would require a complex relationship, which indeed has not be accomplished. Equation 13, much simpler in form, is well suited to provide capillary pressure-saturation relationships for individual rock types.

Experimental data presented by Brown[16] demonstrate the value of Eq. 13 in correlating capillary pressure-liquid saturation data obtained from core samples of a particular geologic formation. He prepared curves known as *J* curves by plotting $J(S_w)$ versus wetting-phase saturation. Figures 11 and 12 are *J* curves for microgranular and coarse-grained limestone cores of the Cretaceous Edwards Formation in the Jourdanton Field of southwest Texas. The data indicate a unique correlation for core samples of differing porosities

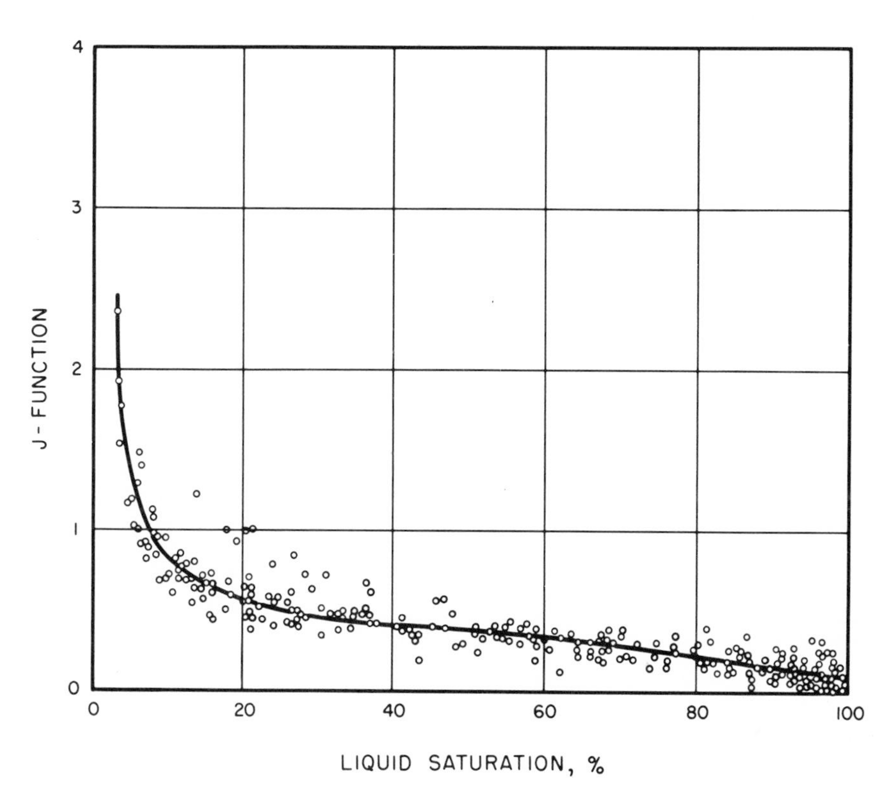

FIG. 11. *J*-curve for microgranular limestone cores, Edwards Formation, Jourdanton Field. (After Brown,[16] Fig. 10, courtesy of AIME.)

and permeabilities but of like rock type in a field. From the *J*-function curve for a particular zone, any one of the parameters P_c, ϕ, and k may be calculated, provided the other quantities in Eq. 13 are known or assumed.

Application of Capillary Pressure Concepts to Oil Finding (Minimum Closure)

The foregoing capillary pressure phenomena are useful in the exploration for oil and gas. It is of vital interest to the petroleum geologist and the engineer to

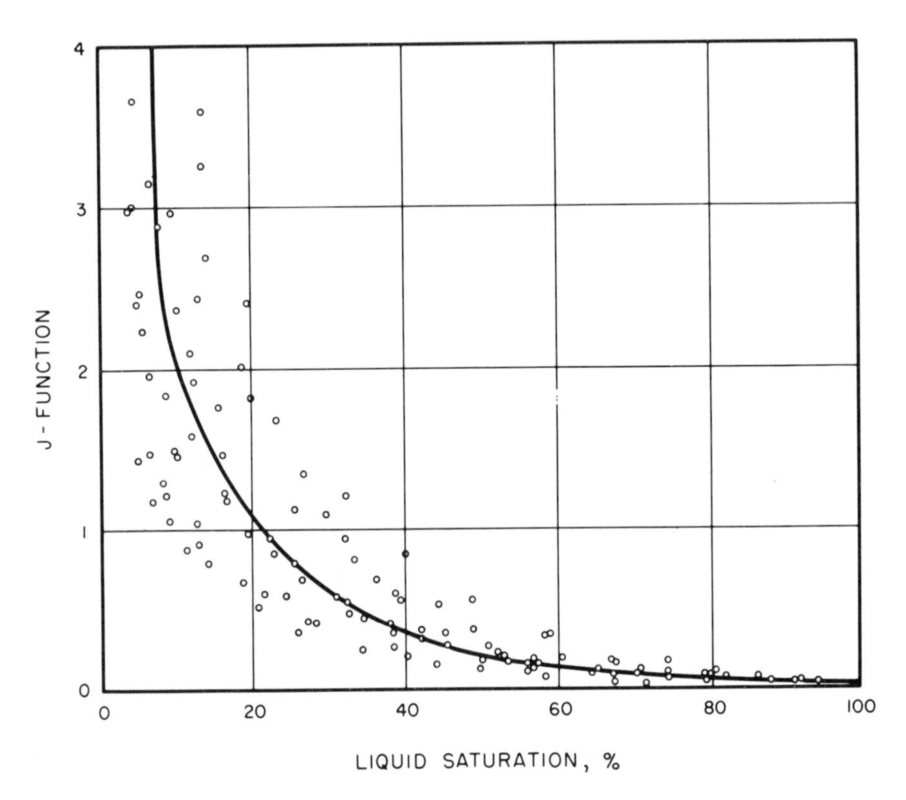

FIG. 12. *J*-curve for coarse-grained limestones cores, Edwards Formation, Jourdanton Field. (After Brown,[16] Fig. 11, courtesy of AIME.)

know the minimum closure required for a given prospective zone to produce water-free oil or gas. This minimum closure corresponds to the thickness of the transition zone. The dashed line in Fig. 7 illustrates the manner in which the thickness of the transition zone (the top of the transition zone is shown in the figure as the lower limit of water-free production) increases with a decrease in the permeability of the zone. In other words, as has been demonstrated previously, the tight zones with low ϕ and k and small pores will experience a greater rise in mobile water above the free water surface than the more permeable zones, which are commonly cavernous or fractured.

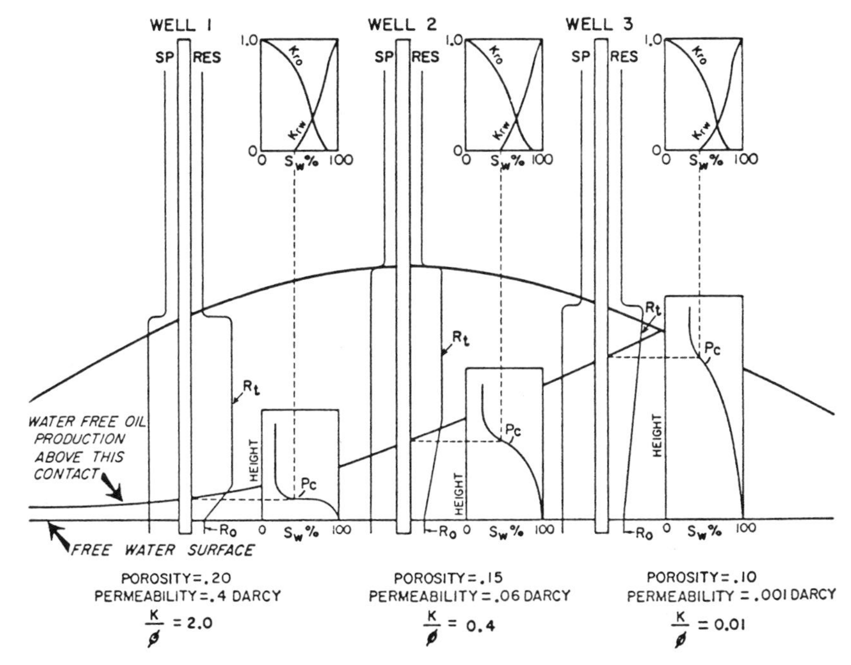

FIG. 13. Tilted oil-water contact with lateral variation in rock characteristics. (After Arps,[17] Fig. 3, courtesy of AAPG.)

Tilted Oil-Water Contact Owing to Lateral Variation in Rock Properties

Capillary pressure curves, along with relative permeability curves, can also be employed to explain tilted water table conditions in the absence of hydrodynamic causes. In the hypothetical example given by Arps[17] in Fig. 13, the porosity and the permeability of the formation decrease from left to right with resulting increase in transition zone thickness, as shown by the capillary pressure curves.

Commercial Water-Free Gas Production and Wet Noneconomic Oil Production

A final example of the principles of capillary pressure and relative permeability is the actual case of a Wyoming field as shown in Fig. 14. The Phosphoria Zone, which is a fractured dolomite, flowed gas with no water initially, whereas the Tensleep Sandstone tested mostly water, even though both zones had apparently the same amount of closure. The combination of high permeability in the fractured Phosphoria and a large density difference between the gas and the water (Eq. 8) yielded a narrow transition zone, much thinner

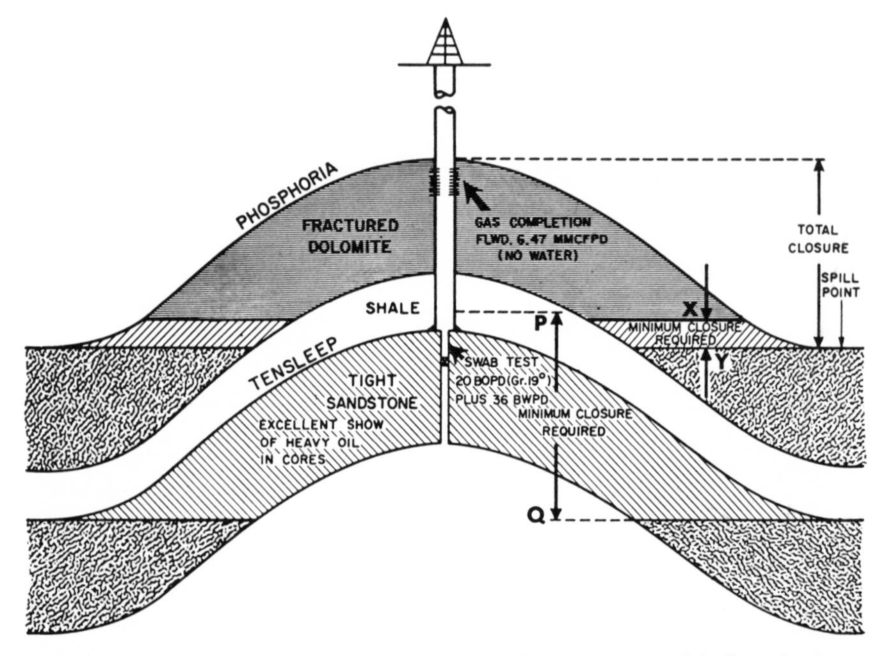

FIG. 14. Water-free gas production on same closure with non-commercial oil production. (After Arps,[17] Fig. 5, courtesy of AAPG.)

than the available closure. On the other hand, the low permeability of the Tensleep Sandstone, coupled with a much lower density difference between the heavy 19° API oil and the fresh water, dictated a minimum closure in excess of that available.

Tank Oil-in-Place by Material Balance

It is apparent from the foregoing discussion that the methods employed for estimating volumetrically oil and gas reserves in carbonate rocks can be rather complicated. This complexity reflects the massive heterogeneity of the zones, in which there are wide variations in porosity, permeability, pore-size distribution, and general geometric configuration of the pore space. It has been observed that elaborate and detailed efforts to measure the inherent properties of individual segments of carbonate zones are sometimes necessary. Inability on the part of the estimator to define the reservoir adequately may lead to gross errors in the estimates. The alternative method of prognosticating in-place values for hydrocarbons by material balance is, therefore, often of critical importance in the analysis. The two methods should be employed jointly whenever possible.

The use of material balance equations to analyze the performance of oil and gas reservoirs was introduced by S. Coleman, H. D. Wilde, Jr., and T. W. Moore of Humble Oil and Refining Company in 1930. The material balance method makes use of the law of conservation of matter with respect to the overall status of the fluids in the reservoir. Two basic assumptions are incorporated in the equations: (1) the original quantities of oil- and gas-in-place in the reservoir are equal to the respective amounts of each fluid produced plus those remaining, and (2) the reservoir volume occupied by the hydrocarbons does not change except as modified by water production and water influx. The second assumption simplifies the equations but assumes further that there is no decrease in porosity with decrease in reservoir pressure. Actually the bulk volume of the rock does decrease slightly, whereas the volume of the solid rock material increases. Both volume changes serve to decrease the porosity slightly. It is not usually necessary to consider the porosity change, however, except possibly in the case of an undersaturated reservoir that is still producing above the bubble point pressure.

The above two assumptions are combined and a final equation is derived in which the quantities of (1) initial oil-in-place, (2) cumulative oil, gas, and water production, (3) size of the original free gas cap, and (4) water influx, are all related at a given reservoir pressure. According to the material balance equation, the sum total of the changes in the volumes occupied in the reservoir by the oil, gas, and water phases during depletion is zero. For example, if the interstitial water volume remains the same, the decrease in volume occupied by the oil in the reservoir equals the increase in the free gas space.

The General Material Balance Equation

The general form of the material balance equation, which has been called the Schilthuis equation,[19] is:

$$N = \frac{N_p[B_t + B_g\,(R_p - R_{si})] - B_w\,(W_e - W_p)}{mB_{oi}\,[(B_g/B_{gi}) - 1] + (B_t - B_{oi})} \tag{14}$$

where N = original oil-in-place, stock tank barrels (STB).

N_p = cumulative oil production, STB.

B_o = oil formation volume factor: volume at reservoir conditions per volume at standard conditions, reservoir bbl/STB (RB/STB).

B_g = gas formation volume factor, barrels of reservoir space (RS) per standard cubic foot of gas, RS/SCF $[B_g = (14.7/520)(1/5.61)(z_R T_R/P_R)]$.

p_R = reservoir pressure, psia.

T_R = reservoir temperature, °R.

z_R = gas deviation factor at reservoir conditions.

R_s = solution gas/oil ratio, standard cubic feet of gas (SCF) per stock tank barrel (STB), SCF/STB.

R_p = net cumulative gas/oil ratio, SCF/STB ($R_p = G_p/N_p$).

G_p = cumulative gas produced, SCF.

B_w = water formation volume factor, RB/STB.

W_e = volume of water encroachment into reservoir during production period, STB.

W_p = cumulative water production during production period, STB.

m = ratio of volume of initial gas cap to volume of initial oil zone, unit space per unit space.

i = subscript indicating initial conditions.

B_t = two-phase formation volume factor: volume of reservoir space occupied by 1 STB of oil plus the gas liberated from the oil in going from initial reservoir pressure p_i to pressure p $[B_t = B_o + B_g(R_{si} - R_s)]$.

In Eq. 14, the pressure-volume-temperature (PVT) data (B_o, B_g, B_w, R_s, etc.) can be determined with reasonable accuracy from laboratory measurements. However, N, m, and W_e are more difficult to evaluate. If the geological and engineering data indicate that there is no free gas-cap or water drive, N is the unknown and can be calculated. If the size of the initial gas cap in relation to the volume of the oil zone (m) can be determined and a fairly long production and pressure history is available, N may again be calculated.

The PVT data employed will represent, at best, an approximation of the true values. Two types of gas evolution from complex hydrocarbon systems are the "flash" and the "differential." In flash liberation the gas is allowed to remain in contact with the oil after it is liberated. The overall system, therefore, maintains a constant composition. In differential liberation the gas is removed from the oil immediately upon liberation, in which case the system composition is continually changing. The investigator is forced to choose between laboratory PVT data based on differential liberation and those based on flash liberation. The production mechanism, however, usually involves a complex combination of the two liberation processes. The difficulty in obtaining representative PVT data is normally secondary to more serious limitations of the general applicability of the material balance equation to reservoir performance. Here the well-known maxim applies that the accuracy of the material balance calculations depends on the quality of the basic data going into the equation. The following comments, made by Muskat[20] early in the history of the material balance method, are still pertinent to its use for reserve estimation:

"The material balance method is by no means a universal tool for estimating reserves. In some cases it is excellent. In others it may be grossly misleading. It is always instructive to try it, if only to find out that it doesn't work, and why. It should be a part of the 'stock in trade' of all reservoir engineers. It will boomerang if applied blindly as a mystic hocus-pocus to evade the admission of ignorance. The algebraic symbolism may impress the 'old timer' and help convince a Corporation Commission, but it will not fool the reservoir. Reservoirs pay little heed to either wishful thinking or libelous misinterpretation. Reservoirs always do what they 'ought' to do. They continually unfold a past with an inevitability that defies all 'man-made' laws. To predict this past while it is still the future is the business of the reservoir engineer. But whether the engineer is clever or stupid, honest or dishonest, right or wrong, the reservoir is always 'right.'"

Derivation of the Material Balance Equation

The reservoir material balance equation is developed by equating the reservoir space occupied by the oil and gas under original conditions to the reservoir space occupied by the oil and gas remaining in the reservoir at a later date at a reduced pressure. Under initial conditions, the hydrocarbon pore volume is equal to the original oil zone volume (NB_{oi}) plus the original gas-cap volume (mNB_{oi}), as shown in Fig. 15.

After discovery, the reservoir pressure in the oil pool decreases as fluids are produced. If all the oil plus its dissolved gas had been allowed to remain in the reservoir at the reduced pressure by producing only free gas from the gas cap, the new volume would be $NB_t = NB_o + N(R_{si} - R_s)B_g$. The term NB_o denotes the new volume in the reservoir of the original tank oil (N), and $N(R_{si} - R_s)B_g$ represents the gas in the reservoir in the free state that was liberated from the tank oil in going from the initial pressure to the lower pressure. If, on the other hand, the pressure decline were the result of oil and gas production from the oil zone only, with no gas produced from the free gas cap, the new volume of the gas cap at the lower pressure would be $mNB_{oi}(B_g/B_{gi})$. In the typical case, however, fluids from both the free gas cap and the oil zone are produced. At the reduced pressure, the reservoir space taken up by the oil produced plus its dissolved gas is represented by $N_pB_t = N_pB_o + N_p(R_{si} - R_s)B_g$. Subtracting this from the original oil yields the quantity $(N - N_p)B_t$, which is the reservoir volume of the oil remaining in the reservoir plus the gas liberated from the oil that has not been produced and is in the free state. The free gas cap has a new volume which equals $mNB_{oi}(B_g/B_{gi})$ (the new volume if no free gas were produced)

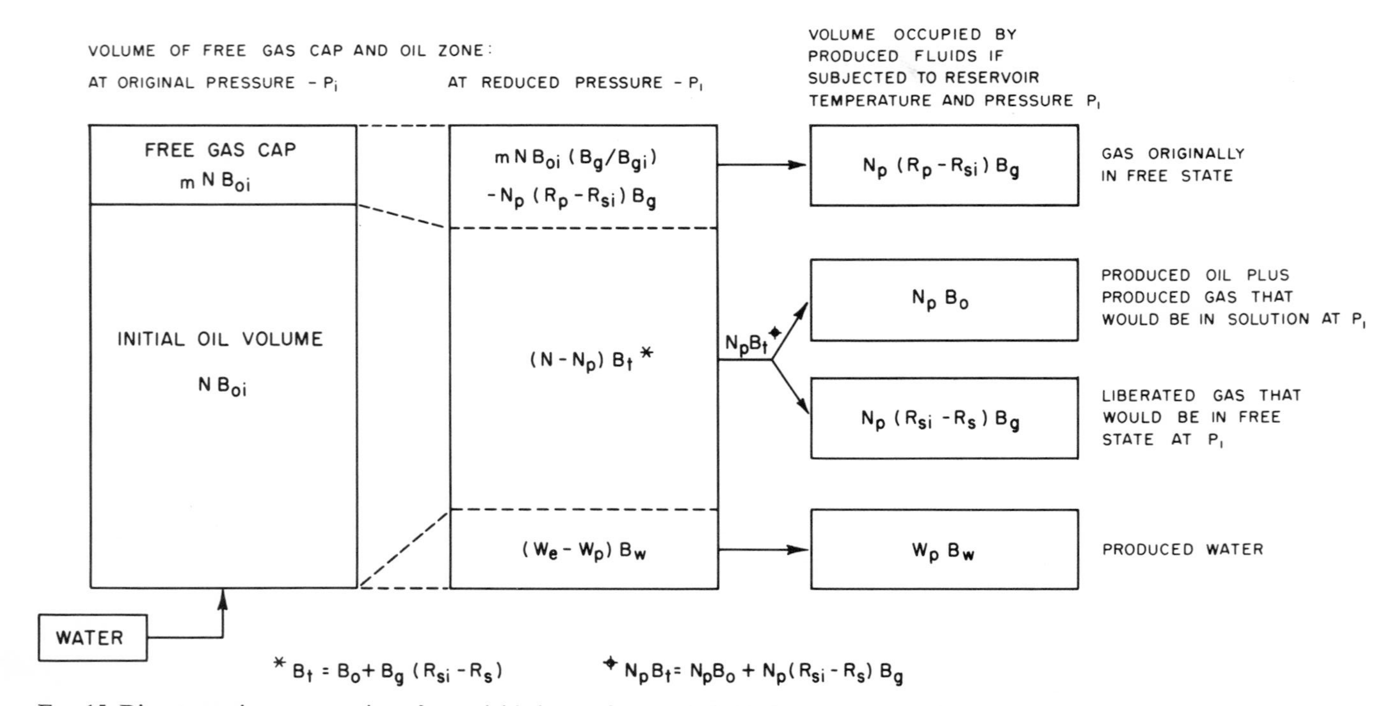

FIG. 15. Diagrammatic representation of material balance of reservoir fluids for general case of oil reservoir having free gas cap and subject to water influx.

minus $N_p(R_p - R_{si})B_g$. The latter quantity is the reservoir volume of the gas produced that was not originally dissolved in the N_p barrels of oil produced. For simplicity, the equation can be stated as follows:

$$\underbrace{NB_{oi}}_{\substack{\text{Original}\\\text{oil}\\\text{volume}}} + \underbrace{mNB_{oi}}_{\substack{\text{Original}\\\text{volume}\\\text{of free}\\\text{gas cap}}} = \underbrace{(N - N_p)B_t}_{\substack{\text{Volume in}\\\text{reservoir of}\\\text{remaining}\\\text{oil plus its}\\\text{liberated gas}}} + \underbrace{mNB_{oi}(B_g/B_{gi}) - N_p(R_p - R_{si})B_g}_{\text{New volume of free gas cap}} + \underbrace{(W_e - W_p)B_w}_{\substack{\text{Water influx}\\\text{minus water}\\\text{production}}}. \tag{15}$$

Because of the complexity of the production process the gas actually produced does not have its origin as represented above. Some of the gas in the quantity $N_p(R_p - R_{si})B_g$ came from the free gas cap. It is convenient and expedient, however, to express this gas in terms of the produced oil (N_p). From a material balance standpoint all the gas is accounted for properly. The term W_e represents the cumulative water influx, if any, into the reservoir during the production of N_p barrels of oil, and W_p is the cumulative water production during the same period.

Solving Eq. 15 for N yields Eq. 14, which is the general equation. If laboratory PVT data are not available, published correlations such as those of Standing[21] and Frick[22] or correlations derived by analogy with similar fluids are aids in approximating fluid behavior. Sometimes the water influx, W_e, can be calculated from methods of Schilthuis,[19] Hurst,[23] and van Everdingen and Hurst.[24] When W_e cannot be calculated or estimated, it is often ignored and N is calculated at several different times during the production period. Normally, increased values for N will result with longer production periods, if water influx is significant. The actual value for N is then estimated by extrapolating to zero time on a plot of calculated values for N versus time.

Sample Problem: Calculation of Initial Oil- and Gas-in-Place, Using Volumetric and Material Balance Methods.

Pressure-Production History

Date	Static reservoir pressure (psia)	Cumulative oil production (bbl)	Cumulative gas production (MCF)	Cumulative water production (bbl)
Initial conditions	2,827	0	0	0
2/14/65	2,535	462,966	363,777	2,017
4/19/66	2,301	732,686	688,288	2,471
3/31/67	2,207	933,796	944,585	3,150

Flash Liberation Data
(Pertains to production through one separator at 75°F and 100 psig)

Pressure (psia)	B_o (bbl/STB)	B_w (bbl/STB)	R_s (SCF/STB)	B_g (bbl RS/SCF)
2,827	1.309	1.005	587	0.000996
2,800	1.305	1.005	585	0.000997
2,600	1.295	1.004	555	0.00103
2,400	1.282	1.004	523	0.00109
2,200	1.269	1.004	485	0.00117
2,000	1.256	1.003	445	0.00130

Reservoir Data

Average porosity	23%
Average permeability	200 md
Average interstitial water saturation	35.5%
Productive area (A)	269 acres
Average net pay thickness (h)	92 ft
Saturation pressure	2,827 psia

The general equation is Eq. 14:

$$N = \frac{N_p[B_t + B_g\,(R_p - R_{si})] - B_w(W_e - W_p)}{mB_{oi}[(B_g/B_{gi}) - 1] + (B_t - B_{oi})}.$$

The data indicate that the reservoir was initially saturated, with no free gas cap, and, therefore, $m = 0$. Inasmuch as water production has been relatively insignificant, W_e is neglected and

$$N = \frac{N_p[B_t + B_g\,(R_p - R_{si})] + B_w W_p}{B_t - B_{oi}}.$$

At 2,535 *psia*:

$$R_p = \frac{363{,}777 \text{ MCF}}{462{,}966 \text{ bbl}} = 785 \text{ SCF/bbl}.$$

Interpolating at 2,535 psi, $B_o = 1.291$, $B_g = 0.00105$, $B_w = 1.005$, and $R_s = 545$ SCF/bbl. $R_{si} = 587$ SCF/bbl at 2,827 psi. Thus:

$$B_t = B_o + B_g(R_{si} - R_s) = 1.291 + 0.00105\,(587 - 545) = 1.335$$

and

$$N = \frac{0.4630\{[1.335 + 0.00105(785 - 587)] + 0.002(1.005)\} \times 10^6}{1.335 - 1.309}$$

$$= \underline{27{,}600{,}000 \text{ STB.}}$$

At 2,301 *psi*:

$$N = \frac{0.7327\{[1.369+0.00113\ (939-587)]+0.0025(1.004)\}\times 10^6}{1.369-1.309}$$

$$= \underline{21{,}600{,}000 \text{ STB}}.$$

At 2,207 *psi*:

$$N = \frac{0.9338\{[1.386+0.00117\ (1012-587)]+0.0032(1.004)\}\times 10^6}{1.386-1.309}$$

$$= \underline{22{,}900{,}000 \text{ STB}}.$$

Volumetric calculation of tank oil-in-place (N):

$$N = \frac{7758\phi(1-S_w)}{B_{oi}} \times h \times A$$

$$= \frac{7758\ (0.23)\ (1-0.355)}{1.309} \times 92 \times 269 = \underline{21{,}800{,}000 \text{ STB}}.$$

The results of the above calculations indicate the advantage of determining N both by material balance and by the volumetric method. With no additional knowledge of reservoir conditions, the calculated N of 27,600,000 STB at a reservoir pressure of 2,535 psia is highly doubtful in light of the other values. The values for tank oil-in-place of 21,600,000 STB and 22,900,000 STB obtained at lower reservoir pressures are supported by the volumetric calculation of 21,800,000 STB. Thus, an estimated N of 22,100,000 STB (average of the three) appears justified on the basis of the available data.

Reserve Estimation by Empirical Methods

The determination of reserves by empirical methods involves the use of performance data derived from other pools as a guide in estimating oil recovery in a specific reservoir. Empirical methods were first used in the forecasting of future production rates in the early days of the oil industry. At that time, empirical techniques were popular because no other methods were available. As the technology of oil and gas production developed, the analytical methods already discussed, as well as others, came into existence. In spite of the emergence of analytical and numerical procedures in reserve estimation, empirical methods have retained a position of prominence for two reasons: (1) Some of the analytical procedures merely estimate tank oil-in-place. A separate estimate of recovery efficiency is, therefore, required for reserve estimation. (2) The analytical methods that compute reserves directly are not universally applicable. In addition, they frequently yield results of a low degree of reliability, so that a separate estimate is desirable.

To predict the performance of the subject pool, empirical methods utilize performance criteria of pools with similar characteristics. In limestone and dolomite reservoirs, this technique of analogy has been used extensively. Chapter 6 examines carbonate reservoir performance. A report entitled *A Statistical Study of Recovery Efficiency*, published by the American Petroleum Institute,[25] is discussed in that chapter. It reveals some interesting relationships concerning rock type, recovery mechanism and efficiency, and important reservoir parameters. In this study a total of 312 case histories of oil reservoirs (226 sandstone and 86 carbonate reservoirs) was examined. Three categories of reservoirs were considered in which the predominant drive mechanism was (1) water drive, (2) solution gas drive without supplemental drives, or (3) solution gas drive with supplementary drives. As would be expected, the study showed that a solution gas-drive mechanism is the least efficient. It also indicated that all drive mechanisms tend to work better in sandstones with intergranular porosity than in carbonates, which commonly have intermediate types of porosity as discussed in Chapter 6. The median values of recovery efficiency, expressed as percentage of initial oil-in-place, for the pool studies were (1) 51.1 and 43.6% for sandstone and carbonate reservoirs, respectively, for the water-drive pools; (2) 21.3 and 17.6%, for the two respective rock types, for solution gas-drive pools without supplementary drive; and (3) 28.4 and 21.8%, respectively, for solution gas-drive pools with supplementary drive. Again referring to the pools studied, it appears that the probability of natural water influx into reservoirs of all rock types rises as the permeability increases. The median permeability of the sandstone reservoirs exceeds by severalfold that of the carbonates.

Conclusions

Two important aspects of reserve estimation in carbonate reservoirs have been pin-pointed in this chapter for detailed discussion, namely, analysis of production decline curves and the volumetric determination of tank oil-in-place.

The problem of decline curve analysis should be approached from the standpoints of raw data validity, techniques of data handling, models of graphic representation, and final interpretation procedures. It is seen that for each well a decision must be made as to what extent the past production history of the well represents future performance. Every effort should be made to estimate remaining reserves and future rates of production by an alternative method, particularly for wells with decline histories that are difficult to interpret.

Capillary pressure phenomena play a critical role in the distribution and the mobility of fluids in carbonate reservoirs. The presence of these capillary forces, combined with gravity forces in the heterogeneous reservoir rock, dictates, at times, rather rigorous methods for determining oil-in-place values. The ability to be ingenious and imaginative and to exercise sound judgment, coupled with experience and a good working knowledge of reservoir mechanics, constitutes the key to reliable reserve estimation.

References and Bibliography

1. Lahee, F. H.: "Oil and Gas Reserves: Their Meaning and Limitations", *Dril. and Prod. Prac.* (1950) 312–315.
2. Eggleston, W. S.: "What Are Petroleum Reserves?" *J. Pet. Tech.* (July, 1962) 719–724.
3. Arps, J. J., "Discussion to W. S. Eggleston: 'What are Petroleum Reserves?' " *J. Pet. Tech.* (July, 1962) 724–725.
4. Sheldon, Dean: "Making Analytical Engineering Appraisals", *Oil and Gas J.* (Oct. 25, 1954) Vol. 53, No. 25, 104–108.
5. Arnold, R. and Anderson, R.: "Preliminary Report on Coalinga Oil District", *U.S. Geol. Survey Bull.* 357 (1908) 79.
6. Lewis, J. O. and Beal, C. H.: "Some New Methods of Estimating the Future Production of Oil Wells", *AIME Bull.* 134 (Feb., 1918) 477.
7. Muskat, Morris: *Physical Principles of Oil Production*, McGraw-Hill Book Co., New York (1949) 316, 341.
8. Arps, J. J.: "Analysis of Decline Curves", *Trans.*, AIME (1944) Vol. 160, 228–247.
9. Chatas, A. T. and Yankie, W. W., Jr.: "Application of Statistics to the Analysis of Production Decline Data", *Trans.*, AIME (1958) Vol. 213, 399–401.
10. McCray, A. W. and Comer, A. G.: "Statistical Basis for Choice Among Hyperbolic Decline Curves and Computer Applications in Calculating Confidence Limits of Reserve Predictions", paper SPE 1930 presented at SPE 42nd Annual Fall Meeting, Houston, Tex. (Oct. 1–4, 1967).
11. Locke, C. D., Schrider, L. A. and Romeo, M. K.: "A Unique Approach to Oil-Production Decline Curve Analysis with Applications", paper SPE 2224 presented at SPE 43rd Annual Fall Meeting in Houston, Tex. (Sept. 29–Oct. 2, 1968).
12. Leverett, M. C.: "Capillary Behavior in Porous Solids", *Trans.*, AIME (1941) Vol. 142, 152–169.
13. Aufricht, W. R. and Koepf, E. H.: "The Interpretation of Capillary Pressure Data from Carbonate Reservoirs", *J. Pet. Tech.* (Oct., 1957) 53–56.
14. Rockwood, S. H., Lair, G. H. and Langford, B. J.: "Reservoir Volumetric Parameters Defined by Capillary Pressure Studies", *Trans.*, AIME (1957) Vol. 210, 252–259.
15. Rose, W. and Bruce, W. A.: "Evaluation of Capillary Character in Petroleum Reservoir Rock", *Trans.*, AIME (1949) Vol. 186, 127–142.
16. Brown, Harry W.: "Capillary Pressure Investigations", *Trans.*, AIME (1951) Vol. 192, 67–74.
17. Arps, J. J.: "Engineering Concepts Useful in Oil Finding", *Bull.*, AAPG (Feb., 1964) Vol. 48, No. 2, 157–163.
18. Hunter, A., Bradford, W. and Allen, D.: "Huntington Beach Oil Field", *Summary of Operations, Calif. Oil Fields* (Jan.–June, 1955) Vol. 41, No. 1, 61–68.
19. Schilthuis, R. J.: "Active Oil and Reservoir Energy", *Trans.*, AIME (1936) Vol. 118, 33–52.
20. *Reservoir Eng. News Letter* (Sept., 1947).
21. Standing, M. B.: *Volumetric and Phase Behavior of Oil Field Hydrocarbon Systems*, Rheinhold Publ. Corp., New York (1952).

22. Frick, T. C. (Ed.): *Petroleum Production Handbook*, Vol. II, McGraw-Hill Book Co., New York (1962).
23. Pirson, S. J.: *Elements of Oil Reservoir Engineering*, 2nd ed., McGraw-Hill Book Co., New York (1958) 608.
24. van Everdingen and Hurst, W.: "The Application of the Laplace Transformation to Flow Problems in Reservoirs", *Trans.*, AIME (1949) Vol. 186, 305–324.
25. American Petroleum Institute: *A Statistical Study of Recovery Efficiency*, Bull. D4 (Oct., 1967).
26. Hall, H. N.: "Compressibility of Reservoir Rocks", *Trans.*, AIME (1936) Vol. 118, 33–52.
27. Craft, B. C. and Hawkins, M. F.: *Applied Petroleum Reservoir Engineering*, Prentice-Hall, Englewood Cliffs, N.J. (1959).

CHAPTER 6

Classification and Performance of Carbonate Reservoirs

HERMAN H. RIEKE, III, ROBERT W. MANNON, GEORGE V. CHILINGAR, AND G. L. LANGNES

Introduction

One of the main objectives of the technology presented in the foregoing chapters has been to provide a basis for reliable predictions of the productive potential and ultimate performance of carbonate reservoirs. Accordingly, a treatise on classification and performance should serve as one end product of the material presented in previous chapters of this book. The areas of petrophysics, fluid flow behavior, formation evaluation, and reserve estimation in carbonates have been examined. What remains is to bring this knowledge and other available information to bear on carbonate reservoir classification and performance; in this chapter an attempt is made to achieve this goal. The major portion of the existing literature on reservoir mechanics and performance is based on studies of sandstone reservoirs. There is less fundamental knowledge about the physical properties and performance characteristics of limestone and dolomite reservoir rocks than of sandstones. The normally more complex heterogeneous depositional environments and post-depositional processes that affect carbonates are underlying causes for the relative paucity of information.[1] A valuable aid in studying carbonate reservoir performance is the discreet application to carbonates of laboratory and field research and development work performed on sandstones.

Even though the oil industry world-wide has a long history of production under many different environments and circumstances, much of the information is not usable. The data concerning many of the pools already depleted or nearing depletion, which might have been used as valuable case histories, are almost entirely worthless. These pools were in their prime before the 1930's when routine pressure measurements and rock and fluid sampling techniques were unknown. In some cases accurate gas/oil ratios and water/oil ratios are not available because the gas and water production was only estimated at best.

It should not be inferred, however, that the problems of predicting carbonate reservoir performance are due solely to the relatively meager

References p. 306.

knowledge of the subject. There are other problems of a different sort. The concept of analogy—that is, the assumption that pools with similar characteristics will have similar performances—is a powerful tool in the hands of the estimator. Nature being as it is, however, these similarities among pools exist only within certain constraints, especially in carbonate rocks. Every pool is unique in some respects and must be treated as such.

Human and political-economic factors, in addition to physical and chemical factors, influence the manner in which the reservoir is developed and operated. A discussion of this topic, however, is not presented in this chapter. Craze[2] has prepared a concise review of the relationship of well spacing, withdrawal rates, and the general development plan to ultimate pool performance and recovery.

Three schemes for the classification of carbonate reservoirs which are presented in this chapter are based on (1) the nature of the contained fluids, (2) the type of energy available to expel the oil and gas, and (3) the type of carbonate reservoir rock pore system. The first two methods are common to all rock types, whereas the third applies only to carbonates.

Classification of Pools Based on Fluid Composition

Hydrocarbon accumulations can be classified by fluid type, using the familiar pressure-temperature phase envelope for a hydrocarbon mixture (Craft and Hawkins[3]). Under this system the reservoir types are (1) dissolved-gas reservoirs, either saturated or undersaturated with gas, and with or without a free gas cap; (2) gas-condensate reservoirs at or above the dew point; and (3) dry-gas reservoirs. It is sometimes helpful to expand the above classification on the basis of the relative quantities of oil, gas, and condensate present in the reservoir. The following classification is modified after Eremenko,[4] whose original classification is presented in Fig. 1.

TYPE I. Highly Undersaturated Oil Reservoir. Saturation pressure is lower, by an order of magnitude or more, than formation pressure with a correspondingly low solution gas/oil ratio. Pools are normally shallow with low-gravity oil. In extreme cases, saturation pressure may approach atmospheric, with well depths of only a few hundred feet and gravities as low as 8–10° API. There is no free gas cap.

TYPE IIa. Moderately Undersaturated Oil Reservoir with No Free Gas Cap. Saturation pressure is somewhat lower than formation pressure.

TYPE IIb. Moderately Undersaturated Oil Reservoir with Initial Free Gas Cap. Saturation pressure is lower than formation pressure. Paradoxically an initial gas cap exists in spite of the undersaturated oil zone. Reservoirs in the Asmari, Bangestan, and Khamie Limestones in Iran are good examples

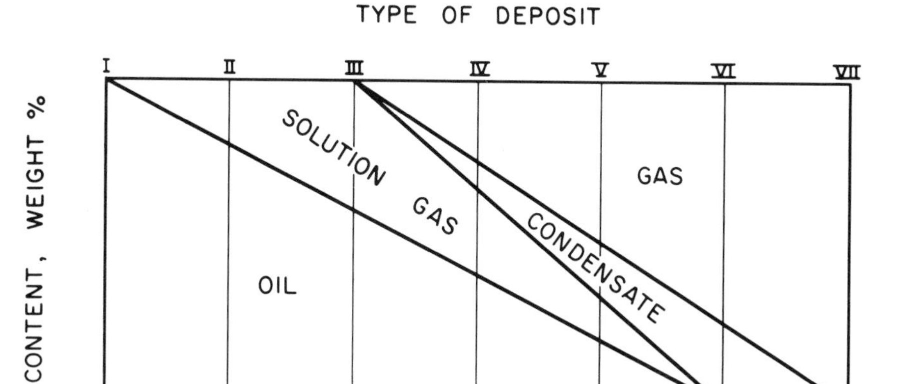

FIG. 1. Eremenko's original classification of hydrocarbon accumulations on the basis of their oil, gas, and condensate contents.[4]

of this type.[5] Possible explanations include changes in the depth of burial and thermal diffusion effects. Standing[6] suggested that possibly circulating formation waters removed some of the gas from the oil after accumulation, owing to the differential solubility of the various hydrocarbons in water. The lighter hydrocarbons are more soluble in the water than the heavier constituents. The importance of this subclass lies in the fact that gas-cap energy is available from inception.

TYPE III. Saturated Oil Reservoir with No Free Gas Cap. Reservoir pressure is initially at the bubble point pressure of the oil. Pressure depletion may result in early formation of the secondary gas cap.

TYPE IV. Saturated Oil Reservoir with Initial Free Gas Cap. Initial reservoir pressure equals saturation pressure. No retrograde condensation occurs in the gas cap with decrease in pressure. Minor amounts of liquid production may consist of natural gas liquids.

TYPE V. Gas-Condensate Reservoir with Black Oil Band. Depending on location of structure, liquid production from individual wells may consist entirely of condensate, a mixture of condensate and black oil, or entirely of black oil.

TYPE VI. Gas-Condensate Reservoir with No Black Oil Band. Initial producing gas/oil ratio is always greater than 3,000 SCF/bbl and normally exceeds 6,000 SCF/bbl; it is a function of reservoir temperature, pressure, and gas composition. Gravity of condensate is in the 40–50° API range. There does not appear to be any correlation between condensate gravity and initial producing gas/oil ratio.

TYPE VII. Dry-Gas Reservoir. Reservoir contains only "dry" gas, and no condensate is formed in the reservoir as the pressure decreases.

The reservoir type (I–VII) is determined by the reservoir pressure and temperature and by the hydrocarbon composition. Changes in the depth of burial of a pool after the oil and gas are trapped may result in a change in reservoir type. A Type III pool, for example, may revert to a Type II when subjected to deeper burial. Conversely, erosion processes that produce shallower burial depths may cause a free gas cap to form (Type III reservoir becomes Type IV). As the pressure decreases, owing to the removal of the overburden, the gas cap will expand and force oil out of the trap at the bottom. In extreme cases, all the oil may be expelled from the trap to form a gas or gas-condensate reservoir (Type VI or VII).

The mechanism of gas expansion in traps with decrease in pressure owing

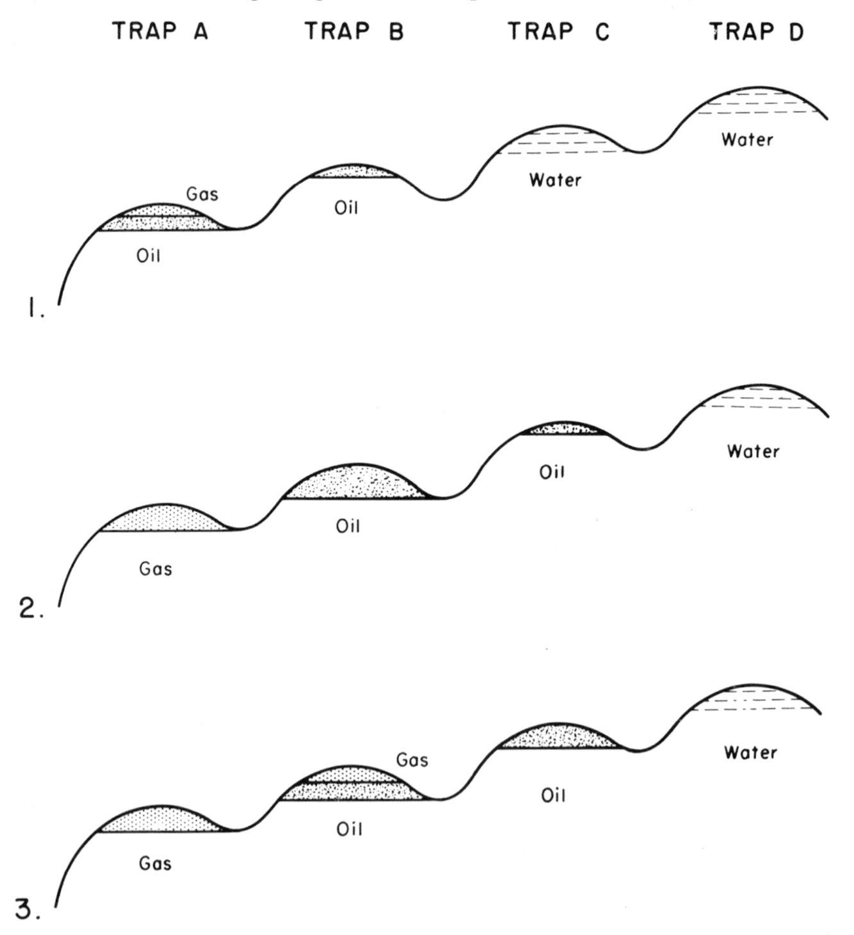

FIG. 2. Theory of differential entrapment: different stages in the migration and accumulation of oil and gas in interconnected traps. (After Gussow,[9] courtesy of AIME.)

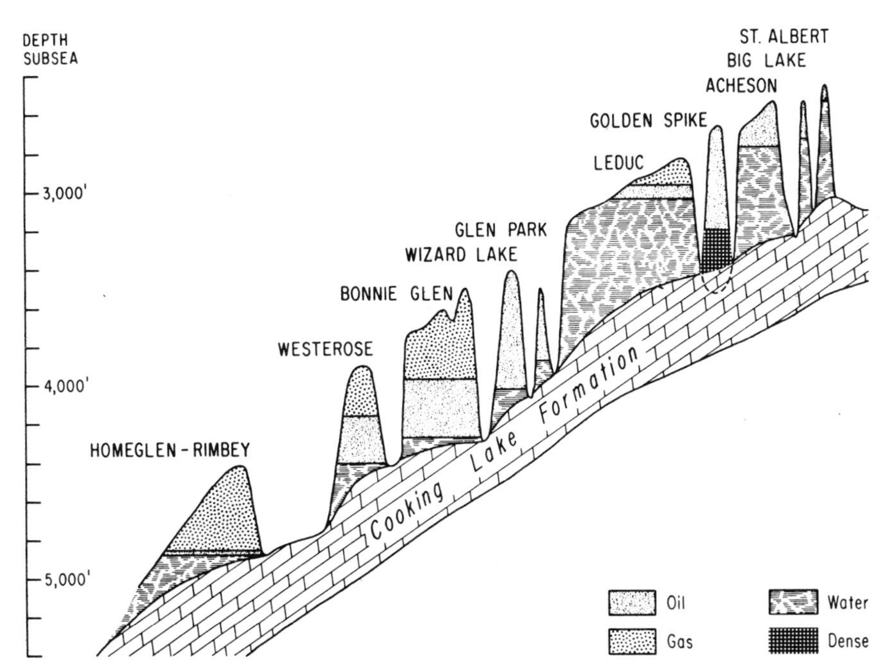

FIG. 3. Profile of the Acheson-Homeglen-Rimbey-Leduc reef trend. (After Horsefield[42] and Hnatiuk and Martinelli,[43] courtesy of Pet. Soc. of CIM.)

to erosion is a well-known concept and has been discussed by Levorsen[7] and by Gussow.[8]

Observations made by both of these authors concern possible explanations for the difference in oil and gas content of adjacent or nearby traps. Gussow's work is of particular interest here because the field examples cited are in carbonate reservoirs. He suggests that the nature of the hydrocarbons trapped in the reservoirs depends on the proximity of the reservoirs to the source beds. As oil migrates across basins and encounters a series of traps, the process of "differential entrapment" becomes operative. The theory states that a trap filled with oil and gas may eventually lose the oil to another trap updip as oil and gas continue to migrate into the area. In Fig. 2, during stage 1, trap A is an oil reservoir with a free gas cap (Type IV). Trap A is filled, and as a result additional oil cannot enter the trap but spills updip to trap B, which contains oil only (Type III). The migrating gas, however, will continue to be trapped in trap A during stage 1. Traps C and D contain only water. In stage 2, trap A is completely filled with gas (Type VI or VII) and all the oil has been displaced to trap B, which is completely filled with oil (Type III). Oil is starting to spill out to trap C (Type III). Trap D still contains only water. In stage 3, trap A remains a gas reservoir and both oil and gas are bypassing it. Trap B now has a gas cap (Type IV) and oil is bypassing it; oil is also expelled as the gas cap expands. Trap C is now filled with oil but still has no gas cap. Trap D is filled with water.

Gussow[9] cited the oil and gas accumulations in the Leduc reefs on the Acheson-Homeglen-Rimbey trend, Alberta, Canada, as a general illustration of this process (Fig. 3). Homeglen-Rimbey is an oil reservoir with a very large gas cap. Westerose South, immediately updip, is all gas. Farther up the trend, Westerose, Bonnie Glen D-38, and Bonnie Glen D-3A are oil pools with free gas caps. Next come Wizard Lake and Glen Park, oil pools without gas caps, and finally Leduc-Woodbend, Yekau Lake, and Acheson, which have gas caps. The presence of gas caps in the last three reservoirs, even though they are updip from the two pools that do not have gas caps, can be explained by the large difference in depth of burial between the two areas, which approximates 1,000 ft. The lower pressure in the shallower reservoirs allowed some of the gas to come out of solution and form a gas cap. The reservoir parameters of these pools are shown in Table 1.

Another example cited by Gussow of differential entrapment in carbonate reservoirs is illustrated by the oil and gas fields in the Swan Hills area in northern Alberta, Canada (Fig. 4). An estimated 4.7 billion bbl of oil-in-place has been trapped in carbonate rocks in this succession of traps. The presence of nonassociated wet gas downdip at the Carson Field can be explained by the theory of differential entrapment. Oil originally trapped here has been flushed updip, as has some of the oil at Carson Creek North, which has a free gas cap. All the traps farther updip (Judy Creek South, Judy Creek, South Swan Hills, Swan Hills, and the C Pool of Swan Hills North) are filled with oil and have no free gas caps. The Virginia Hill Field is also all oil but may have been isolated from the other pools during the time of the migration of the fluids.

The effects of the process of differential entrapment, as reflected in the distribution of oils of varying gravities in the oil pools of the Swan Hills area, are also cited by Gussow. Where gravity effects predominate in the individual pools, the higher API gravity oil is at the top and the lower API gravity oil at the bottom. As the oil is displaced from a trap by migrating gas, the lower API gravity oil at the bottom is displaced initially updip to eventually accumulate in traps updip. Accordingly, there may be a decrease in API

TABLE 1. Comparison of Reservoir Parameters for Acheson-Homeglen-Rimbey D-3 Reef Trend

(After Hnatiuk and Martinelli,[43] courtesy of the Pet. Soc. of CIM.)

Pool	Depth (ft)	Porosity (%)		Connate water (%)	Permeability (md)		Area (acres)	Maximum total pay (ft)	Oil-in-place (million bbl)	Gas-in-place (Bcf)
		Oil zone	Gas cap		Horiz.	Vert.				
Acheson	5,076	9.1	7.5	10	3,100	1,500	3,640	234	149	10
Leduc-Woodbend	5,344	8.0	7.4	15	1,000	5–100	21,600	232	308	420
Glen Park	6,304	9.6	...	7.6	1,604	105	433	421	28	...
Wizard Lake	6,458	9.4	...	7	700	n.a.	3,250	646	380	...
Bonnie Glen	6,995	9.4	8.3	6	350	n.a.	8,800	711	625	430
Westerose	7,233	9.34	7.95	7	1,934	195	1,757	589	158	117
Westerose South	7,740	...	8.6	10	1,200	60	10,720	673	0	1,915
Homeglen-Rimbey	7,934	7.6	7.8	10	250	72	14,053	553	110	1,285

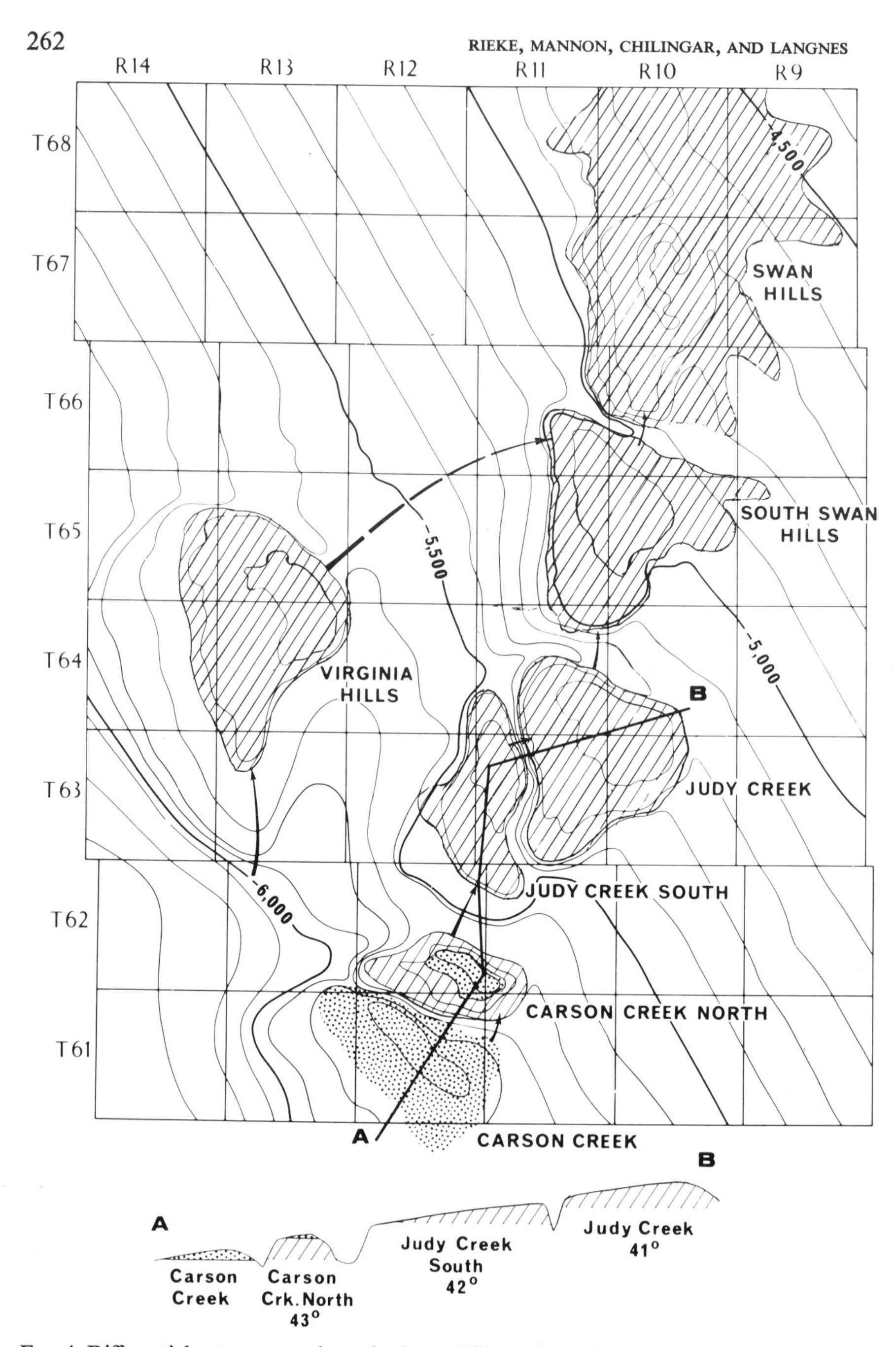

FIG. 4. Differential entrapment along the Swan Hills reef trend, northern Alberta, Canada, showing oil and gas accumulations in the Swan Hills carbonate reefs and shoals of the Upper Devonian Beaverhill Lake. Spill points and migration paths are shown. (After Gussow,[9] courtesy of AIME.)

gravity along a migration path toward the updip edge of the basin. The oil in the Swan Hills trend ranges from 43° API at Carson Creek North to 39° API at Swan Hills North.

The API gravity of Illinois Basin crudes can be characterized by individual frequency distribution plots. All gravity data were obtained from the numerous producing Illinois oil fields.[10] Histograms constructed for the crudes from three formations: Cypress Formation (Miss., Chesterian), Aux Vases Sandstone (Miss., Chesterian), and the members of the Ste. Genevieve Formation (Miss., Valmeyeran), consisting of Ohara Limestone, Rosiclare Limestone, and the McClosky Limestone, are shown in Fig. 5. The API gravity frequency distributions from each pay zone are generally similar in that they are all unimodal with the mode lying between 36 and 39° API. With respect to texture, the McClosky Limestone is highly oolitic and very much like a homogeneous, clean sandstone. The API gravity histogram of the McClosky pay appears to be nearly symmetrical, possibly indicating regional migration and mixing of the crude oils, whereas the other pay zones are quite irregular. The McClosky is also continuously permeable over great distances. In contrast to the Illinois Basin crudes are the gravity histograms characterizing the Viking Sandstone (Cret.), Frobisher Limestone (Miss.), and the Devonian D-1, D-2, and D-3 zones in western Canada (Fig. 6). These distributions show a much wider API gravity range than the crudes in Illinois. The spread of gravity values in the Devonian carbonates illustrates the frequency distribution of differential entrapment (Fig. 6). Similar histograms can be easily constructed for other carbonate formations. Such data could be useful in the correlation and prediction of other reservoir properties and in the determination of crude origin.

The differential entrapment mechanism proposed by Gussow appears to be a distinct possibility in instances such as those cited above. In the case of aquifers having good porosity and permeability over a wide area, so that regional migration of fluids is possible, the theory is certainly plausible. It is difficult, however, to envision, as noted by Dickey, [11] "differential entrapment as a valid explanation for oil accumulations in the common instance of oil fields with lenticular sands and limited aquifers."

The classifications of oil reservoirs on the basis of fluid content is useful for several reasons. First, a reservoir is readily identifiable in its initial state on this basis. Also, some interesting theories, based on the fluid content, can be advanced regarding the events leading up to its formation, as discussed above. In addition, information on the nature of the initial fluids-in-place in a reservoir brings to the mind of the experienced geologist or petroleum engineer certain preliminary concepts of how the reservoir may behave initially and how it should be exploited. For example, if an oil reservoir is highly undersaturated, the initial production period above the bubble point

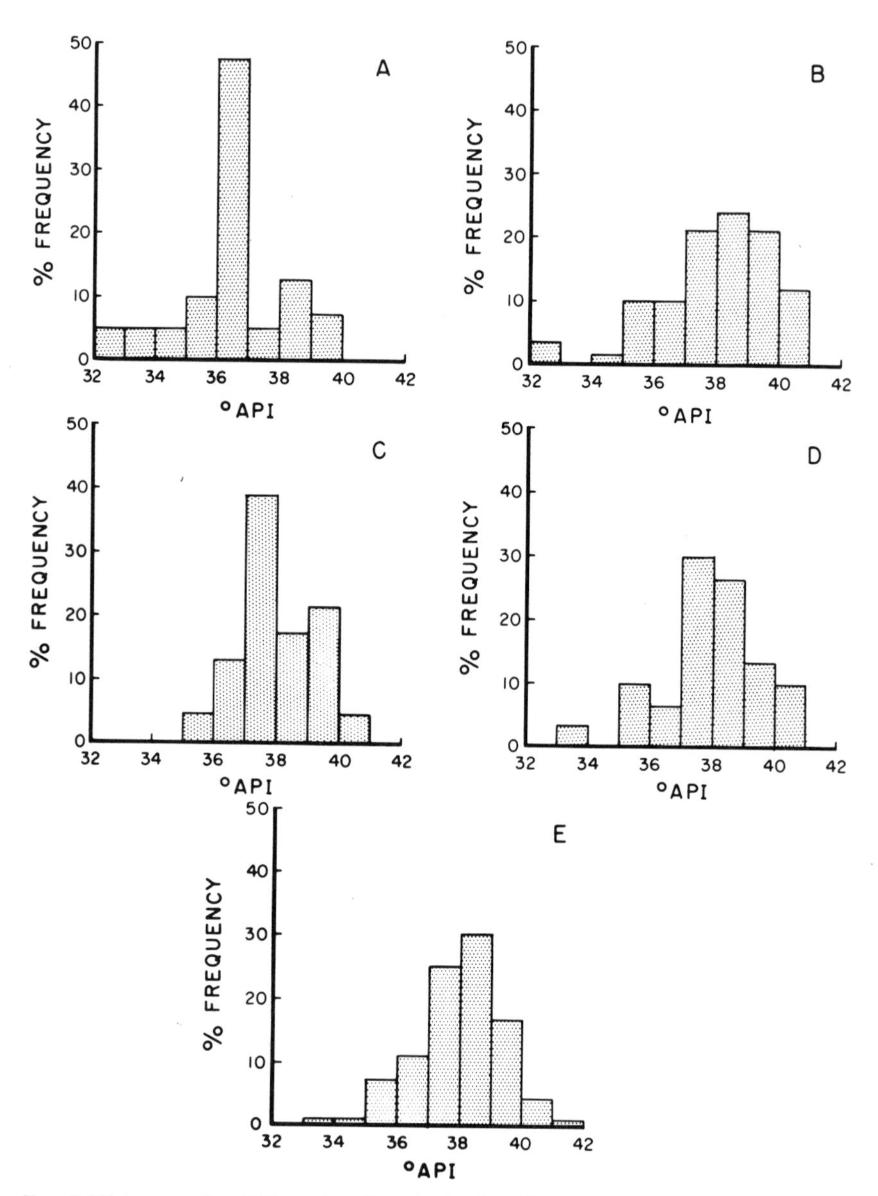

FIG. 5. Histograms for API gravity of crudes in the Illinois Basin. (A) Cypress Sandstone, (B) Aux Vases Sandstone, (C) Ohara Limestone, (D) Rosiclare Limestone, and (E) McClosky Limestone.

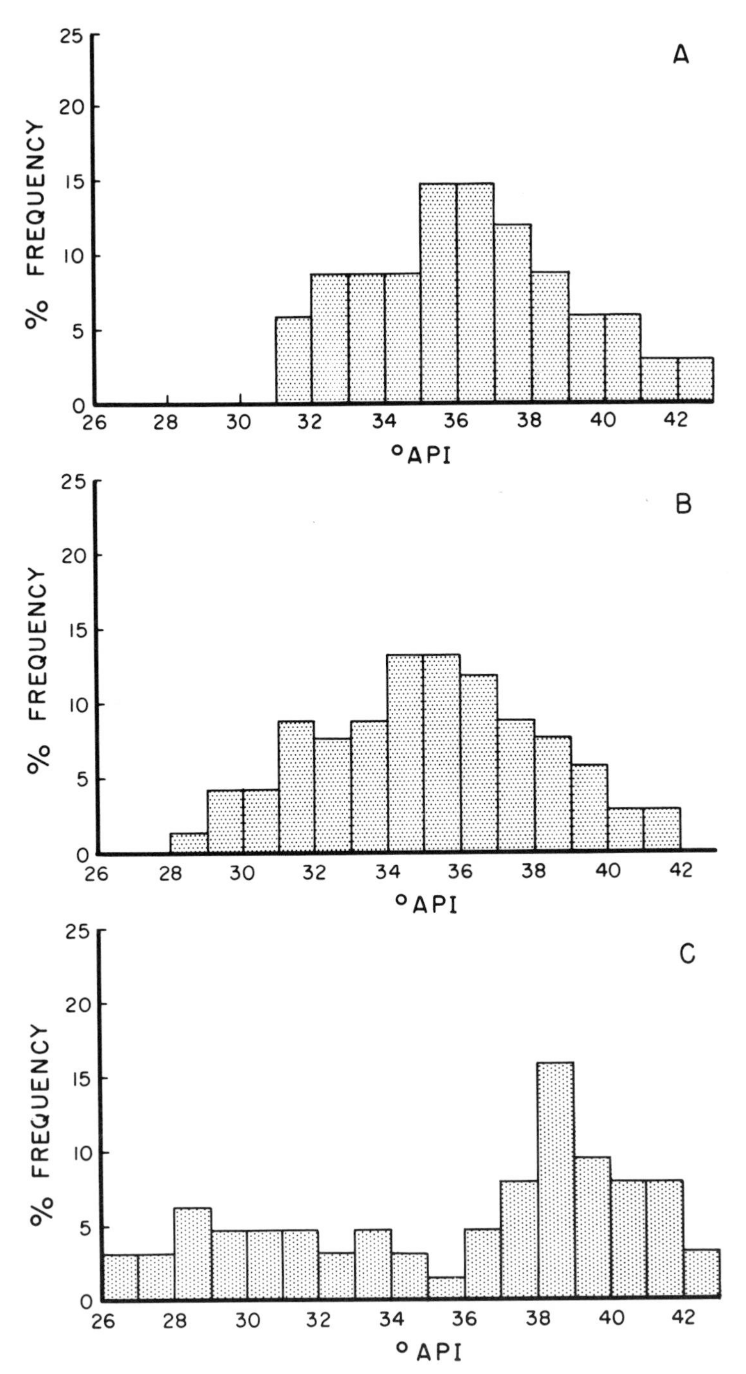

FIG. 6. Histograms for API gravity of crudes in western Canada. (A) Viking Sandstone, (B) Frobisher Limestone, and (C) Devonian D-1, and D-3 limestone pays.

pressure will be fairly predictable. There will be a sharp decline in oil production rates and an only moderate increase, if any, in the producing gas/oil ratio as the oil is produced under conditions of oil expansion. In addition, the wells can be produced at capacity during the initial period without danger of harming the reservoir. If the pool has an initial free gas cap, however, individual well rates, at least in wells near the gas-oil interface, should be controlled to prevent gas coning. Immediate steps should also be taken to attempt to conserve gas energy as well as to provide for expansion of the gas cap if possible and, in any case, guard against gas-cap shrinkage. In the case of a gas reservoir it is, of course, important to know whether or not it is a gas-condensate reservoir.

Classification of Oil Pools Based on Drive Mechanism

Whereas the above categorization of pools by fluid content is helpful, it is apparent that further classification by producing mechanism is imperative for a careful study of the technology of oil and gas recovery. In many cases, however, the performance of the pool, throughout its productive life, cannot be explained by a clearly defined production mechanism; a combination of two or more production mechanisms is usually operative. Classification of producing mechanisms, however, permits a stepwise examination of the predominating factors that influence reservoir behavior, either individually or in combination.

The potential energy sources available to move oil and gas to the wellbore include (1) gravitational energy of the oil, acting over the vertical distance of the productive column, (2) energy of compression of the free gas in the gas cap or within the oil-producing section, (3) energy of compression of the solution gas dissolved in the oil or the water, (4) energy of compression of the oil and water in the producing section of the reservoir, (5) energy of compression of the waters peripheral to the production zone, (6) energy of capillary pressure effects, and (7) energy of the compression of the rock itself. All these forces are active during the productive life of a reservoir.

The predominant producing mechanism operating to produce the oil and gas reflects the relative influence each energy source has on reservoir behavior. The major drive mechanisms are (1) solution gas drive, (2) gas-cap expansion drive, (3) water drive, and (4) gravity drainage. Each drive mechanism when effective in a pool will give rise to a certain characteristic form of reservoir behavior. Although in practice most pools behave in a manner that represents a combination ("mix") of two or more drive mechanisms, each mechanism will be described in the context of a single-drive pool. The common characteristics of each drive mechanism can be discussed for reservoirs having intergranular porosity and then placed in the context of the observed behavior of carbonate reservoirs with other types of porosity. The interest is centered on the record of performance: variation of oil, gas, and water production rates, gas/oil and water/oil ratios, and reservoir pressure with time. Movement of the water-oil contact and creation or expansion of a free gas cap are also of great importance. In addition, individual well performance is of concern.

Solution Gas Drive

In solution gas drive, also called depletion drive, dissolved gas drive, or internal gas drive, the major source of energy to produce the oil and gas from a pool comes from the evolution of dissolved gas from the oil with decline in pressure. No initial free gas cap exists, and the free gas phase formed remains within the oil-producing section. The reservoir is sealed off to a large extent from communication with contiguous water zones by faults or permeability pinchout. As a result, with declining pressure, the water influx into the reservoir is minor.

Typical solution gas-drive performance is shown in Figs. 7 and 8. Initially, there is no free gas phase and the instantaneous producing gas/oil ratio is equal to the original solution gas/oil ratio. Except in cases of undersaturated

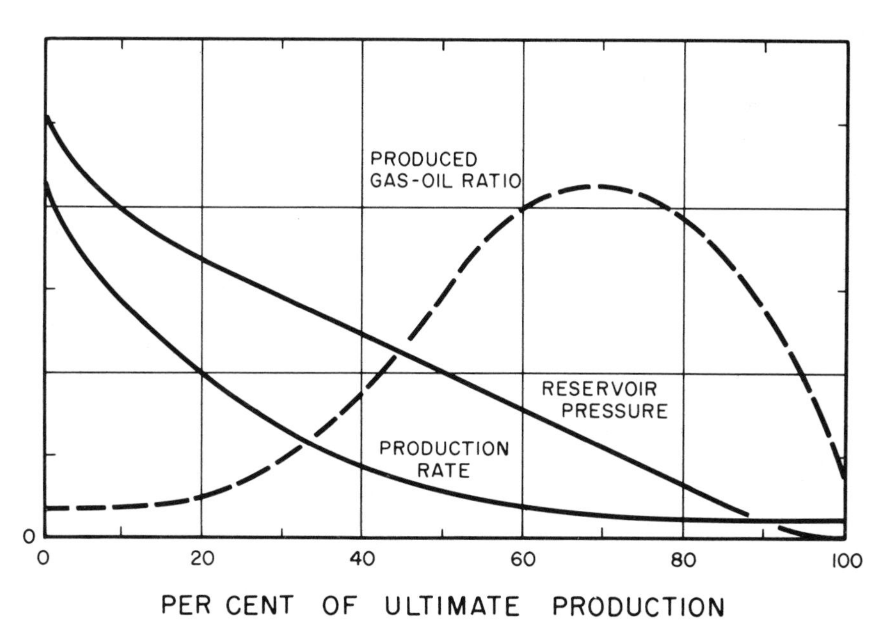

FIG. 7. Generalized performance of a solution gas-drive pool. (After Torrey,[46] courtesy of Prentice-Hall, Inc.)

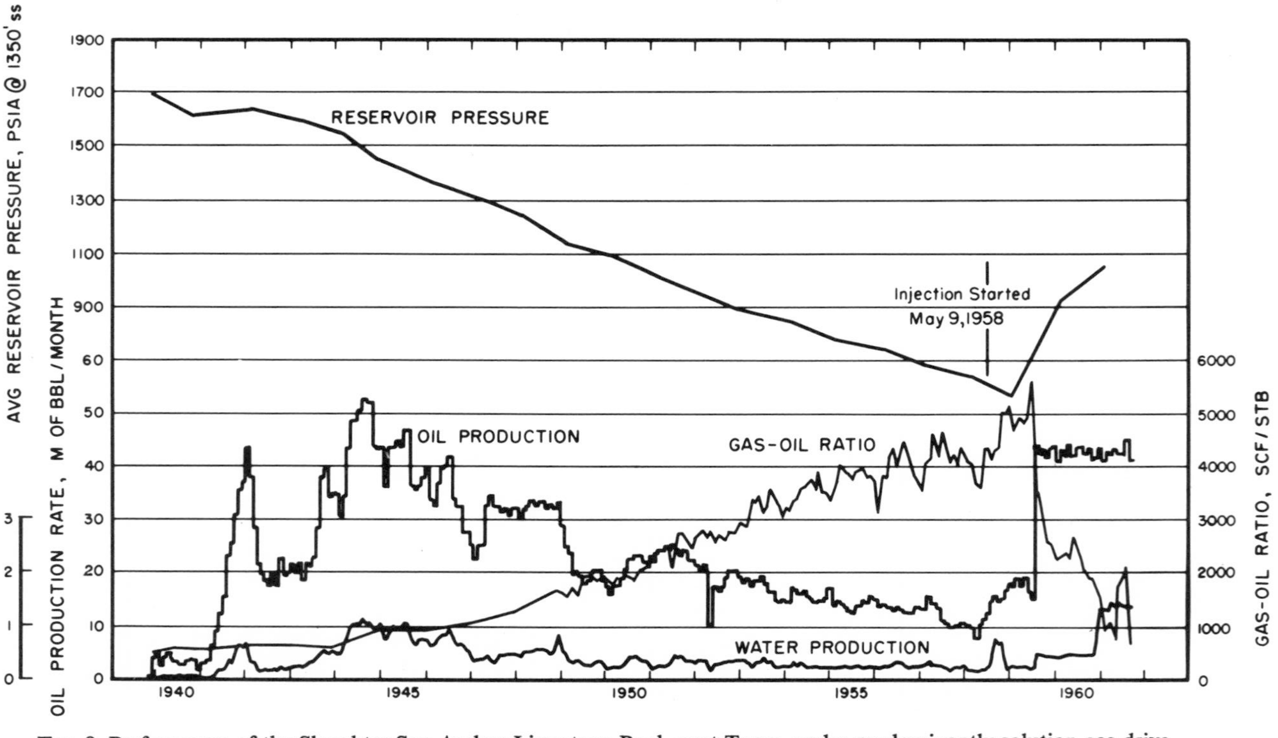

FIG. 8. Performance of the Slaughter San Andres Limestone Pool, west Texas, under predominantly solution gas drive. (After Sessions,[47] courtesy of AIME.)

reservoirs, a finite gas saturation quickly develops and continues to increase as depletion proceeds. At the time the gas saturation reaches the equilibrium value of 5–10%, the gas phase has sufficient mobility and free gas is flowing to the wellbore along with the oil. This results in a rather abrupt increase in the producing gas/oil ratio. The gas/oil ratio continues to rise with increased gas saturation, reflecting the rapid increase in gas flow rate and the attendant decrease in oil production rate. At a gas saturation of 20–30%, the flow of oil becomes negligible, and the gas/oil ratio reaches a peak and then declines as the reservoir reaches the latter stages of depletion.

In undersaturated reservoirs, where the initial reservoir pressure is substantially above the saturation pressure, as mentioned above, the production mechanism is oil expansion. Under these conditions, the producing gas/oil ratio will remain at a low level during the time that the reservoir pressure is above bubble point pressure. It will approximate the solution gas/oil ratio and ideally should actually decrease slightly as the pressure falls, even though this is rarely observed in the field. The peak gas/oil ratio before it begins to decline, reflecting ultimate reservoir depletion, will normally be 5–10 times as great as the solution gas/oil ratio.

In purely solution gas-drive carbonate pools with intergranular porosity, reservoir pressure depends primarily on cumulative oil recovery. Neither reservoir pressure nor ultimate oil recovery is sensitive to oil production rate, unless the production rate affects the producing gas/oil ratio. A rapidly increasing gas/oil ratio, after equilibrium gas saturation is reached, is characteristic of solution gas-drive pools in general. Reducing the production rate will not serve to increase the ultimate oil recovery appreciably. An exception to this rule is found when excessive drawdowns at individual wells lead to extensive transient effects on the reservoir. Any tendency for the reservoir to exhibit significant gravity drainage or water influx, or to form a secondary gas cap, however, may make ultimate recovery sensitive to production rate.

Solution gas-drive performance is closely related to a number of physical parameters. The ratio of reservoir oil viscosity to reservoir gas viscosity (μ_o/μ_g), solution gas/oil ratio, formation volume factor, interstitial water saturation, and oil and gas permeability relationships largely control performance. The close interrelationship among these parameters is indicated by the fact that a change in one factor results in a change in one or more of the others. Some general and meaningful observations can be made, however, regarding the effect of altering the value of a single factor. As oil viscosity increases, there is a corresponding rise in the instantaneous producing gas/oil ratio because of greater gas bypassing, which results in lower solution gas-drive efficiency and lower oil recovery. Also, as the amount of gas available to be in solution decreases, the oil recovery declines. Muskat,[12] however, found that doubling the solution gas/oil ratio resulted in only a 10%

increase in ultimate recovery. The greater oil shrinkage at higher solution gas/oil ratios serves to dampen somewhat the effect of increased oil solubility on oil recovery, but the shrinkage effect is of only minor importance. An increase in the crude oil gravity (°API) as an overall characteristic of the fluid system likewise results in an increase in ultimate recovery. Again the effect is dampened at the higher gravity ranges owing to the greater oil shrinkage, and the ultimate recovery will actually decrease with an increase in oil gravity in the 40–50° API range.

Gas-Cap Drive

Oil pools with initial free gas caps are subject to a gas drive, which is external to the oil zone and separate from the solution gas drive-mechanism. The oil expulsion mechanism is typically a combination of solution gas drive within the oil column plus the added benefit of gas permeating and diffusing into the oil zone from the gas cap. The idealized performance of a gas-cap-drive reservoir is presented in Fig. 9. The decline in production rate and reservoir pressure is not as rapid as in solution gas-drive reservoirs. The gas/oil ratio performance is more favorable. Gas-cap-drive reservoirs, however, are more sensitive to production rate than are solution gas-drive pools.

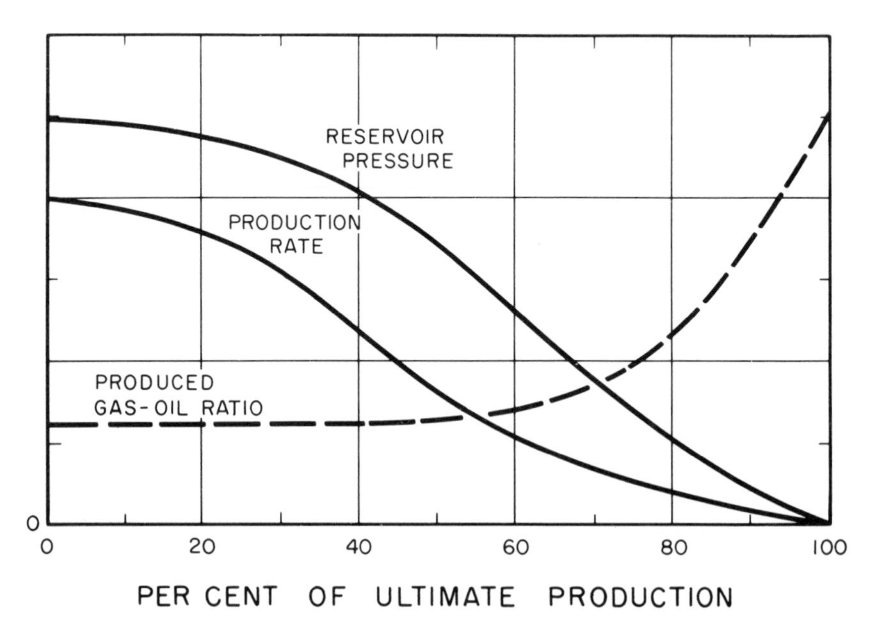

FIG. 9. Idealized performance of a gas-cap-drive reservoir. (After Torrey,[46] courtesy of Prentice-Hall, Inc.)

Wells producing from intervals close to the gas cap must be produced at low rates to prevent gas coning or recompleted to exclude these upper intervals. The overall gas/oil ratio performance largely reflects such procedures. The performance of the Goldsmith San Andres Dolomite Pool in west Texas (Fig. 10) early in its history typifies gas-drive performance with a gradual increase in gas/oil ratio. The oil production is curtailed, and no decline is evident.

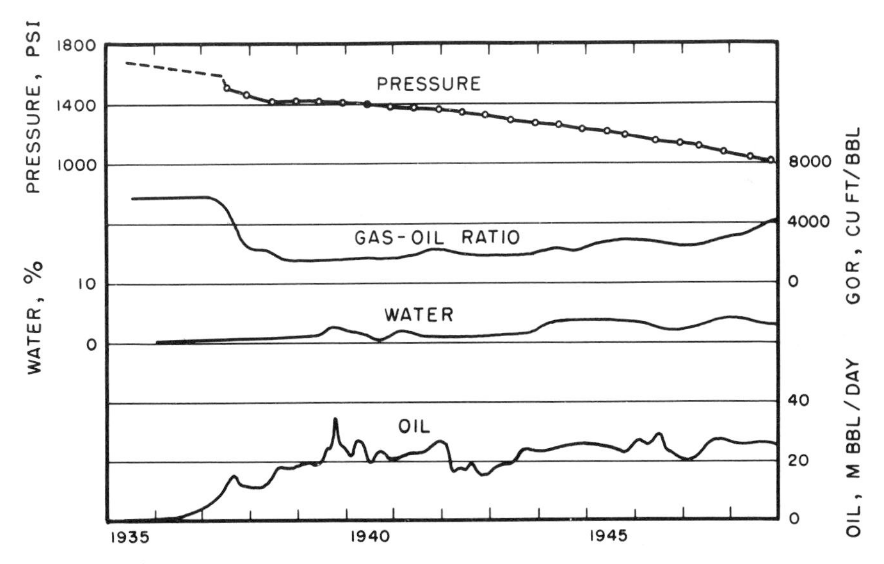

FIG. 10. Typical gas-cap-drive performance, Goldsmith San Andres Dolomite Pool in west Texas. (After Craze,[44] courtesy of AIME.)

Some gravity segregation of oil and gas takes place in virtually every gas-cap-drive reservoir. More pronounced fluid segregation will promote the expansion of the gas cap and downdip movement of the oil, with resultant higher oil recoveries. The size of the gas cap will also affect oil recovery. Normally, the thicker the gas cap, the greater is the ultimate recovery. Notable exceptions are carbonate pools in the Acheson-Homeglen-Rimbey reef trend, Alberta, Canada, which have large gas caps underlain by thin oil bands. The estimated ultimate oil recoveries under primary production are often very low (5–10%) owing to excessive gas and water coning problems. A comparison of the performance of these pools indicates that in the case of thicker oil bands and correspondingly thinner gas caps the oil recoveries are higher.

Water Drive

A reservoir of high permeability, such as a fissured or cavernous limestone, in contact with an aquifer of broad areal extent will normally have an active water drive. The degree to which the reservoir withdrawals are replaced by water determines the efficiency of the water-drive mechanism. In complete water-drive systems, which are not common, substantially all the fluid withdrawals are replaced by intruding water. Some excellent examples of

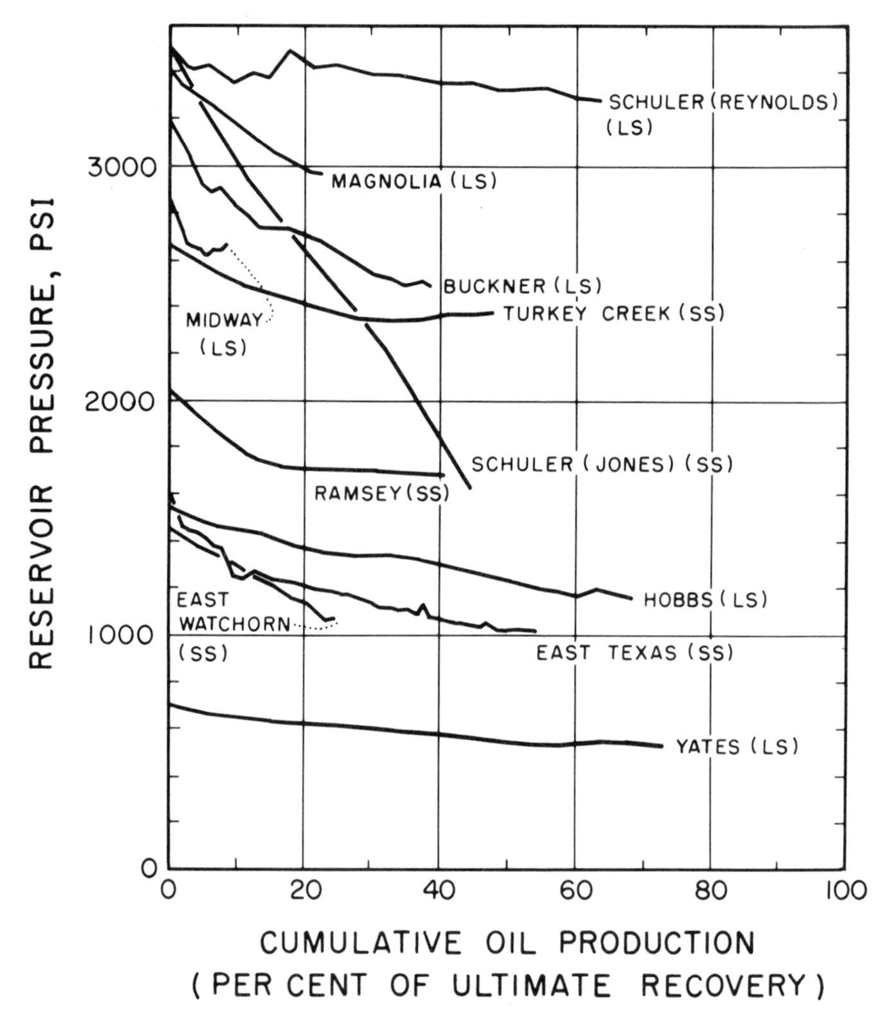

FIG. 11. Pressure-production performance of some water-drive pools. (After Elliott,[13] courtesy of AIME.)

complete water-drive pools in carbonate rocks are the Arbuckle Limestone fields in Arkansas and Kansas.

If the reservoir is initally undersaturated, the natural maintenance of the pressure by water influx may result in oil production above the bubble point pressure for an extended period. During this time a small portion of the reservoir voidage resulting from production (withdrawals) is replaced by expanding oil. Later in the life of the reservoir a free gas phase may form, which will provide part of the energy for the oil expulsion. The existence ot the free gas will depend largely on the rate of withdrawal of fluids.

In all types of water-drive reservoirs, including complete water-drive systems, an initial pressure decline provides the necessary pressure differential at the reservoir boundary to induce water movement into the reservoir. This initial rapid decline preceding water influx is illustrated in general by Fig. 11. The Schuler (Reynolds), Magnolia, Buckner, and Midway Fields are Smackover Limestone pools, and the Hobbs and Yates Pools produce from a Permian limestone. The remaining pools are sandstone reservoirs. All of the pools, except the solution gas-drive Schuler Jones Sand Pool, which is included for comparison purposes, are subject to at least substantial, if not complete, water drive.[13]

To summarize water-drive performance, the producing zone first of all is

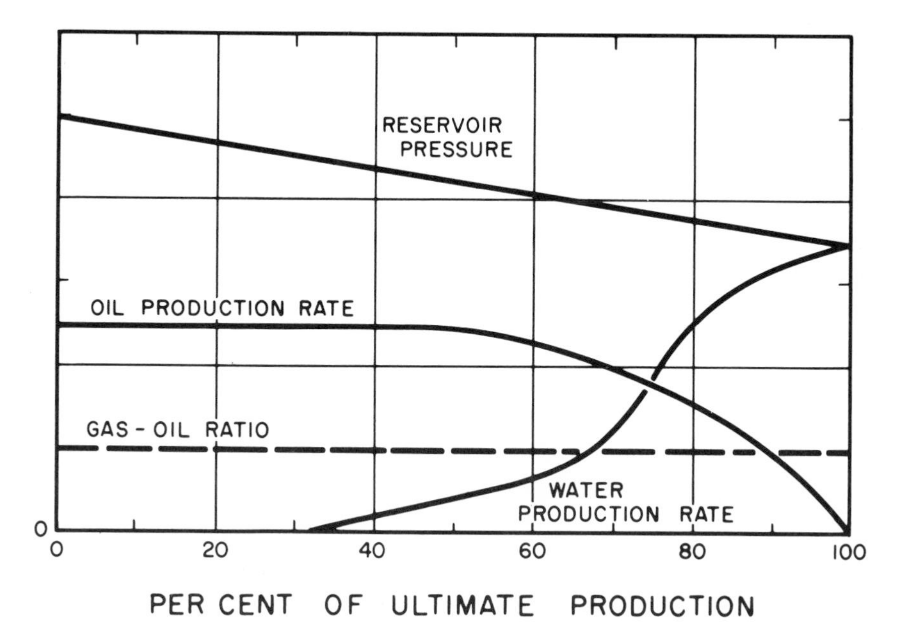

FIG. 12. Idealized water-drive performance. (After Torrey,[46] courtesy of Prentice-Hall, Inc.)

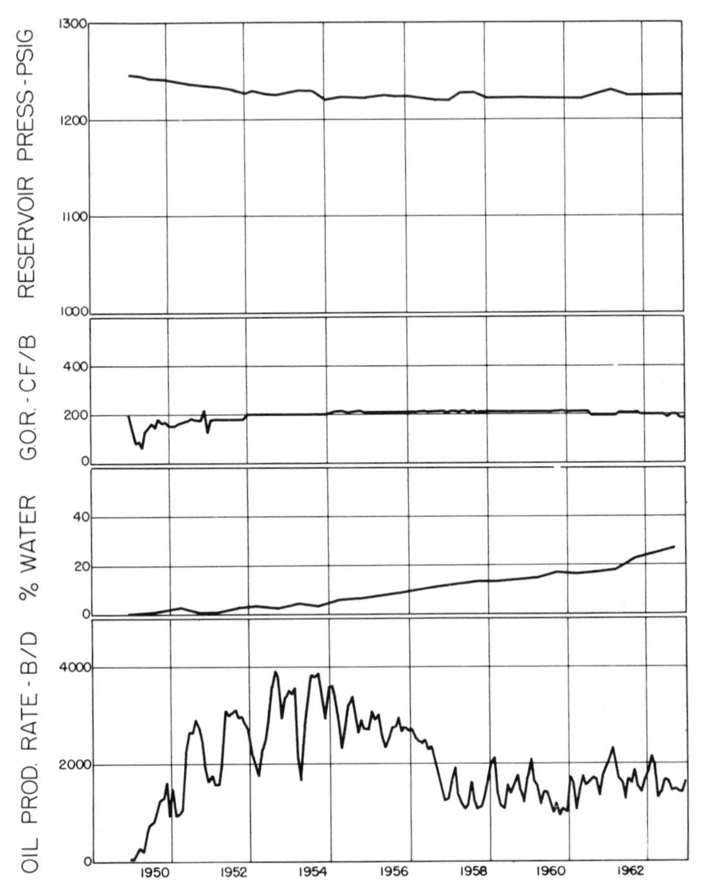

FIG. 13. Performance of a water-drive pool, Excelsior D-2 reef pool, Excelsior Field, Alberta, Canada. (Data from Alberta Oil and Gas Conservation Commission.)

in contact with a broad aquifer, normally of high permeability. Decrease in production capacity is minor until water begins to be produced. The produced gas/oil ratio is substantially constant. Figures 12 and 13 illustrate this type of performance. Recovery factor depends on reservoir rock characteristics such as pore size and fracture distribution, values for mobility ratio ($k_w\mu_o/k_o\mu_w$), and pool geometry. The rate at which the pool is produced may also affect recovery, particularly if the pool is subject to only partial water drive. Reservoir withdrawal rates greatly in excess of the rate of water influx can lead to performance similar to that of solution gas-drive pools. Free gas saturations in the reservoir can develop in the more permeable sections to the extent that incoming water will bypass tighter sections to move preferentially into areas of high gas saturation, with a resulting loss in recovery.

Gravity Drainage

In oil pools subject to gravity drainage, the gravity segregation of fluids during the primary production process is clearly evident in the production history. Oil migrates downdip to maintain down-structure oil saturation at a high level, and free gas accumulates high in the structure. If a primary gas cap exists, it will expand as a result of the segregation process. A pool without a primary gas cap will soon form a secondary gas cap. Early in the life of the pool, the gas/oil ratios of the structurally high wells will increase rapidly. A program of shutting-in wells with high gas/oil ratios and of overall discreet control of individual well rates will provide for maximum benefit from gravitational fluid movement. Figure 14 indicates the two cases of the generalized performance of a gravity-drainage pool with and without such control. The oil gravity, permeability of the zone, and formation dip dictate

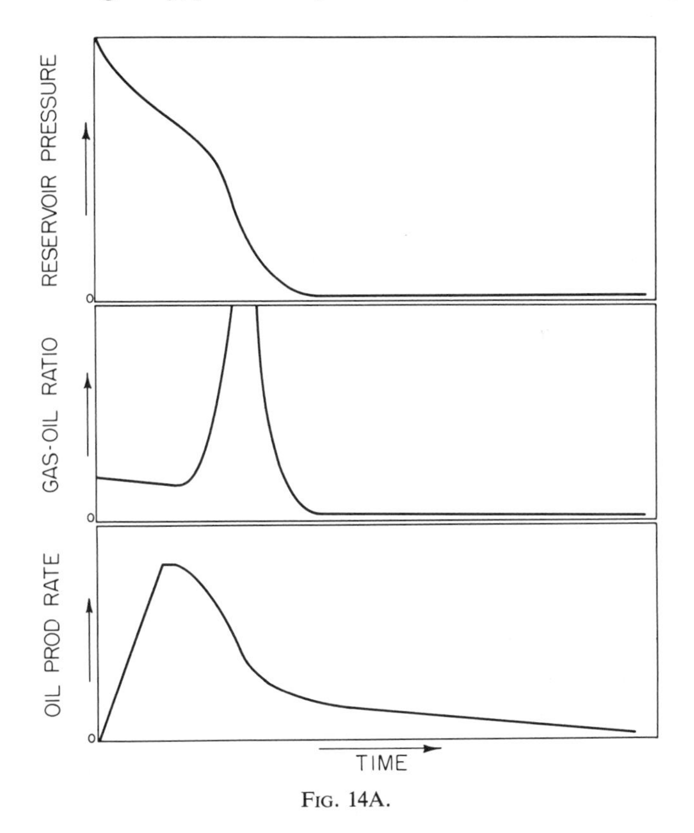

FIG. 14A.

FIG. 14. Generalized performance of a gravity-drainage pool. A—produced without control, B—produced under control by shutting in high-gas/oil-ratio wells. (Courtesy of E. C. Babson.)

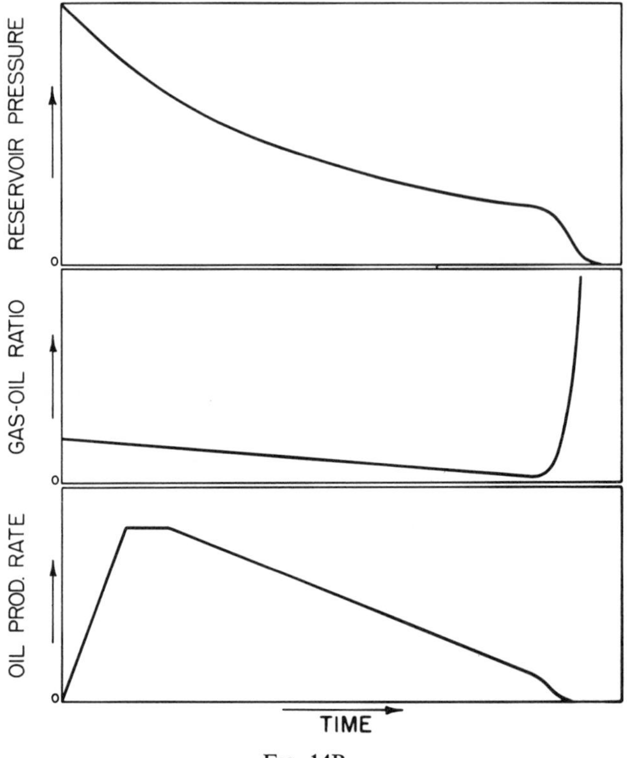

FIG. 14B.

the magnitude of the gravity drainage. The combination of low viscosity and low specific gravity (high API gravity), along with the high zone permeability and steeply dipping beds, accentuates the down-structure oil movement. Typically, in gravity-drainage pools the water influx is minor and the down-dip wells produce at the lowest gas/oil ratios and have the highest oil recovery. In cases of strict gravity drainage, a major portion of the recovery occurs after complete pressure depletion, when gravity is the primary dynamic force moving the oil to the wellbore.

Combination-Drive Reservoirs

Some reservoirs may produce under the influence of one or more of the drives discussed above. Throughout the life of a pool a sequence such as solution gas drive–gravity drainage–water drive may develop, or two different drives may be competing simultaneously (e.g., water-drive reservoir with a primary gas cap). Combination-drive reservoirs may be further classified as open or closed, as proposed by E. C. Babson.[14]

Open Combination-Drive Reservoirs

The performance of open combination-drive reservoirs is affected by the method of operation and by individual well and pool rates. This type of reservoir is in open communication with the aquifer. Sources of energy include water drive, gravity drainage, and gas expansion. In uncontrolled operations, a gas-expansion stage may occur first, accompanied by a rising gas/oil ratio and declining pressures. A gravity-drainage stage may come next at a reduced reservoir pressure. The effects of a moderate water influx will frequently be evident in the latter stages of the life of the pool. Curtailing production rate will cause a buildup in reservoir pressure as water influx exceeds fluid withdrawals. The performance of the D-3 Pool, Redwater Field in Alberta, Canada, which is a typical open combination-drive pool, demonstrates this effect in Fig. 15.

In order to facilitate gravity drainage, the rate of water influx into the reservoir can be reduced by maintaining high water-withdrawal rates from

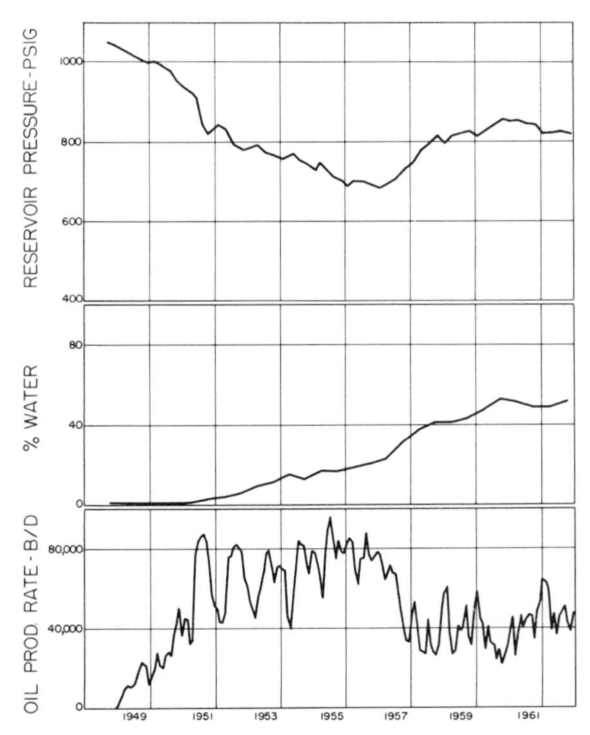

FIG. 15. Performance of an open combination-drive pool, Redwater D-3 reef pool, Redwater Field, Alberta, Canada. (Data from Alberta Oil and Gas Conservation Commission.)

the high-water/oil-ratio wells close to the water table and controlling the rates of the wells situated higher on the structure. On the other hand, to take full advantage of the water influx, wells producing at high water cuts should be shut in and rates on other wells carefully regulated to reduce problems of nonuniform water influx and water coning.

The combined effect of water drive and an expanding gas cap can serve as a very effective oil displacement mechanism. The Leduc D-3 Pool discussed later in this chapter is currently under both water and gas-cap injection, and flushing efficiencies in the gas cap and water zone appear to be about 80%.

When a pool with an initial free gas cap is subject to an active water drive, steps must be taken to prevent oil movement into the primary gas cap. Oil encroaching into the gas cap, even though subsequently displaced, will leave a residual oil saturation, causing a substantial loss in ultimate recovery. Bruce[15] cited the case of the Mount Holly Pool in Arkansas, producing from the Smackover Oolite Limestone. An oil recovery of only 30–40% is indicated because of oil lost in the shrinking primary gas cap. The recovery rate of the other Smackover pools in the area should be at least 60%.

Closed Combination-Drive Pools

There is no significant water encroachment in closed combination-drive reservoirs owing to the faulting and/or facies changes so common in carbonate rocks. A gas-expansion stage is usually followed by gravity drainage. Reservoir pressure is low, gas/oil ratios are falling, and updip wells produce very little fluid of any kind during the secondary stage. The role that gravity drainage plays depends, of course, on the net effect of the dip and permeability of the zone and the oil gravity, in addition to the individual well rates. The control of production rates to allow for effective gravity drainage is normally more critical than in the case of strict gravity-drainage reservoirs because of a reduced tendency for downward oil migration. The generalized performance of a closed combination-drive pool produced without control is presented in Fig. 16. A good example of this type of reservoir is the Golden Spike Field in Alberta, Canada.

Classification of Carbonate Reservoirs based on Type of Pore System

Classification on the basis of fluid type or drive mechanism applies to both carbonate and sandstone oil reservoirs. A system of classification for carbonate reservoirs based on the configuration of the void space of the host rock is by its very nature unique to carbonate rocks. A classification of this type is important because the fundamental difference between the performance of sandstone reservoirs and that of carbonate reservoirs resides in

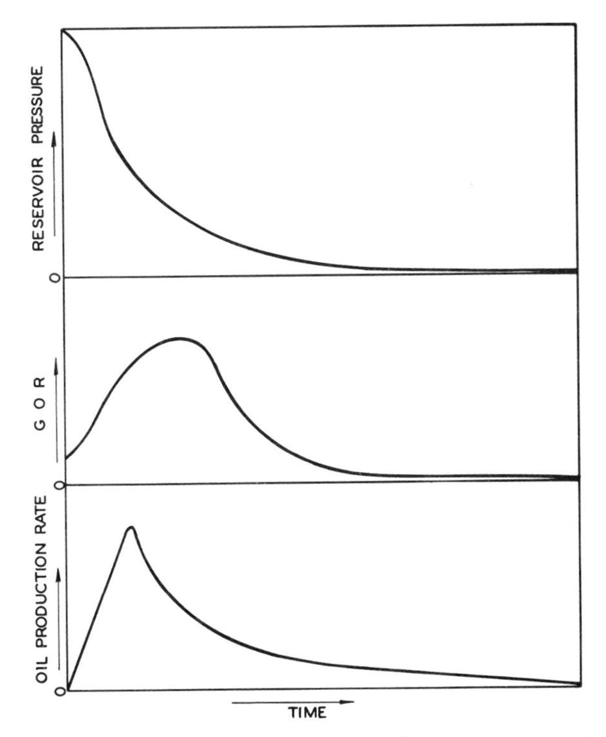

FIG. 16. Generalized performance of a closed combination-drive pool produced without control. (Courtesy of E. C. Babson.)

the unique porosity of limestones and dolomites and its influence on hydrocarbon movement.

Among the early investigators of the reservoir performance of carbonate formations were Bulnes and Fitting.[16] They recognized three distinct classes of porosity in carbonates and termed them intergranular, intermediate, and foramenular. Carbonate rock with the intergranular type of porosity is like sandstone and has mainly primary voids. These carbonate rocks are composed of calcareous fragments, shells, crystals, or oolites. The geometry of the pore space is determined by the size, shape, and packing of the solid fragments themselves. Intermediate porosity displays, in addition to the intergranular openings, fissures, solution caverns, and other openings induced by solution and fracturing that have sizes and shapes independent of the geometrical properties of the constituent fragments, grains, and crystals. The volume of the fissures, fractures, vugs, and other voids is frequently much greater than the initial intergranular pore space. Consequently, the intermediate type of porosity system is much more complex than the intergranular system. The foramenular class of porosity, which Bulnes and Fitting did not dwell upon, is characterized by void spaces in an otherwise nonporous continuous solid, such as a fractured reservoir with essentially no matrix porosity.

There is no basic difference in performance between sandstone and carbonate reservoirs having intergranular porosity. Carbonate pools with intermediate porosity, however, deviate in performance over a wide range from sandstone pools with comparable fluid content and drive mechanism. In order to demonstrate this difference, it is helpful to mention some recent investigations of fluid flow in porous media in general.

Originally, it was widely believed that porosity and effective permeability were sufficient to characterize reservoir rock performance. Barfield *et al.*[17] and Mortada and Nabor[18] have shown, however, that directional permeability is also a pertinent characteristic. Klute,[19] Fatt,[20] and Goodknight *et al.*[21] have proposed a fourth parameter: storage porosity, which consists of the regions of porosity that contribute significantly to the pore volume of the reservoir but little to the flow capacity. On a microscopic basis, in sandstones with intergranular porosity, for instance, these regions consist of "dead-end" or "storage" pores. The variation of sandstone pool performance because of directional permeability and storage porosity, however, falls normally within a relatively narrow range as compared to carbonate reservoirs of intermediate porosity, which show a wider variation in pool performance. Vugular or reef carbonates illustrate the importance of directional permeability, as they usually exhibit fair to good vertical permeabilities and their performance may be controlled more by movements in the vertical direction than in the horizontal. The dead-end or storage pores in sandstones correspond to the large volumes of low-permeability matrix porosity in naturally fractured carbonate reservoirs (the fracture system porosity and the matrix porosity). Although the fracture system generally accounts for only a small percentage of the total porosity, it serves as a path for oil movement to the wells. The matrix, which contains most of the oil, normally has low permeability and yields oil and gas to the fractures reluctantly. In carbonates, the directional permeability and the double-porosity effects are more exaggerated than in sandstones and accordingly carbonate pools will vary in performance to a much greater extent.

Warren and Root[22] developed an idealized model to study the characteristic behavior of reservoirs with intermediate porosity. Their reservoir model described unsteady-state flow mathematically and indicated that two additional parameters are sufficient to distinguish the flow behavior in an intermediate-porosity rock from fluid flow in homogeneous media. One of the parameters is related to the degree of heterogeneity of the system, and the other is a measure of the fluid capacitance of the secondary porosity. These parameters can be evaluated from pressure buildup tests which take the form, as shown in Chapter 3 (p. 89), of parallel segments. Stratified reservoirs, however, yield a similar type of buildup curves, and therefore unambiguous interpretations are difficult.

Foramenular porosity, the third class mentioned by Bulnes and Fitting[16] as occurring in carbonates, is somewhat rare. This type of porosity, in which an essentially nonporous matrix is intersected by fractures, can be found in the Dean-Wolfcamp pay in west Texas and in the San Andres Formation in the Chaveroo and Cato Fields in southeastern New Mexico.[23] The Mississippi-Solid Formation, which is a dense, massive, and cherty limestone in the S. W. Lacey Field, Kingfisher County, Oklahoma, has also been placed in a "pore space in fractures" type of porosity.[24]

It appears that a classification of carbonate oil reservoirs based on type of pore system and having at least a generalized correlation with reservoir performance is needed. The following classification is an attempt to achieve this end. It should be stated at the outset, however, that no classification, no matter how comprehensive, simple, or, on the other hand, complex, will ever be developed that is entirely satisfactory. The classification system proposed here consists of three categories: (1) intergranular-intercyrstalline porosity, (2) vugular-solution porosity, and (3) fracture-matrix porosity.

Intergranular-Intercrystalline Porosity

As stated previously, carbonate pools having intergranular porosity are comparable to unfractured sandstone reservoirs in behavior. The spatial distribution and the general configuration of the pores are more orderly than in systems falling into categories 2 and 3. Commonly, the basic building blocks of the system or the allochemical constituents[25] are predominantly oolites, or well-rounded and reasonably equant intraclasts or pellets or other similar transported calcareous constituents. Some or all of the calcareous grains may be fossils that are current-sorted, broken, and abraded as a result of transportation. Sedentary fossils making up a reef, however, normally would not form an intergranular porosity. In category 1, the size and the shape of the allochems have a direct bearing on the nature of the void space. The sedimentary units may also be crystals precipitated from solution to form an intercrystalline porosity. Dolomitization may also give rise to inter-rhombohedral porosity.

Vugular-Solution Porosity

Vugular-solution porosity systems are inherently more complex than either intergranular-intercrystalline or fracture-matrix porosity systems. The original pore structure has usually been altered by the formation of solution cavities and channels. Reef pore structures usually belong in this category; reefs have been formed by sedentary fossils that have been subjected to only slight movement, if any, after the death of the organisms. Directional permeability may also be very pronounced. Although fractures may exist, they do not dominate the mechanics of flow. The performance of carbonate pools with vugular-solution porosity may closely resemble sandstone pool performance, but commonly differs considerably.

Fracture-Matrix Porosity

In reservoirs having fracture-matrix porosity, the double-porosity system may be well developed. Strong directional permeability may also exist. The matrix is usually of low permeability and contains most of the oil. The fractures contribute a minor portion of the total hydrocarbon pore space but allow the reservoir to produce at economic rates. Pool performance is markedly different from that of sandstone reservoirs. The size of the matrix "blocks" varies from a few inches to several feet.

Determination of Degree of Heterogeneity of Carbonate Pore Systems from Laboratory Gas-Drive Tests

In Chapter 2, Jodry explained in detail the use of capillary pressure data to obtain a better description of carbonate pore systems. Stewart *et al.*[26] have performed model flow tests on limestone cores that also provide valuable information on carbonate pore structure. The results of their laboratory work are especially pertinent here in light of the foregoing discussion of carbonate rock classification.

In their laboratory investigation solution gas-drive and external gas-drive tests were conducted on large limestone cores. The results of the flow tests indicated some divergence between the k_g/k_o curves calculated from the external gas-drive tests and the k_g/k_o curves calculated from the solution gas-drive tests. Stewart *et al.* found the extent of this divergence to be indicative of the carbonate rock type. In the case of intergranular limestones, the solution and external gas-drive k_g/k_o data were in close agreement. For rock samples having vugular-solution porosity or fracture-matrix porosity the respective k_g/k_o curves diverged to varying degrees, depending on the nature of the multiple-porosity development. The flow tests also indicated that dead-end or storage pores exist in some intergranular limestones.

Figures 17, 18, and 19, presented by Stewart *et al.*,[26] indicate gas-drive performance for limestone cores possessing intergranular-intercrystalline, vugular-solution, and fracture-matrix porosities. The relationship between the external-drive and solution-drive k_g/k_o curves for the different rock types

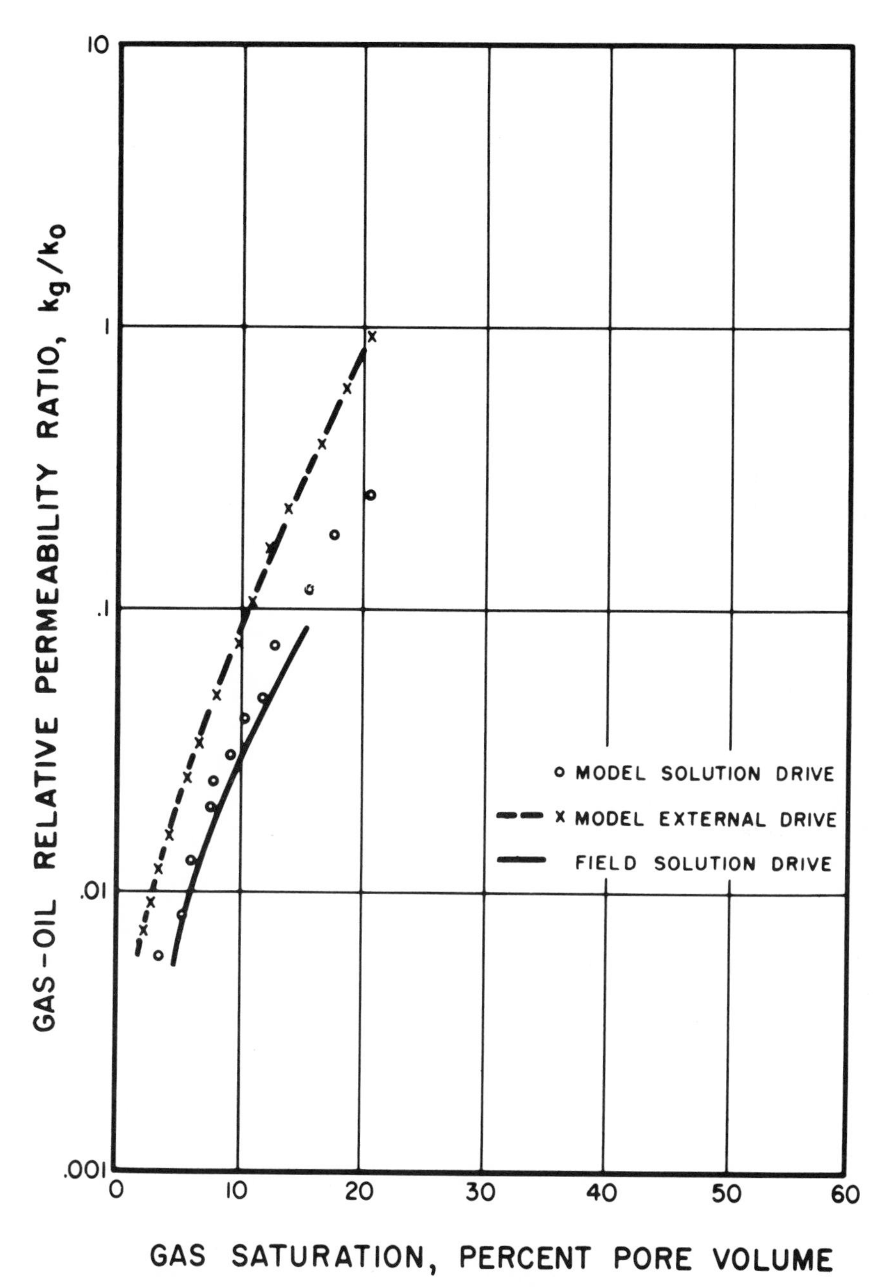

FIG. 17. Comparison of gas-drive performance of limestone having intergranular porosity; laboratory external and solution gas-drive tests and field solution gas-drive data. (After Stewart *et al.*,[26] courtesy of AIME.)

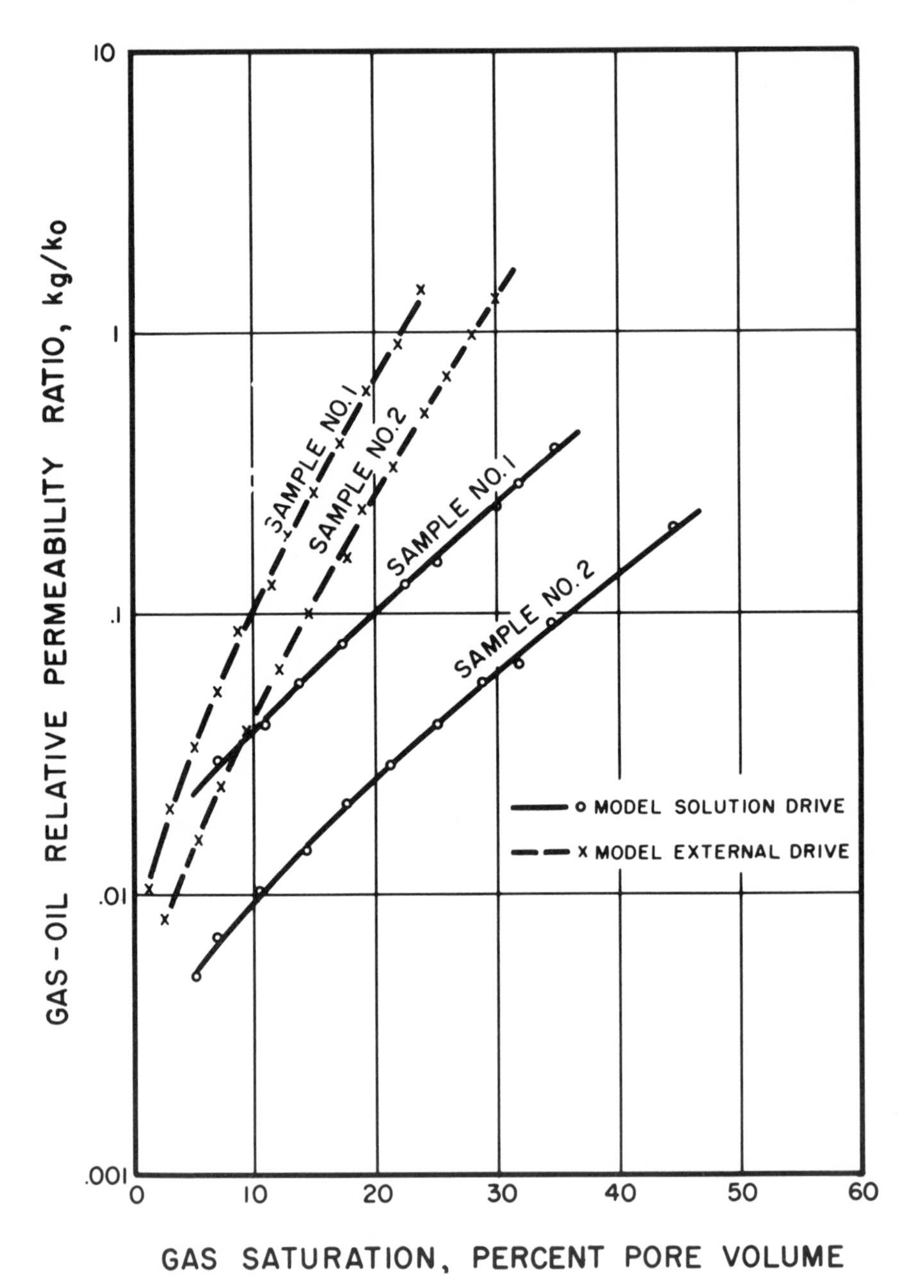

FIG. 18. Comparison of gas-drive performance on two samples of reef limestone; laboratory external and solution gas-drive tests. (After Stewart *et al.*,[26] courtesy of AIME.)

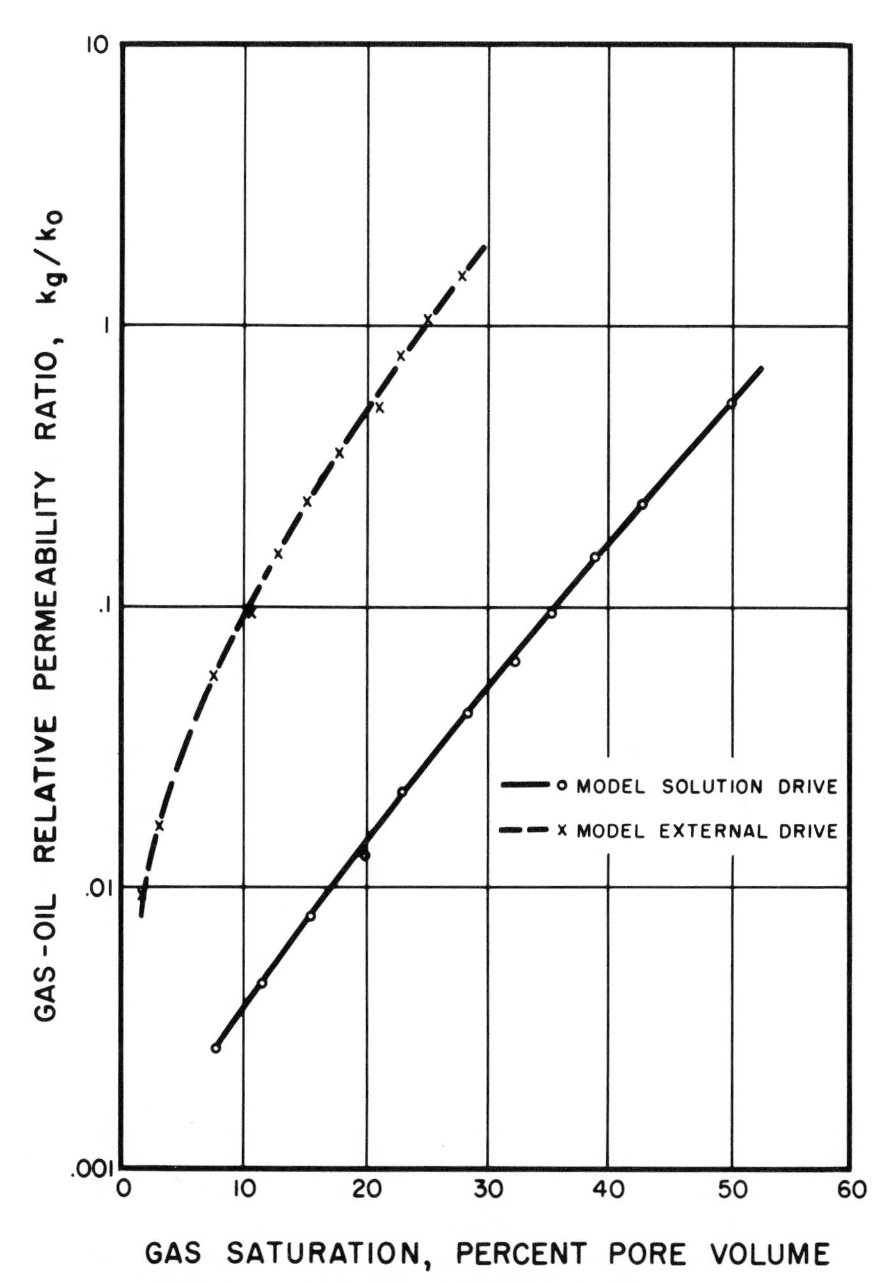

FIG. 19. Comparison of gas-drive performance in fractured limestone; external and solution gas-drive tests. (After Stewart *et al.*,[26] courtesy of AIME.)

typifies the qualitative correlation between gas-drive performance and type of pore system. As shown in Fig. 17, for an intergranular system there is close agreement between the two k_g/k_o curves. Figure 18, for two samples having vugular-solution prosity, indicates substantial deviation; and, finally, Fig. 19 demonstrates the greatest divergence, for a fractured limestone.

In a later study, Stewart *et al.*[27] considered the results of the laboratory gas-drive tests just described in the light of the formation of gas bubbles during the depletion process. In the solution gas-drive mechanism the displacement operation is dependent on the formation of gas bubbles within the pores of the rock itself, whereas in the external-drive mechanism the gas is injected from an outside source. They explained that in the case of fracture-matrix porosity, with external drive, gas tends to channel through portions of the fracture system, resulting in very inefficient oil displacement. Under conditions of solution drive, the gas will also channel through the fissures and larger pore throats to some extent as the bubbles unite to form a continuous free gas phase. At the same time, however, individual gas bubbles will continue to emerge from isolated pores and to displace oil until they join the continuous gas phase. As a result, solution gas drive will be more efficient than external gas drive in vugular-solution and fracture-matrix porosity rocks.

Stewart *et al.*[27] also found that differences ranging up to twofold in oil recovery could be obtained by varying the test conditions. The variations in test results were substantial for the vugular-solution and fracture-matrix porosity types of rocks and only minor for rocks having intergranular porosity. For limestones having nonuniform porosity, the laboratory gas-drive results were found to be affected by (1) rate of pressure decline, (2) original bubble point pressure of the gas-oil solution, (3) oil viscosity, and (4) gas solubility characteristics. The greater the rate of pressure decline, the larger the number of gas bubbles formed. Laboratory tests indicated that an increase in rate of pressure decline by a factor of 10 results in the formation of 10 times as many bubbles. With a greater number of bubbles formed, solution gas-drive recovery will increase in rocks having nonuniform porosity. The reasoning is that the recovery of oil from the storage pores of rocks with nonuniform porosity will occur only when gas bubbles form in the storage pores themselves.

Laboratory solution gas-drive tests in carbonate rocks usually demonstrate more favorable recoveries than those anticipated in actual practice, because fewer bubbles form at the rates of pressure decline which exist in the field. It is not known whether significant improvements in oil recovery can be achieved within the limits of pressure decline rate possible in the field. Increases in oil recovery by this means may be feasible under the proper conditions. This question is but one aspect of the overall problem of production rate and its effect on recovery, which is discussed in the next section.

The available field evidence does not support broad conclusions at this time as to whether or not increased oil recovery would result from high withdrawal rates in carbonate reservoirs under solution gas drive.

Stewart *et al.*[26] also conducted similar gas-drive experiments on sandstone cores. The relative permeability relationships were identical under both external- and solution-drive conditions in all cases. It appears that in rocks having uniform porosity the pores act as fluid conductors as well as fluid storage spaces. It is not necessary, therefore, for gas bubbles to form in the pore spaces themselves, as the gas evolved upstream is able to enter essentially all the individual pores to achieve the oil displacement. A close examination of the pore structure of the limestone sample, the k_g/k_o curves of which are presented in Fig. 17, indicated intergranular porosity; however, many of the passages between the original fragments have been sealed off by secondary cementation with resulting formation of storage pores. As a result, the external-drive k_g/k_o curve is slightly less favorable than the solution-drive curve, as shown in Fig. 17.

Discussion of Some Theoretical and Practical Aspects of Carbonate Reservoir Performance

Solution Gas-Drive and Gas-Cap-Drive Reservoirs

It has been observed that the performance of carbonate reservoirs under solution gas drive or gas-cap drive varies over a wide range, depending on the nature of the producing zone. The general field k_g/k_o behavior of pools having porosities of the intergranular, vugular-solution, and fracture-matrix types is presented in Chapter 3. The performance of reservoirs with intergranular-intercrystalline porosity resembles closely that of sandstone reservoirs with similar k_g/k_o curves and ultimate oil recovery. Reservoirs with vugular-solution porosity may have lower ultimate recoveries owing to less favorable and more unpredictable k_g/k_o behavior. Finally, reservoirs having fracture-matrix porosity have proved to be the most difficult to evaluate because of very erratic performance; these reservoirs generally have low ultimate recoveries owing to low-permeability host rock, even though the existence of fractures greatly increases recoveries. Carbonate pools with nonuniform porosity typically have k_g/k_o curves with low equilibrium gas saturations, sometimes approaching zero. As indicated above, the low recoveries are due in part to the inefficient displacement of the oil contained in the pores by the solution gas bubbles and the gas coming from outside the oil zone. In addition, it appears that gravity segregation of fluids in the secondary channels, particularly the fractures, may play an important role in the production process. In a highly fractured reservoir, the fracture system may, at least at low withdrawal rates, act as an effective oil and gas separator. On emitting from the matrix, the oil and gas separate, with the gas migrating upward to form a secondary gas cap in some cases. Fluid segregation in the fractures may have a pronounced effect on pool performance.

Pirson,[28] in considering this problem from a theoretical standpoint, assumed a highly fractured reservoir model with a low matrix and high fracture permeability, both horizontally and vertically. Under these conditions the nature of the fluid segregation and its effect on ultimate oil recovery were examined. Using an average k_g/k_o curve for a group of dolomite reservoirs with an equlibrium gas saturation of about 7%, Pirson calculated the theoretical reservoir performance for various types of fluid segregation. For simplicity, the production process was viewed as a succession of stages of depletion separated by shut-in periods during which static and capillary equilibrium is reached in both the matrix and the fracture systems. As a result, a readjustment of fluid saturations takes place in the porous systems during the shut-in periods.

Several degrees of fluid segregation were considered in the theoretical calculations: (1) no segregation of oil and gas in a simple depletion performance; (2) segregation from the upper $\frac{3}{4}$ of the reservoir, with enough oil draining downdip to resaturate the lower $\frac{1}{4}$ of the zone during the shut-in periods to 90% liquid saturation; (3) comparable to case 2, but with the oil draining from the upper $\frac{1}{2}$ of the reservoir to resaturate the lower $\frac{1}{2}$ of the zone; and (4) comparable with cases 2 and 3, but with oil coming from the upper $\frac{1}{4}$ of the reservoir to replenish the lower $\frac{3}{4}$ of the zone. As would be expected, there is a decrease in calculated ultimate oil recovery and an attendant increase in peak gas/oil ratio in going from case 1 to case 4. The respective ultimate oil recoveries and peak gas/oil ratios, progressing from case 1 to case 4, are 27.0% and 7,500 cu ft/bbl, 20.8% and 12,000 cu ft/bbl, 14.5% and 37,000 cu ft/bbl, and 9.6% and 83,000 cu ft/bbl. In view of the continuous fracture system assumed, a k_g/k_o curve with an equilibrium gas saturation approaching zero would be more realistic. Under this circumstance, Pirson's calculated ultimate recoveries would be reduced by $\frac{1}{4}$ to $\frac{1}{2}$. Most of the k_g/k_o curves for carbonate reservoirs suggest lower values for equilibrium gas saturation than those in sandstones.

Another phenomenon, mentioned by Pirson,[28] Elkins and Skov,[29] and Stewart *et al.*[26] is the possible reduction in oil recovery in fractured reservoirs caused by the capillary end effect, which develops at the efflux end of a core in the laboratory or at the end of a matrix block in the reservoir. As described by Hadley and Handy[30] in a theoretical and experimental study, it is caused by the discontinuity in capillary pressure when the flowing fluids leave the porous medium and abruptly enter a region with no capillary pressure. The capillary pressure discontinuity tends to decrease the rate of efflux of the preferentially wetting phase, in this case the oil, from the core or matrix blocks, compared to the rate of efflux of the nonwetting phase or the gas phase. Accordingly, oil tends to accumulate near the edges of the blocks.

Laboratory experiments indicate that the end effect becomes less important at higher flow rates. Figure 20 shows a comparison of the observed and the

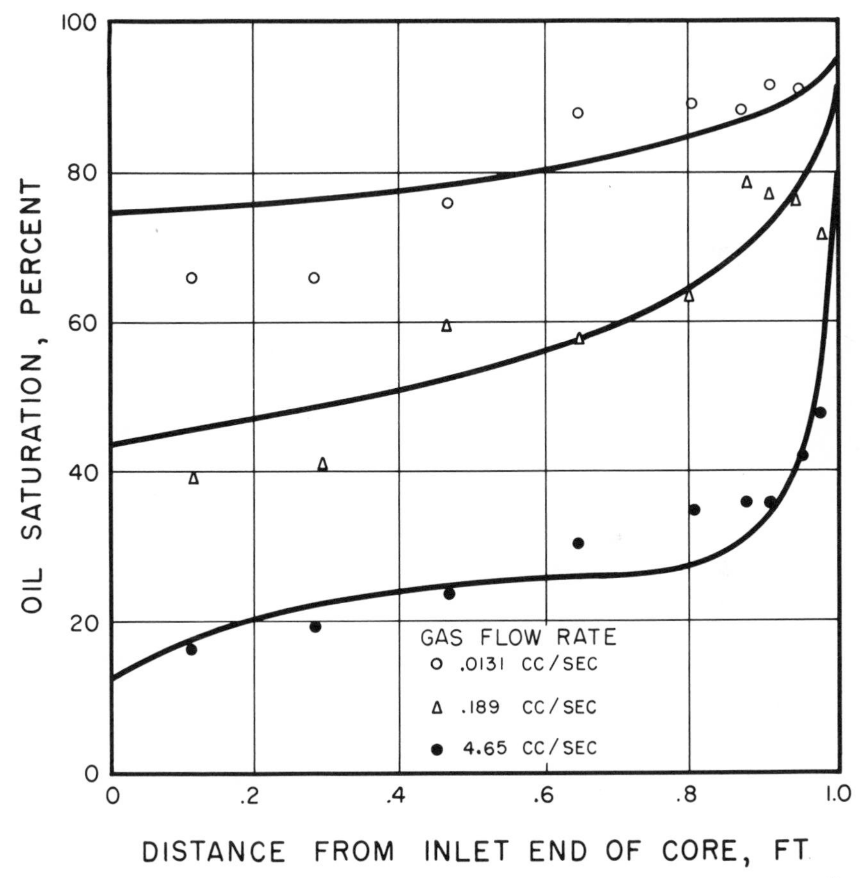

FIG. 20. Comparison of observed and calculated fluid distribution in a core for a gas-oil system at various flow rates. (After Hadley and Handy.[30])

calculated saturation distribution in a core for a gas-oil system at various flow rates under conditions approaching steady state. At higher flow rates, the oil recovery increases and the oil saturation buildup becomes more localized toward the outlet end of the core. The overall effect of the capillary end effect in a fractured reservoir is to decrease oil recovery and increase the average gas/oil ratio during the life of the pool. At the flow rates experienced in the field, the end effect is normally unimportant and would probably be significant only in extensively fractured zones in which the dimensions of the matrix blocks are of the order of several inches rather than several feet.[29]

A possible solution to the problems of gravity segregation and end effect is to make an effort to prevent them from taking place. This is done by producing the wells at high drawdowns, provided high producing rates are compatible with water influx and market demand considerations. If sufficiently high horizontal pressure gradients are established, the oil and gas flowing from the matrix blocks move more directly to the producing wells and fluid segregation and end effects are reduced.

Jones-Parra and Reytor[31] investigated the effect of fluid segregation in the fracture system on production performance and ultimate recovery through the use of a mathematical model: an idealized network consisting of a high-permeability fracture system exhibiting complete segregation of fluids and a low-permeability matrix with no gravity segregation. Their purely mathematical treatment tends to support the contention that higher withdrawal rates at increased gas/oil ratios may enhance oil recovery in some instances. (See Fig. 26, p. 116.)

Oil Recovery Owing to Water Invasion in Fractured Reservoirs

During the water influx into a fractured reservoir, oil displacement may be the result of (1) the flow of water under naturally imposed pressure gradients (viscous forces) and (2) imbibition, which is the spontaneous movement of water into the matrix under capillary forces. The environmental conditions of fractured carbonate reservoirs are conducive to the predominance of capillary forces over viscous forces. As a result, the tendency of water to channel through the more permeable stratum is offset by the tendency of water to imbibe into the tight matrix and displace the oil into fractures. Numerous investigators have examined imbibition behavior by mathematical methods and in the laboratory.[32–37]

Graham and Richardson,[33] for example, investigated in the laboratory the role of water imbibition in the recovery of oil from a fracture-matrix porous system. Their results are particularly interesting as they pertain to fractured carbonate fields. In a fractured zone the process is described as a condition of water imbibing from the fracture system into the matrix with simultaneous countercurrent movement of the oil from the matrix into the fractures. In the process, the rate of imbibition is directly proportional to the interfacial tension and the square root of the permeability, and is dependent on the wettability, the fluid viscosities, and the rock characteristics.

Discussion of Carbonate Reservoir Field Performance (Examples)

Asmari Pools in Iran

The Asmari reservoirs in southern Iran provide good field examples of the producing mechanisms discussed in the preceding sections. The producing zone is the Asmari Limestone of Lower Miocene to Oligocene age. Characteristically, the reservoir rock is a fine- to coarse-grained, hard, compact limestone with evidence of some recrystallization and dolomitization. It generally has a low porosity and permeability. The reservoir rock is folded into elongated anticlines and is abundantly fractured into an elaborate pattern of separate matrix blocks.

Andresen *et al.*[38] have analyzed the Asmari reservoirs and their performance. During the depletion of a typical Asmari reservoir, the mechanisms of gas-cap drive, undersaturated oil expansion, solution gas drive, gravity drainage, and imbibition displacement are all in operation at one time or another. Figure 21 is a schematic diagram of a typical Asmari reservoir, showing the distribution of fluids during production. At normal drawdown pressures, the free gas that flows from the matrix is not produced with the oil. Instead, it migrates freely through the fissures into the gas cap. The gas-oil level moves downward and the water-oil level upward under the action of dynamic and capillary forces.

The high-relief Asmari reservoirs have extremely thick oil columns with free gas gaps as indicated in Fig. 21. The oil columns consist of four sections: (1) the secondary gas cap, (2) the gassing zone, (3) the oil expansion zone, and (4) the water-invaded zone, as reported by Andresen *et al.*[38] The secondary gas-cap zone is bounded by the original and the current gas-oil levels in the fissure system. Owing to the low permeability of the matrix, there is no significant segregation of fluids in the matrix blocks themselves. The gassing zone has the current gas-oil level as its upper boundary, and the lower boundary is the level at which the reservoir oil is at saturation pressure. Located within the gassing zone is the equilibrium gas saturation level. Above this position in the zone, the gas evolved from solution is mobile and flows from the matrix blocks to the fractures and then migrates vertically through the fissure system to the gas cap. The free gas below the level of equilibrium gas saturation is immobile and is not produced from the matrix blocks. The oil expansion zone extends from the saturation pressure level to the current oil-water level. The water-invaded zone is below. Water displaces oil in this zone primarily by imbibition.

The phenomenon of convection also occurs in Asmari reservoirs. At initial conditions, the reservoir is in a state of equilibrium. This equilibrium is disrupted by the production process. According to Sibley,[39] saturation pressure increases with depth in most Asmari reservoirs at a rate of 4–5

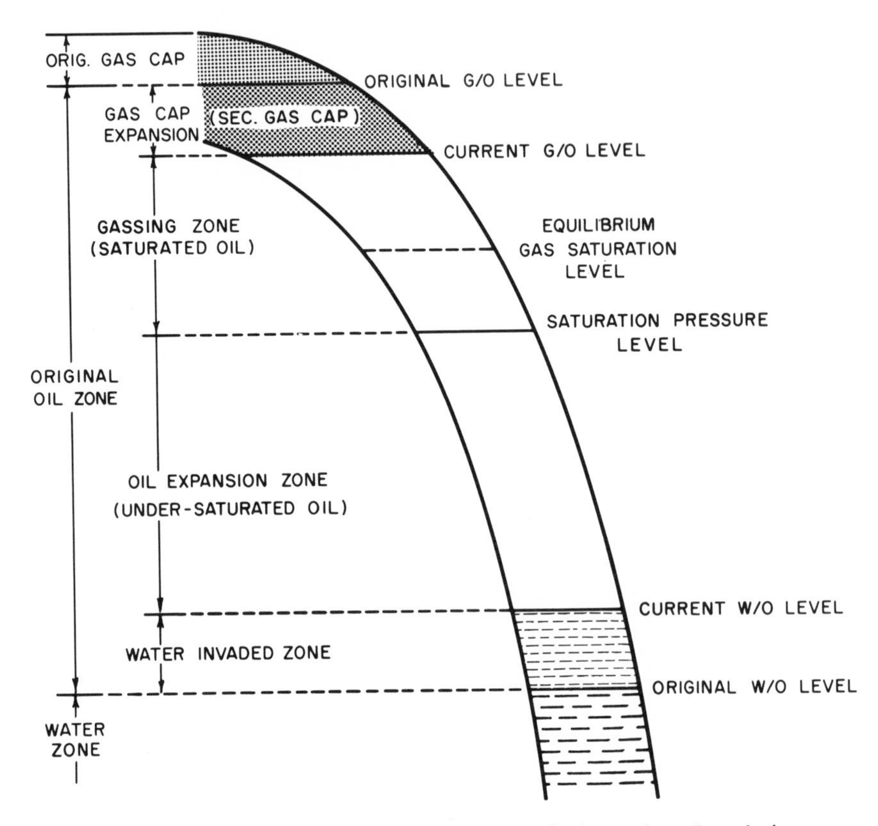

FIG. 21. Fluid distribution in Asmari Limestone reservoirs in southern Iran during production. (After Andresen *et al.*[38])

psi/100 ft, and the solution gas/oil ratio correspondingly shows an increase of 0.8 SCF/STB/100 ft of depth. As a result, the density of the reservoir oil decreases slightly with depth, which provides for convection in the highly permeable fissure system. The above description illustrates the complexity of production mechanisms in highly fractured high-relief reservoirs. The analysis of such reservoirs can be extremely difficult.

Kirkuk Pool in Iraq

The Kirkuk oil field in Iraq is another classic example of a complex reservoir system. Free water movement, pronounced gas segregation, and oil convection all occur in an extensively fractured and vuggy limestone.[40] The degree of fracturing and vugginess is highest at the crest of the anticlinal structure, where convection is correspondingly greatest. Temperature profiles of wells located on the crest of the structure indicate that convection is substantial.

In Fig. 22, the temperature profile of a well drilled on the crest of the structure is presented; the well has been idle for a long time. From the top of the fractured section of the oil zone to the water table, a distance of over 400 ft, there is virtually no change in temperature.

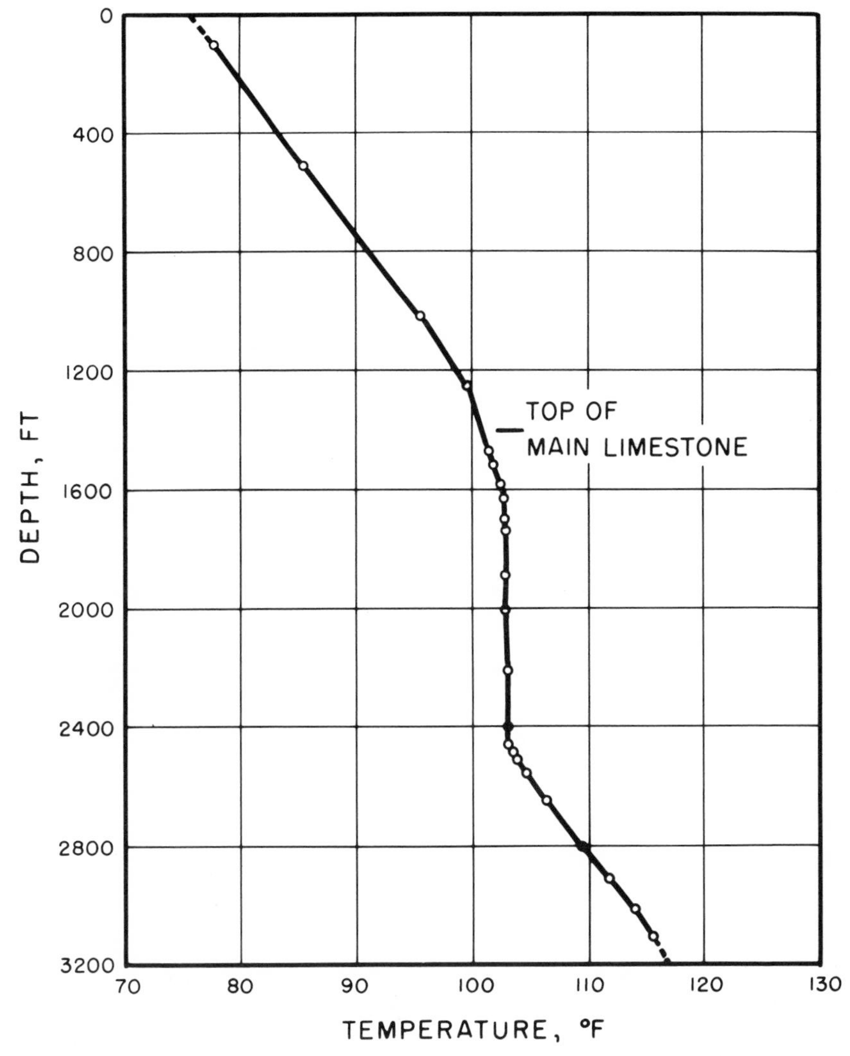

FIG. 22. Temperature profile of well in Kirkuk Field. (After Freeman and Natanson.[40])

At Kirkuk imbibition is a major driving mechanism. Freeman and Natanson[40] have described two types of imbibition taking place in the Kirkuk reservoir. When the matrix block is totally immersed in water, countercurrent and direct flow types of imbibition should ideally yield the same ultimate recovery, even though there may be some trapping of oil droplets in the water-filled fracture under countercurrent flow conditions. In any given time interval, however, the direct flow conditions will yield more oil if this imbibition process acts over a larger area. The reverse may also be true. Aronofsky *et al.*,[37] using a simple abstract model, have examined the effect of water influx rate on the imbibition process. Their treatment is confined to countercurrent imbibition.

San Andres Field, Mexico—an Extremely Undersaturated Oil Reservoir

The San Andres Limestone reservoir of the San Andres Field[41] (Gulf of Mexico, state of Veracruz, Mexico) is an example of a highly undersaturated oil reservoir. The original reservoir pressure was 466.1 kg/cm^2 (6,620 psi) and the bubble point pressure was 177.5 kg/cm^2 (2,525 psi). The San Andres zone is a chalky, oolitic limestone exhibiting intergranular porosity with occasional vugs and fissures; the average permeability is 2.5 md. The performance of the pool, as shown in Fig. 23, has demonstrated a high degree of undersaturation with a rapid decrease in production rate and an essentially constant gas/oil ratio of 110 m^3/m^3 (615 cu ft/bbl) and a water cut of 1–2%. The increase in gas/oil ratio in the middle of 1960 reflects a pilot gas injection test, which was started in December, 1959, and terminated in the latter part of 1960 because of an unfavorable response to the gas injection in the surrounding wells. By March, 1960, the reservoir pressure had declined sharply to 285.4 kg/cm^2 (4,060 psi) with a cumulative recovery of 5% of the initial oil-in-place. Garaicochea Petrirena *et al.*[41] calculated a recovery of 7.4% to bubble point pressure under an oil expansion producing mechanism. By the middle of 1962, 8.4% of the oil-in-place had been produced, which should have lowered the reservoir pressure to slightly below bubble point pressure.

Leduc D-3 Reservoir, Combination of Gas-Cap and Water Drives

The Leduc D-3 Pool in Central Alberta, Canada, is a remarkable example of a pool subject to active concurrent gas-cap and water drive. The producing D-3 reservoir (Upper Devonian) is a dolomitized organic reef deposit of high permeability (approximately 1,000 md in the horizontal direction and substantially less in the vertical). The producing zone is of the vugular-solution porosity type with some fissuring. Average porosity is estimated at

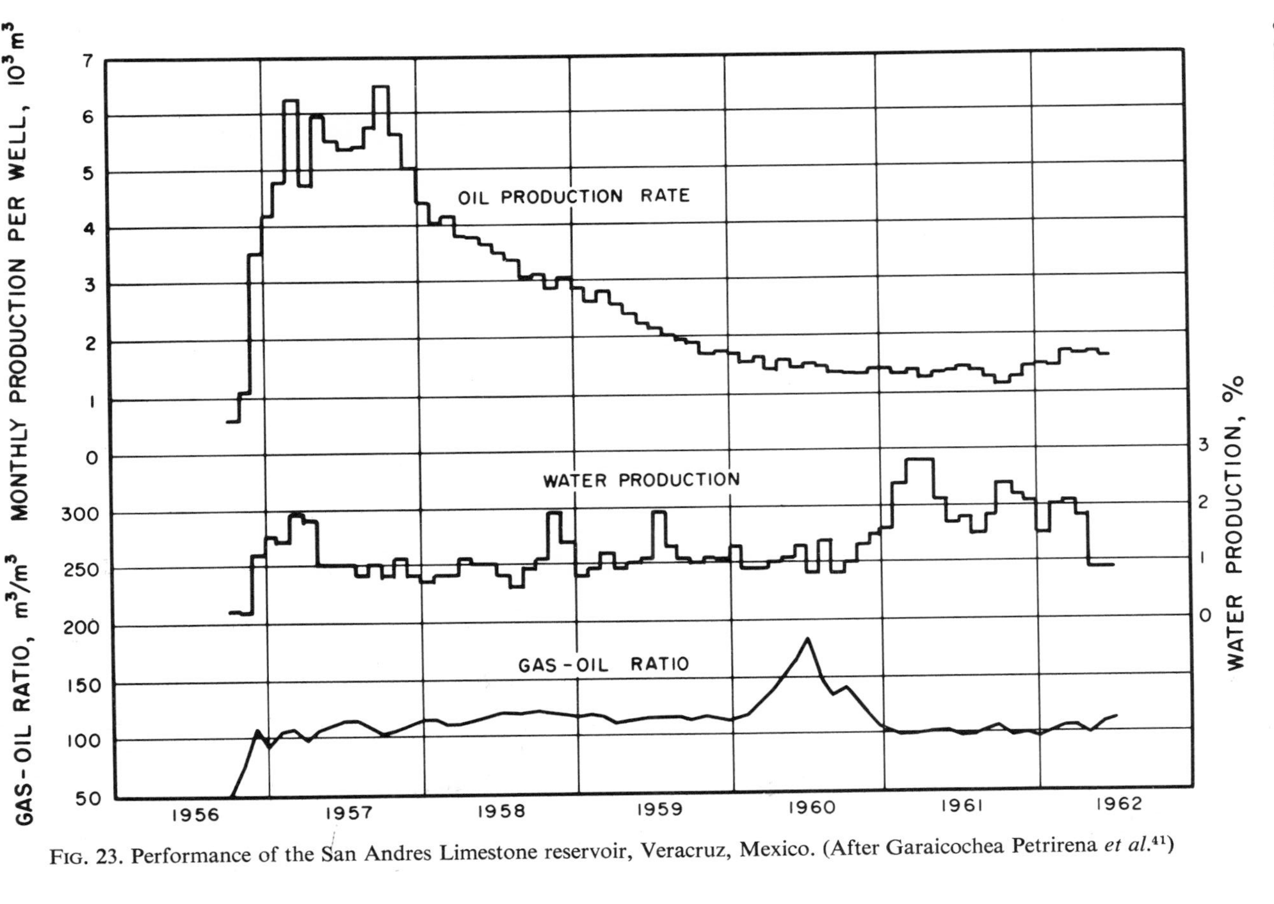

FIG. 23. Performance of the San Andres Limestone reservoir, Veracruz, Mexico. (After Garaicochea Petrirena *et al.*[41])

8%. The relatively thin oil zone of 30–38 ft is overlain by a large gas cap, ranging up to 158 ft in thickness, and underlain by a broad aquifer composed of more than 900 ft of permeable reef deposit. According to Horsefield,[42] the estimated original volume of the gas cap is 553,140,000 reservoir bbl, whereas the volume of the oil zone is 401,841,000 reservoir bbl. From the outset of production, there has been a continuous and fairly uniform down-structure movement of the gas-oil contact, along with an up-structure movement of the oil-water contact, as shown by Fig. 24. The indicated elevation of the gas-oil contact early in the history of the pool was due to a well that blew out of control for 6 months. To minimize gas and water coning, the wells were normally completed in a 5-ft interval in the middle portion of the oil zone. As a result, water production has been minor and the gas/oil ratio has been equal to approximately 600 SCF/bbl.

Volumetric balance studies[42] indicate that by 1954 water drive had contributed 50%, gas-cap drive 40%, and solution gas drive 10% to the replacement of reservoir oil. The extent to which gravity drainage has aided in the

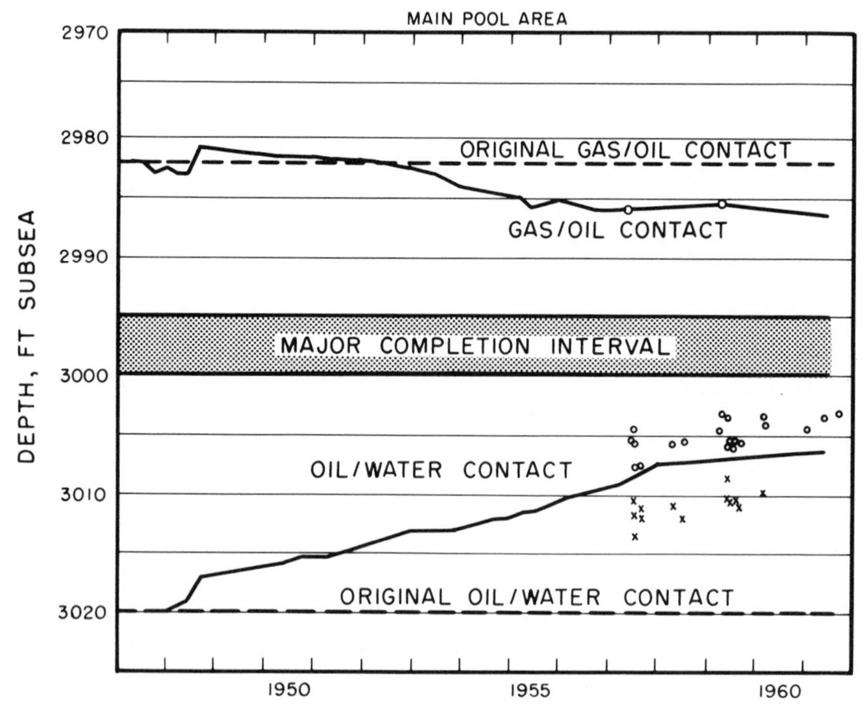

FIG. 24. History of gas-oil and oil-water contact, main pool area, Leduc D-3 reef pool, Alberta, Canada. (After Horsefield,[42] courtesy of AIME.)

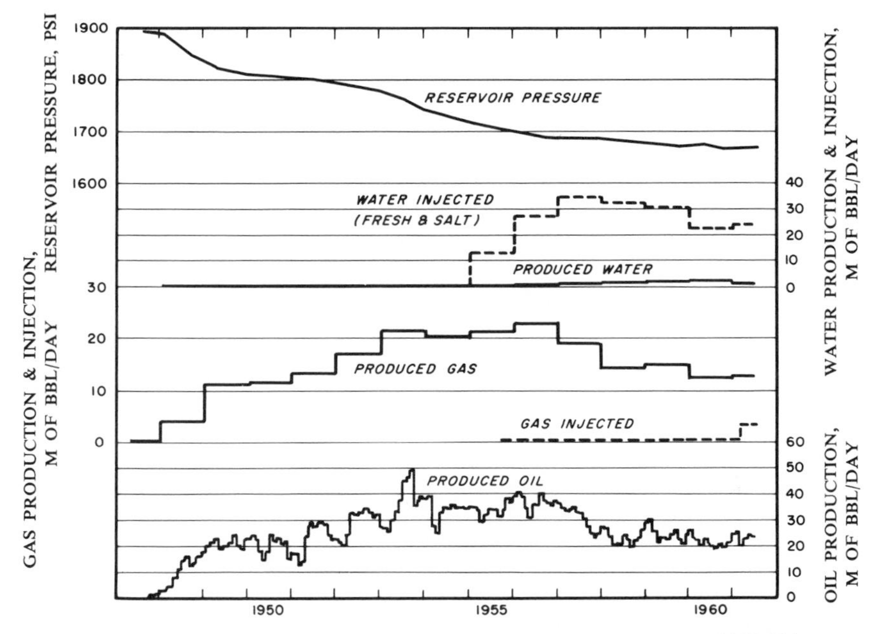

FIG. 25. Leduc D-3 reef pool performance. (After Horsefield,[42] courtesy of AIME.)

expansion of the gas cap is not reported, but it is considerable owing to the substantial vertical permeability and the high oil gravity (39° API). The pool is under simultaneous water and gas-cap injection to ensure that a minimum "oil sandwich" is left at depletion. Figure 25 shows the pool performance. Material balance studies based on the movement of the gas-oil and water-oil contacts have indicated that the flushing efficiency in the gas cap and in the water influx zone are of the order of 80% of the hydrocarbon pore volume (HCPV). This high estimate of oil recovery is supported in part by the fact that the average oil saturation of cores cut in the water-invaded region in deepened wells was found to be 13.6% HCPV.

Recent Statistical Studies on Oil Recovery Efficiency

Recently the American Petroleum Institute[45] presented a statistical study of oil recovery efficiency, which is also discussed in Chapters 3 and 5. The study was based on case histories from 226 sandstone and 86 carbonate reservoirs, classified by their predominant drive mechanism: solution gas drive without supplemental drive, solution gas drive with supplemental drive, gas-cap drive, water drive, and gravity drainage. It is by far the most comprehensive and exhaustive study of oil pool performance ever published and affords the opportunity for meaningful comparisons between sandstone and carbonate reservoirs under the various drive mechanisms mentioned. The results of the study can be summarized as follows:

Tables 2, 3, 4, 5, and 6 show the range of values for the reservoir parameters under the five different producing mechanisms. The parameters for the solution gas- and water-drive pools have been grouped on the basis of the type of reservoir rock: sandstone or limestone, dolomite, and other (chert, etc.). The gas-cap drive and gravity-drainage pool data were more limited. The sandstone and carbonate reservoirs have been combined in the gas-cap-drive analysis. The gravity-drainage parameters are presented for sandstones only. Seventy-seven sandstone and 21 carbonate pools producing under essentially a solution gas-drive mechanism were studied. Eighty-one pools (60 sandstone and 21 carbonate) had solution gas drive as the predominant recovery mechanism, supplemented by water drive, gravity drainage, and, in a few cases, partial gas or water injection. Seventy-two sandstone and 39 carbonate water-drive pools were examined. Only 14 gas-cap-drive and 6 gravity-drainage pools were included. The value of these data is primarily for comparison purposes. As expected, the solution gas-drive pools have the lowest recovery efficiencies and the pools under water drive and gravity drainage show the highest recoveries. The sandstone reservoirs exhibit higher oil recoveries than the carbonate pools, and also have higher porosities and much greater permeabilities. The carbonates, however, are much more likely to be fractured and fissured, a factor which will not always be reflected accurately in the stated values for porosity and permeability. In this connection, it should be borne in mind that fractures may form during the core recovery process. The wider range of recovery efficiencies in carbonates corresponds to the inherently greater complexity of carbonate rocks.

In Table 2, the median value for percentage recovery of initial oil-in-place for solution gas drive without supplementary drive is 21.3% for sandstone pools, compared to 17.6% for carbonate reservoirs. The median values of 154 and 88 bbl/acre-ft for sandstones and carbonates, respectively, show, however, that the actual recoveries in sandstones are almost twice as high as the carbonate values, partly owing to the much higher sandstone porosities. Supplemental drives also appear to have benefited sandstone reservoirs more than carbonate pools in the case of the solution gas-drive reservoirs examined in this study (Table 3). The median values show that the effect of supplementation by additional drive mechanisms is about $\frac{1}{3}$ for sandstones and $\frac{1}{4}$ for carbonates. When gas-cap-drive is the predominant mechanism, the average recovery efficiency is higher by about $\frac{2}{3}$ than the comparable value for the solution gas-drive mechanism (Table 4).

The median values for ultimate recovery (bbl/acre-ft) by water drive in sandstones exceed those for limestones and dolomites by a factor of 3 and,

TABLE 2. Range of Values for Reservoir Parameters with Solution Gas Drive as Predominant Drive Mechanism without Supplemental Drives. (After Arps *et. al.*,[45] courtesy of API.)

Parameters*	Sand and sandstone			Limestone, dolomite, and other		
	Minimum	Median	Maximum	Minimum	Median	Maximum
Rock						
k, darcys	0.006	0.051	0.940	0.001	0.016	0.252
ϕ, fraction	0.115	0.188	0.299	0.042	0.135	0.200
S_w, fraction	0.150	0.300	0.500	0.163	0.250	0.350
Fluids						
g_o, °API	20	35	49	32	40	50.2
μ_{ob}, cp	0.3	0.8	6	0.2	0.4	1.5
R_{sb}, cf/bbl	60	565	1,680	302	640	1,867
B_{obd}, ratio	1.050	1.310	1.900	1.200	1.346	2.067
B_{obf}, ratio	1.050	1.297	1.740	1.200	1.402	2.350
B_{oad}, ratio	1.000	1.090	1.400	1.060	1.120	1.420
Environment						
h_n, ft	3.4	32.2	772	3.9	27	425
α, deg	0–5	5–15	>45	0–5	0–5	5–15
D_{bs}, ft	1,500	5,380	11,500	3,100	6,300	10,500
T, °F	79	150	260	107	174	209
p_b, psig	639	1,750	4,403	1,280	2,383	3,578
p_a, psig	10	150	1,000	50	200	1,300
Ultimate recovery						
BAF, bbl/AF	47	154	534	20	88	187
RE, %	9.5	21.3	46.0	15.5	17.6	20.7
S_{gr}, fraction	0.130	0.229	0.382	0.169	0.267	0.447

* Nomenclature for Tables 2, 3, 4, 5, and 6; g_o=stock-tank oil gravity (°API); μ_o=reservoir oil viscosity (cp); R_s=solution gas/oil ratio (SCF/bbl); B_o=oil formation volume factor (reservoir bbl/STB); h_n=net pay thickness (ft); α=formation dip (degrees); D_{bs}=depth below the surface (ft); RE=recovery efficiency (% of initial stock tank oil-in-place); S_{gr}=residual gas saturation (fraction of pore space); S_{or}=residual oil saturation (fraction of pore space). Subscripts: i=initial conditions; b=bubble point conditions; a=abandonment conditions; d=differential liberation; f=flash liberation; k=arithmetic average of absolute permeability, d.

in terms of recovery efficiency, by 17% (Table 5). Water drive seems to be about 2.5 times more efficient than unsupplemented solution gas drive in both sandstone and carbonate reservoirs.

Data for the water-drive cases in sandstones and the solution gas-drive cases in both sandstones and carbonate rocks were subjected to regression analysis. For the water-drive reservoirs the following regression equation for the recovery efficiency (RE, in % of initial oil-in-place) was found:

$$RE = 54.898\left[\frac{\phi(1-S_w)}{B_{oi}}\right]^{0.0422}\left(\frac{k\mu_{wi}}{\mu_{oi}}\right)^{0.0770}\left(S_w\right)^{-0.1903}\left(\frac{p_i}{p_a}\right)^{-0.2159} \quad (1)$$

where μ_{wi} = initial water viscosity (cp), μ_{oi} = initial oil viscosity (cp), k = arithmetic average of absolute permeability (d), S_w = interstitial water content (fraction of total pore space), ϕ = porosity (fraction), B_{oi} = initial oil formation volume factor (reservoir bbl/bbl tank oil), p_i = initial reservoir pressure (psig), and p_a = reservoir pressure at abandonment (psig).

The API subcommittee established the following regression equation for recovery efficiency below the bubble point in solution gas-drive reservoirs:

$$RE = 41.815\left[\frac{\phi(1-S_w)}{B_{ob}}\right]^{0.1611}\left(\frac{k}{\mu_{ob}}\right)^{0.0979}\left(S_w\right)^{0.3722}\left(\frac{p_b}{p_a}\right)^{0.1744} \quad (2)$$

where B_{ob} = oil formation volume factor at bubble point pressure (reservoir bbl/bbl tank oil), μ_{ob} = oil viscosity at bubble point pressure (cp), and p_b = bubble point pressure (psig).

An examination of Eqs. 1 and 2 shows the presence of four common variables: (1) oil-in-place under initial conditions for the water-drive reservoirs and at bubble point pressure for the solution gas-drive reservoirs; (2) mobility factor—at initial conditions for water-drive pools and at bubble point pressure for solution gas-drive pools; (3) interstitial water saturation; and (4) pressure drop ratio—initial/abandonment pressure ratio for water-drive reservoirs and bubble point/abandonment pressure ratio for solution gas-drive reservoirs. It is interesting to note that recovery efficiencies increase or decrease for both mechanisms when the oil-in-place and the mobility factor variables, respectively, are either high or low. On the other hand, low values for water saturation (S_w) and pressure drop rates result in higher recovery efficiencies in water-drive reservoirs, and lower recovery efficiencies in solution gas-drive reservoirs. Even though high S_w values serve to increase recovery efficiency in solution gas-drive reservoirs, the actual recoveries in barrels per acre-foot are usually less.

Equations (1) and (2) can be of value to the geologist and the petroleum engineer in the estimation of ultimate recovery from carbonate reservoirs in instances where the data required for more detailed studies are lacking.

TABLE 3. Range of Values for Reservoir Parameters with Solution Gas Drive as Predominant Drive Mechanism and with Supplemental Drives. (After Arps *et. al.*,[45] courtesy of API.)

Parameters	Sand and sandstone			Limestone, dolomite, and other		
	Minimum	Median	Maximum	Minimum	Median	Maximum
Rock						
k, darcys	0.010	0.216	2.500	0.002	0.019	0.867
ϕ, fraction	0.120	0.210	0.359	0.033	0.133	0.248
S_w, fraction	0.100	0.310	0.579	0.035	0.250	0.600
Fluids						
g_o, °API	15.5	36	46	22	38	46
μ_{ob}, cp	0.4	0.9	20	0.3	0.8	2
R_{sb}, cf/bbl	10	390	1,010	60	615	1,325
B_{obd}, ratio	1.010	1.230	1.580	1.050	1.328	1.682
B_{obf}, ratio	1.015	1.230	1.845	1.050	1.310	1.680
B_{oad}, ratio	1.000	1.050	1.220	1.020	1.110	1.500
Environment						
h_n, ft	4	30	714	8	31	154
α, deg	0–5	0–5	>45	0–5	0–5	5–15
D_{bs}, ft	300	4,237	10,280	2,800	6,000	10,530
T, °F	77	146	260	88	128	225
p_b, psig	5	1,360	4,275	530	1,830	2,935
p_a, psig	5	100	800	40	200	1,550
Ultimate recovery						
BAF, bbl/AF	109	227	820	32	120	464
RE, %	13.1	28.4	57.9	9.0	21.8	48.1
S_{gr}, fraction	0.077	0.255	0.435	0.112	0.260	0.426

Note for Tables 2, 3, and 4: Data for gas-drive pools do not include any oil produced above bubble point.

TABLE 4. Range of Values for Reservoir Parameters with Gas-Cap Drive as Predominant Drive Mechanism

(After Arps *et al.*,[45] courtesy of API.)

Parameters	All rock types combined		
	Minimum	Median	Maximum
Rock			
k, darcys	0.047	0.600	1.966
ϕ, fraction	0.086	0.225	0.358
S_w, fraction	0.150	0.262	0.430
Fluids			
g_o, °API	34	40	43
μ_{ob}, cp	0.3	0.6	2.3
R_{sb}, cf/bbl	226	703	1,335
B_{obd}, ratio	1.116	1.374	1.675
B_{obf}, ratio	1.116	1.350	1.631
B_{oad}, ratio	1.040	1 159	1.490
Environment			
h_n, ft	7	15	35
α, deg	0–5	0–5	15–45
D_{bs}, ft	3,300	5,500	7,675
T, °F	108	175	200
p_b, psig	854	2,213	3,583
p_a, psig	0	500	2,900
Ultimate recovery			
BAF, bbl/AF	68	289	864
RE, %	15.8	32.5	67.0
S_{gr}, fraction	0.223	0.271	0.517

In all cases, these equations are helpful for comparison purposes. To facilitate the use of Eqs. 1 and 2, Arps[48] has developed the nomographs shown in Figs. 26 and 27.

In Appendix C, the writers have assembled data on reservoir characteristics and performance for selected carbonate reservoirs throughout the world.

TABLE 5. Range of Values for Reservoir Parameters with Water-Drive as Predominant Drive Mechanism

(After Arps *et al.*,[45] courtesy of API.)

Parameters	Sand and sandstone			Limestone, dolomite, and other		
	Minimum	Median	Maximum	Minimum	Median	Maximum
Rock						
k, darcys	0.011	0.568	4.000	0.010	0.127	1.600
ϕ, fraction	0.111	0.256	0.350	0.022	0.154	0.300
S_w, fraction	0.052	0.250	0.470	0.033	0.180	0.500
Fluids						
g_o, °API	15.5	35.3	50	15	37	54
μ_{oi}, cp	0.2	1.0	500	0.2	0.7	142
μ_w, cp	0.24	0.46	0.95	...	...	...
B_{oif}, ratio	0.997	1.238	2.950	...	...	...
B_{obf}, ratio	1.008	1.259	2.950	1.110	1.321	1.933
B_{oaf}, ratio	1.004	1.223	1.970	...	...	...
$\rho_w - \rho_o$, g/cc	0.054	0.241	0.490	...	...	...
Environment						
h_n, ft	6.5	17.5	160	9	50.2	185
α, deg	0–5	0–5	15–45	...	...	...
D_{bs}, ft	1,400	6,260	12,400	2,210	6,790	13,100
T, °F	84	163	270	90	182	226
p_i, psig	450	2,775	6,788	700	3,200	5,668
p_b, psig	52	1,815	5,400	30	1,805	3,821
p_a, psig	100	1,970	5,010	...	...	...
Ultimate recovery						
BAF, bbl/AF	155	571	1,641	6	172	1,422
RE, %	27.8	51.1	86.7	6.3	43.6	80.5
S_{or}, fraction	0.114	0.327	0.635	0.247	0.421	0.908

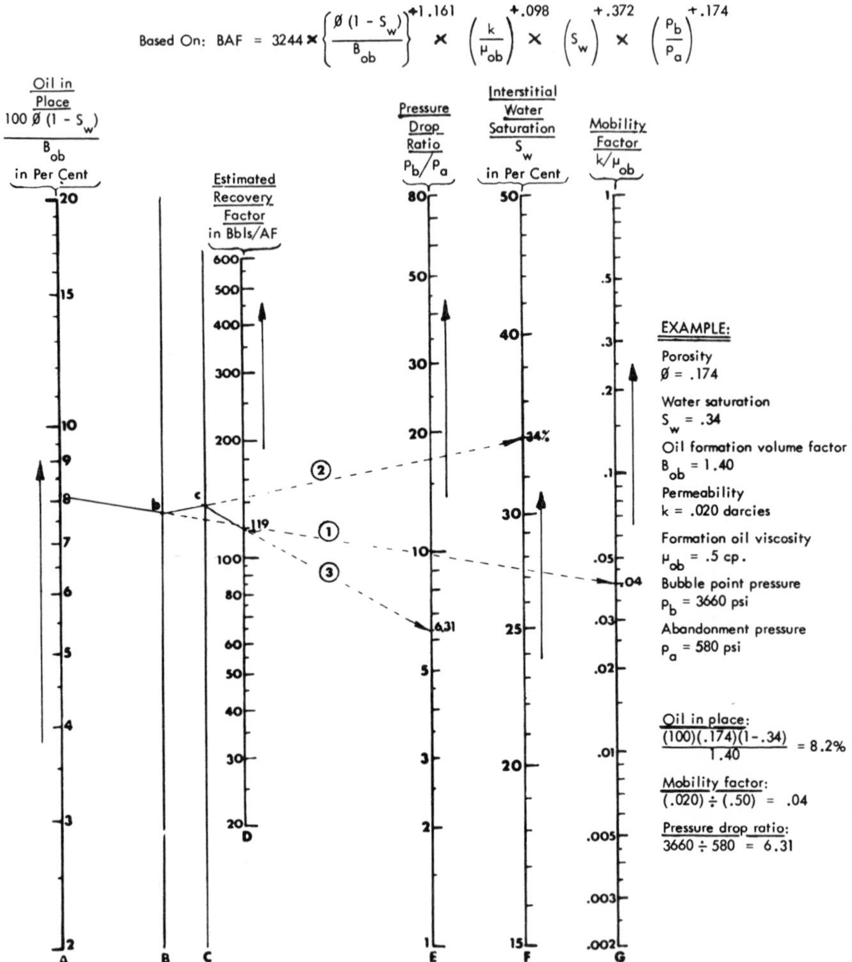

FIG. 26. Nomograph for estimated recovery factor (bbl/acre-ft) by solution gas drive (below bubble point) in sandstones and carbonates. (After Arps,[48] courtesy of AIME.)

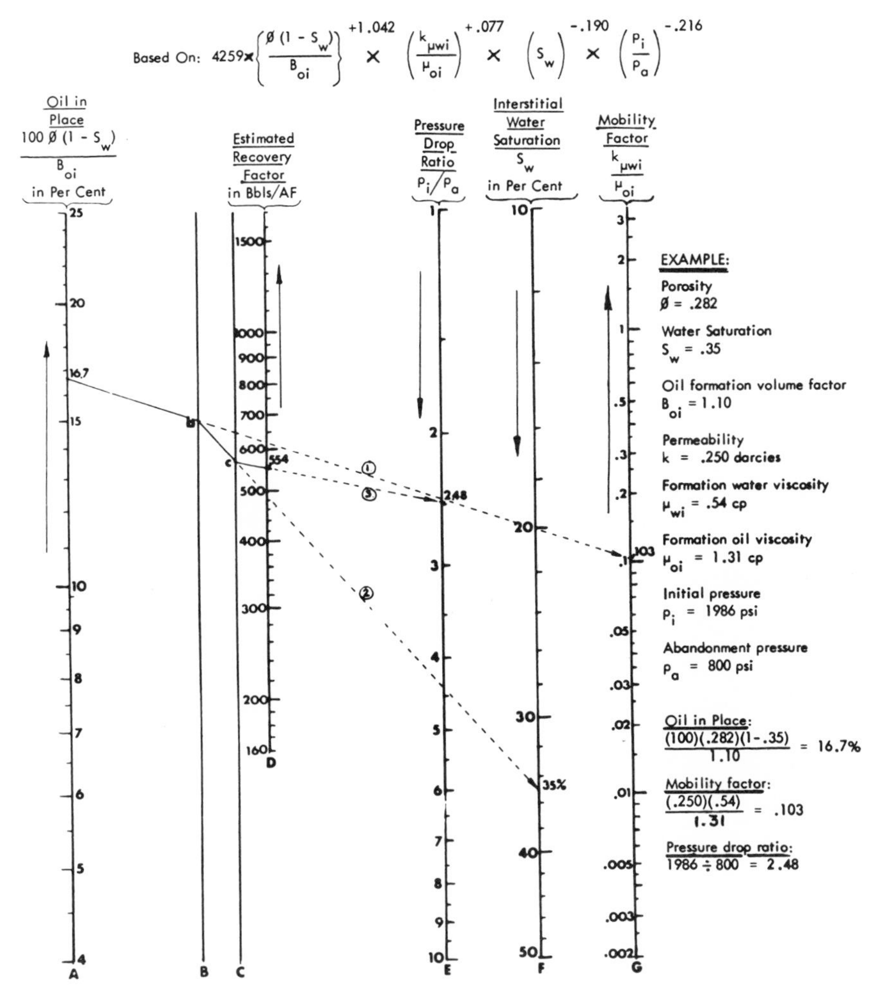

1. Connect 16.7% on oil in place scale A with mobility factor .103 on scale G and find intercept b on scale B.
2. Connect point b on scale B with water saturation 35% on scale F and find intercept c on scale C.
3. Connect point c on scale C with pressure drop ratio 2.48 on scale E and find estimated recovery factor 554 barrels per acre foot at intercept with scale D.

FIG. 27. Nomograph for estimated recovery factor (bbl/acre-ft) by water drive in sandstones. (After Arps,[48] courtesy of AIME.)

TABLE 6. Range of Values for Reservoir Parameters with Gravity Drainage as Predominant Drive Mechanism

(After Arps *et al.*,[45] courtesy of API.)

Parameters	Sand and sandstone only		
	Minimum	Median	Maximum
Rock			
k, darcys	0.305	1.285	2.000
ϕ, fraction	0.194	0.329	0.350
S_w, fraction	0.030	0.293	0.400
Fluids			
g_o, °API	15	22.5	38.5
μ_{ob}, cp	0.7	4	8
R_{sbd}, cf/bbl	96	200	735
B_{obd}, ratio	1.070	1.106	1.383
B_{obf}, ratio	1.070	1.106	1.340
B_{oad}, ratio	1.030	1.040	1.100
Environment			
h_n, ft	40	71	200
α, deg	0–5	5–15	15–45
D_{bs}, ft	1,170	2,400	6,500
T, °F	100	100	132
p_b, psig	497	1,044	2,670
p_a, psig	0	20	50
Ultimate recovery			
BAF, bbl/AF	250	696	1,124
RE, %	16	57.2	63.8
S_{gr}, fraction	0.151	0.377	0.654

Acknowledgments

The writers are greatly indebted to E. C. Babson for his contribution to the chapter. They are also grateful to Mrs. O'Detta Hawkins and Mrs. Ardis Meyer for typing the manuscript, and to Mr. Cliff Mathieson for drafting the diagrams.

References

1. Chilingar, G. V., Bissell, H. J. and Wolf, K. H.: "Diagenesis of Carbonate Rocks", in: G. Larsen and G. V. Chilingar (Editors), *Diagenesis in Sediments*, Elsevier Publ. Co., Amsterdam (1967) 179–322.
2. Craze, R. C.: "Development Plan for Oil Reservoirs", Chapter 33, in: T. C. Frick (Editor), *Petroleum Production Handbook*, McGraw-Hill Book Co., New York (1962).

3. Craft, B. C. and Hawkins, M. F.: *Applied Petroleum Reservoir Engineering*, Prentice-Hall, Englewood Cliffs, N.J. (1959) 62.
4. Eremenko, N. A.: *Geology of Oil and Gas*, lzd. Nedra, Moscow (1968) 389 pp.
5. Andresen, K. H., Baker, R. I. and Raoofi, J.: "Development of Methods for Analysis of Iranian Asmari Reservoirs", *Proc.*, Sixth World Pet. Cong. (1963) Sec. II, 13–27.
6. Standing, Marshall: Personal communication (Nov. 25, 1969).
7. Levorsen, A. I.: *Geology of Petroleum*, 2nd ed., W. H. Freeman and Co., San Francisco (1967) 573.
8. Gussow, W. C.: "Differential Entrapment of Oil and Gas: a Fundamental Principle", *Bull.*, AAPG (May, 1954) 816–853.
9. Gussow, W. C.: "Migration of Reservoir Fluids", *J. Pet. Tech.* (Apr., 1968) Vol. 20, No. 4, 353–363.
10. Bell, A. H., Mast, R. F., Oros, M. O., Sherman, C. W. and Van Den Berg, J.: *Petroleum Industry in Illinois*, 1959, Ill. Geol. Survey Bull. 88, 127 pp.
11. Dickey, P. A.: "Discussion to W. C. Gussow: 'Migration of Reservoir Fluids' ", *J. Pet. Tech.* (Apr., 1968) 364.
12. Muskat, Morris: *Physical Principles of Oil Production*, McGraw-Hill Book Co., New York (1949) 417–421.
13. Elliott, G. R.: "Behavior and Control of Natural Water-Drive Reservoirs", *Trans.*, AIME (1946) Vol. 165, 205.
14. Babson, E. C., *SPE Distinguished Lecture Series* (1965).
15. Bruce, W. A., "A Study of the Smackover Limestone Formation and the Reservoir Behavior of Its Oil and Condensate Pools", *Trans.*, AIME (1944) Vol. 155, 88–119.
16. Bulnes, A. C. and Fitting, R. U., Jr.: "An Introductory Discussion of the Reservoir Performance of Limestone Formations", *Trans.*, AIME (1945) Vol. 160, 179–201.
17. Barfield, E. C., Jordan, J. K. and Moore, W. D.: "An Analysis of Large-Scale Flooding in the Fractured Spraberry Trend Area Reservoir", *J. Pet. Tech.* (Apr., 1959) 15.
18. Mortada, M. and Nabor, G. W.: "An Approximate Method for Determining Areal Sweep Efficiency and Flow Capacity in Formations with Anisotropic Permeability", *Trans.*, AIME (1961) Vol. 222, 277.
19. Klute, C. H.: "The Permeability of Unoriented Polymer Films", *J. Polymer Sci.* (1959) Vol. 41, 307.
20. Fatt, I.: "A Demonstration of the Effect of 'Dead-End' Volume on Pressure Transients in Porous Media", *Trans.*, AIME (1959) Vol. 216, 449.
21. Goodknight, R. C., Klikoff, W. A. and Fatt, I.: "Non-Steady State Flow and Diffusion in Porous Media Containing Dead-End Pore Volume", *J. Phys. Chem.* (1960) Vol. 64, 1162.
22. Warren, J. E. and Root, P. J.: "The Behavior of Naturally Fractured Reservoirs", *Trans.*, AIME (1963) Vol. 228, 245–255.
23. Groves, D. L. and Abernathy, B. F.: "Early Analysis of Fractured Reservoirs Compared to Later Performance", paper SPE 2259 presented at SPE Annual Fall Meeting, Houston, Tex. (Sept. 29–Oct. 2, 1968).
24. Morris, E. E. and Tracy, G. W.: "Determination of the Pore Volume in a Naturally Fractured Reservoir", paper SPE 1185 presented at SPE Annual Fall Meeting, Denver, Colo. (Oct. 3–6, 1965).
25. Folk, R. L.: "Practical Petrographic Classification of Limestones", *Bull.*, AAPG (Jan., 1959) Vol. 43, No. 1, 1–38.
26. Stewart, C. R., Craig, F. F., Jr. and Morse, R. A.: "Determination of Limestone Flow Performance Characteristics by Model Flow Tests", *Trans.*, AIME (1953) Vol. 198, 93–102.
27. Stewart, C. R., Hunt, E. B., Jr., Schneider, F. N., Geffen, T. M. and Berry, V. J., Jr.: "The Rate of Bubble Formation in Oil Recovery by Solution Gas Drives in Limestones", *Trans.*, AIME (1954) Vol. 201, 294–301.

28. Pirson, S. J.: "Performance of Fractured Reservoirs", *Bull.*, AAPG (Feb., 1953) Vol. 37, No. 2, 232–244.
29. Elkins, L. F. and Skov, A. M.: "Cycle Waterflooding the Spraberry Utilizes 'End Effects' to Increase Oil Production Rate", *Trans.*, AIME (1963) Vol. 228, I, 877-884.
30. Hadley, G. F. and Handy, L. L.: "A Theoretical and Experimental Study of the Steady State Capillary End Effect", paper SPE 707-G presented at SPE Annual Fall Meeting, Los Angeles, Calif. (Oct. 14–17, 1956).
31. Jones-Parra, Juan and Reytor, R. S.: "Effect of Gas-Oil Rates on the Behavior of Fractured Limestone Reservoirs", *Trans.*, AIME (1959) Vol. 216, 395–397.
32. Blair, P. M.: "Calculation of Oil Displacement by Countercurrent Water Imbibition", *Soc. Pet. Eng. J.* (Sept., 1964) 195–202.
33. Graham, J. W. and Richardson, J. G.: "Theory and Application of Imbibition Phenomena in Recovery of Oil", *J. Pet. Tech.* (Feb., 1959) 65–69.
34. Handy, L. L.: "Determination of Effective Capillary Pressures for Porous Media from Imbibition Data", *Trans.*, AIME (1960) Vol. 219, 75–80.
35. Mattax, C. C. and Kyte, J. R.: "Imbibition Oil Recovery from Fractured, Water-Drive Reservoir", *Soc. Pet. Eng. J.* (June, 1962) 177–184.
36. Parsons, R. W. and Chaney, P. R.: "Imbibition Model Studies on Water-Wet Carbonate Rocks", *Soc. Pet. Eng. J.* (Mar., 1966) 26–34.
37. Aronofsky, J. S., Massé, L. and Natanson, S. G.: "A Model for the Mechanism of Oil Recovery from the Porous Matrix Due to Water Invasion in Fractured Reservoirs", *Trans.*, AIME (1958) Vol. 213, 17–19.
38. Andresen, K. H., Baker, R. I. and Raoofi, J.: "Development of Methods for Analysis of Iranian Asmari Reservoirs", *Proc.*, Sixth World Pet. Cong. (1963) Sec. II, 13–27.
39. Sibley, W. P.: "Handling of Reservoir Fluid Properties within a Fissure-System Reservoir Simulation Model", paper SPE 2374 presented at SPE Second Regional Technical Symp., Dhahran, Saudi Arabia (1969).
40. Freeman, H. A. and Natanson, S. G.: "Recovery Problems in a Fracture-Pore System: Kirkuk Field", *Proc.*, Fifth World Pet. Cong. (1959) Sec. II, 297–317.
41. Garaicochea Petrirena, F., Perez-Rosales, C. and Ortiz de Maria, M. J.: "An Example of an Extremely Undersaturated and Low Permeability Oil Reservoir—San Andres Field, Mexico", *Proc.*, Sixth World Pet. Cong. (1963) 117–128.
42. Horsefield, R.: "Performance of the Leduc D-3 Reservoir", with Supplement, *SPE Pet. Trans. Reprint Series* (1962) No. 4, 65–72.
43. Hnatiuk, J. and Martinelli, J. W.: "The Relationship of the Westerose D-3 Pool to Other Pools on the Common Aquifer", *J. Can. Pet. Tech.* (Apr.–May, 1967) 43–49.
44. Craze, R. C.: "Performance of Limestone Reservoirs", *Trans.*, AIME (1950) Vol. 189, 287–294.
45. Arps, J. J., Brons, F., van Everdingen, A. F., Buchwald, R. W. and Smith, A. E.: *A Statistical Study of Recovery Efficiency*, API Bull. D14 (Oct., 1967) 33 pp.
46. Torrey, P. D.: "Evaluation of Secondary Recovery Prospects", Paper No. 9, in: *Economics of Petroleum Exploration, Development and Property Evaluation*, Southwest Legal Foundation, Prentice-Hall, Englewood Cliffs, N.J. (1961).
47. Sessions, R. E.: "Small Propane Slug Proving Success in Slaughter Field Lease", *J. Pet. Tech.* (Jan., 1963) 31–36.
48. Arps, J. J.: "Reasons for Differences in Recovery Efficiency", paper SPE 2608 presented at SPE Symp. on Petroleum Economics and Valuation, Dallas, Tex. (1968) 77–82.
49. Arps, J. J. and Roberts, T. G.: "The Effect of Relative Permeability Ratio, the Oil Gravity, and the Solution Gas-Oil Ratio on the Primary Recovery for a Depletion Drive Reservoir", *Trans.*, AIME (1955) Vol. 204, 120–127.
50. Keller, W. O. and Morse, R. A.: "Some Examples of Fluid Flow Mechanism in Limestone Reservoirs", *Trans.*, AIME (1949) Vol. 186, 224–234.
51. Elkins, L. E.: *Drill. and Prod. Prac.* (1946) 160.

CHAPTER 7

Stimulation of Carbonate Reservoirs

A. R. HENDRICKSON

Introduction

The use of acid to stimulate or otherwise improve oil production from carbonate reservoirs was first theorized in 1895 by Herman Frasch, chief chemist in Standard Oil Company's Solar Refinery in Lima, Ohio. He and J. J. Van Dyke, general manager of the refinery, were issued patents covering the use of hydrochloric and sulfuric acid for this purpose and actually conducted several "well treatments" with acid. Although these treatments showed limited success, the process failed to arouse general interest because of the severe corrosion problems encountered. The next attempts to employ acid were made by subsidiaries of the Gulf Oil Company between 1926 and 1930. These attempts consisted of the use of inhibited hydrochloric acid by Gypsy Oil Company to remove carbonate scale from wells in the Glenpool Field in Oklahoma, and the use of raw hydrochloric acid by Eastern Gulf Oil Company to increase production from the "Corniferous" Formation in Lee County, Kentucky. Neither of these projects produced outstanding results, and work was abandoned.

In 1932 arsenic inhibitors were discovered, and The Dow Chemical Company and Pure Oil Company used them with hydrochloric acid to treat Pure's Fox No. 6 well in Isabella County, Michigan, which was producing from a limestone formation. Results of this treatment were outstanding. It was soon followed by several others in neighboring wells, which produced even more spectacular results, and the acidizing industry was born. Since that time, the amount of acid used in well treating has grown to more than 100 million gallons applied in tens of thousands of wells each year.

Over the years, many changes and innovations have been made in the acidizing process. These continue to improve the effectiveness of acid treatments for the stimulation of production and injection wells in both sandstone and carbonate formations. Treating techniques have changed considerably as better equipment and new technology have been developed. New additives to alter the chemical and physical characteristics of acid have evolved. With all these, the basic concept of acidizing still remains that of dissolving rock

References p. 337.

or flow obstructions in the well or reservoir. Of more importance than how much rock is dissolved are the factors of where the acid reaction occurs and how far it can penetrate from the wellbore itself. As a result of research and development advancements, the "art of acidizing" has become a scientific, practical technology.

Although stimulation of carbonate reservoir production is not confined exclusively to acidizing, this method is the primary one used. In this chapter, therefore, the technology and application of acidizing as it has been developed in present treating techniques are discussed. Most of the wells producing from carbonates have benefited or will benefit from an acid treatment at some point in their production life—as an aid in either stimulating or restoring flow capabilities.

A wide variety of acid formulations and additives is available for use with specific problems encountered. The interrelationship of the various factors involved can be a useful guide to optimum acidizing treatments. An indiscriminate use of any acid or additive can be wasteful and, under some conditions, even harmful. The design and application of a treatment, therefore, depend on a thorough understanding of individual well characteristics and production history, as well as of the acidizing materials and techniques.

Acidizing Chemicals and Reactions

In spite of the dozens of formulations available, only four major types of acid are used in well treatments: (1) hydrochloric, (2) hydrofluoric, and the organic acids (3) formic and (4) acetic. Other acids, such as chloroacetic, proprionic, and sulfamic, have had limited experimental and commercial application. Hydrochloric acid constitutes over 90% of all acid types used today; however, the amount of the organic acids being used has grown considerably, particularly in the deeper reservoirs.

Compounds such as acetic anhydride and allyl chloride have also been used. These materials, when pumped into the well, convert to acids which, in turn, can dissolve a portion of the carbonate formation.

There are two primary requirements for an acid to be acceptable: (1) it must react with the carbonates or other minerals and deposits to form soluble products; (2) it must be capable of being inhibited to prevent excessive reaction with metal parts of the well. Important, too, are safety in handling and reasonable basic costs. The four major types of acids, of course, meet all these requirements.

Types of Acid

Hydrochloric acid, the type most commonly used, is ordinarily supplied in concentrations of 32–36%. In well treatments, its normal strength has been

15% by weight. Use of higher concentration, however, has grown in popularity, particularly since the early 1960's. One thousand gallons of 28% hydrochloric acid will dissolve 3,670 lb (about 22 cu ft) of limestone ($CaCO_3$). Approximately 4,100 lb of $CaCl_2$ is produced in the reaction, along with 80 gal of water and 13,200 SCF of carbon dioxide. A 35% calcium chloride brine solution, saturated with carbon dioxide and commonly referred to as "spent acid," is produced as a result of this reaction. A similar reaction occurs with dolomite [$CaMg(CO_3)_2$]. Regardless of the acid strength used, the reaction is the same, dissolving equivalent amounts of carbonate rock. For example, 10,000 gal of a 3% acid solution will dissolve the same amount of rock as 1,000 gal of 28% acid. The main difference between the two solutions is their reaction rates or spending times and, of course, their physical volumes. Although the lower-strength acid solutions have greater equivalent volumes, their reaction times and the penetration into the reservoir from the wellbore are considerably less than those of the higher-strength solutions. Because of the economy and efficiency in using hydrochloric acid, most of the developments and innovations in acidizing technology since its commercialization have been based on this acid.

The two organic acids, formic and acetic, are used less frequently in acidizing carbonates. They are suitable primarily for wells with high temperatures or for conditions where prolonged reaction times are desired. Because of limited ionization ($K_i = 1.77 \times 10^{-4}$ for formic acid and $K_i = 1.75 \times 10^{-5}$ for acetic acid), their reaction rates with carbonates are considerably slower than that of the highly ionized hydrochloric acid (0.8 or more). The longer reaction times make these organic acids more readily adaptable to use in wells with high bottom-hole temperatures (above 250°F). Inhibition of formic and acetic acid corrosion of the metal parts of a well is much more effective than that of hydrochloric acid corrosion, especially at these high temperatures.

The normal strength of formic acid is limited to 9 or 10%. Above this concentration one of the reaction products, calcium formate, can precipitate from its "spent acid" because of its limited solubility.

Acetic acid solutions are normally limited to 15% or less because of the solubility limits of calcium acetate. Hydrochloric acid solutions can be blended with either formic or acetic acid to provide formulations having extended reaction times. Such mixtures are used to obtain more dissolving power per gallon of acid solution. The formic and acetic acids also can be blended together. Table 1 illustrates some of the more common acid strengths and blends.

Hydrofluoric acid is used in a blend with hydrochloric or formic acid in well treating. Such blends are primarily employed to react with silicate minerals in sandstone reservoirs to form soluble complex fluorosilicates. In

TABLE 1. Different Acidizing Solutions

Concentration, %, and type of acid	Pounds of $CaCO_3$ equivalent to 1,000 gal of acid	Relative reaction time*
7.5 HCl	890	0.7
15 HCl	1,840	1.0
28 HCl	3,670	6.0
36 HCl	4,860	12.0
10 Formic	910	5.0
10 Acetic	710	12.0
15 Acetic	1,065	18.0
7.5 Formic, 14 HCl	2,420	6.0
10 Acetic, 14 HCl	2,380	12.0
8 Formic, 14 Acetic	1,700	18.0

* Approximate time for acid reaction to be completed ("spent") to an equivalent strength of 1.5% HCl solution. Comparative values will vary depending on reaction conditions.

carbonates, their application must be carefully controlled because of possible precipitation of reaction products such as calcium fluorides and complex fluorosilicates, which have a very limited solubility. For reaction with silicates such as the clays in drilling muds or natural clays, the concentration blends are usually 2–10% hydrofluoric acid with 5–26% hydrochloric acid.

Hydrofluoric acid in trace quantities, about 0.25%, is used at times in hydrochloric acid for the treatment of dolomites. This small concentration speeds up the hydrochloric acid reaction with the dolomite. It has very little effect, however, in relatively pure limestones.

Acid Reaction Rates

The kinetics of reaction of an acid solution with limestone or dolomite are rather complex. The reaction may most closely follow first-order kinetics, that is, the rate is proportional to the concentration of one of the reacting substances. Actually, however, the reaction appears to be so sophisticated as not to follow a given kinetic order. Because of the many factors involved, first-order kinetics serve as a reasonable guide for correlating the various aspects of acidizing technology. The purpose of studies of the reaction itself has been to establish a better understanding of what governs acid spending time. Spending time can be useful in its relationship to acid penetration of the rock in establishing improved flow paths within the reservoir.

A knowledge of reaction factors, along with an understanding of the role of the reservoir properties, can form a guideline to optimum choice and evaluation of acid type and volume requirements. Such information aids in knowing how and where acid can be used most effectively. The major reaction factors are pressure, temperature, acid type, acid concentration, flow velocity, area/volume ratio, formation composition (chemical and physical), and reaction products.

These factors have been the subject of much reported research study. Details of these investigations are available in the published literature; therefore, only a brief general review of the findings is presented here.

Pressure

The reaction rate between the acid and carbonate rock decreases with increasing pressure (Fig. 1). The effect is most noticeable at the low-pressure ranges, where evolution of the CO_2 in gaseous form is only partially suppressed. At higher pressures (over 1,000 psi), the effect of pressure changes

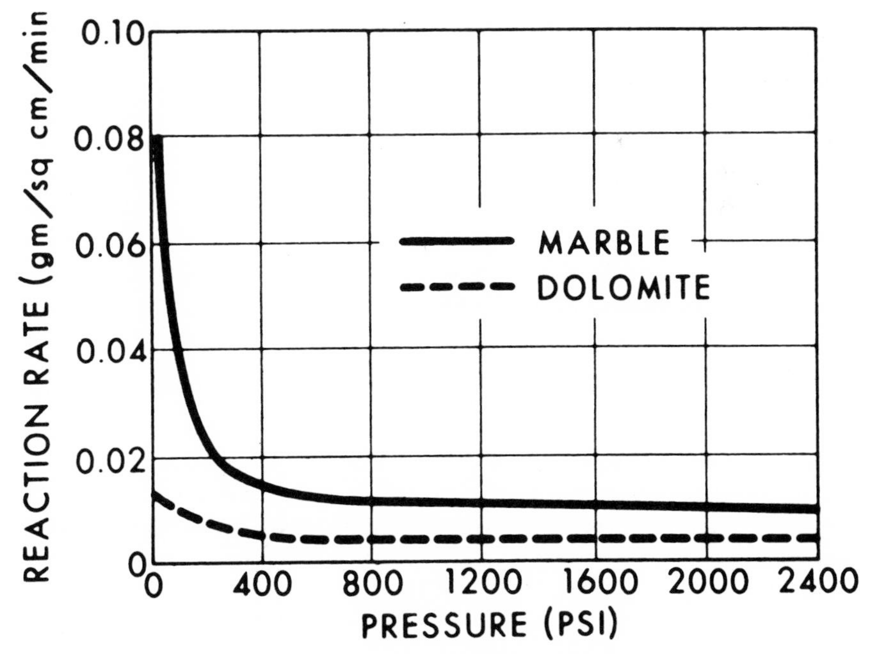

FIG. 1. Effect of pressure on reaction rate of hydrochloric acid. Increasing pressure up to 500 psi tends to decrease reaction rate. Above this pressure, little effect is noted. (Courtesy of Dowell Division of the Dow Chemical Co.)

is negligible and the CO_2 is maintained in solution or in a more liquid-like form. Although only a portion of the CO_2 reaction product is actually soluble in the spent (completely reacted) acid, the pressures used in treatment are normally sufficient to suppress the turbulent conditions caused by gas evolution. This turbulence, in turn, results in the increased reaction at low pressures.

It is also of interest to note that at lower pressures there is a great difference in acid reaction rate between limestones and dolomites (a level of 10 or so). At the normal treating pressures, however, a difference level of 1.5 or 2.0 has been observed. This can be attributed to the more turbulent effect of gaseous CO_2 on the acid reaction with limestone than on that with dolomite at the lower pressures. The turbulent effect at high pressures is much less.

Temperature

Acid reaction rate increases directly with temperature. A general effect is a rate at 140°–160°F double that at 70°–80°F. Likewise, a triple rate exists at 200°–230°F, etc. It is recognized that the actual temperature governing the acid reaction is affected by the injection temperature of the acid and the reservoir temperature (major factors) and by the heat liberated by reaction (minor factor).

Acid Concentration

The initial reaction rate of hydrochloric acid with carbonate rock is nearly proportional to its concentration up to 15–20%. Above 20%, however, the increase is less, reaching a maximum at 24–25% (Fig. 2). Above this level, the rate is actually decreased because of the reduced acid strength and the effect of the dissolved reaction products, that is, calcium or magnesium chlorides. (See the discussions of common ion effect and retardation.)

Flow Velocity

Increased flow velocity of acid along a carbonate rock surface acts to increase the reaction rate. The velocity effect is more pronounced in small-diameter channels or in the case of narrow fracture widths, as illustrated by the following equation:

$$R_r = [(28.5(v/W)^{0.8} + 184)] \times 10^{-6} \qquad (1)$$

where reaction rate R_r (lb/sq ft/sec) is for 15% HCl with $CaCO_3$ at 80°F and 1,100 psi, v = acid flow velocity (ft/sec), and W = fracture width (in.).

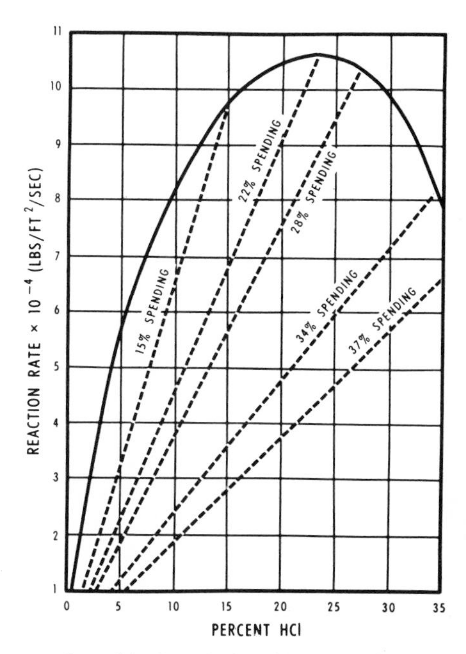

FIG. 2. Effect of concentration of hydrochloric acid on reaction and spending rates. Reactivity reaches a maximum at a concentration of 24–25%. (Courtesy of Dowell Division of the Dow Chemical Co.)

The flow velocity v in a treatment is related to the injection rate and flow geometry of the acid:

$$v = 17.2q_i/d^2 \qquad \text{(cylindrical channels)} \qquad (2)$$

$$= 1.15q_i/hW \qquad \text{(linear fractures)} \qquad (3)$$

$$= 0.18q_i/r_fW \qquad \text{(radial fractures)} \qquad (4)$$

where q_i = acid injection rate (bbl/min), d = channel diameter (in.), W = fracture width (in.), h = fracture height (ft), and r_f = fracture radius (ft).

Flow velocity is discussed further in the section on treatment design in regard to acid spending time and reservoir penetration.

Area/volume ratio

Area/volume ratio, that is, the reactive surface area of rock in contact with a given volume of acid, is inversely proportional to the pore channel diameter or fracture width. The importance of this factor is illustrated by the following wide range of ratios that may be encountered:

Condition		Ratio, sq ft/gal
Rock matrix:	10% porosity, 10-md permeability	28,000 : 1
Fracture:	0.001-in. width	3,200 : 1
	0.100 ,,	32 : 1
	0.250 ,,	13 : 1
Channel:	0.001-in. diameter	6,400 : 1
	0.100 ,,	64 : 1
	0.25 ,,	26 : 1
Wellbore:	2.0-in. diameter	3.2 : 1
	4.0 ,,	1.6 : 1
	6.0 ,,	1.1 : 1

With an extremely large surface area of rock matrix having small-diameter channels and narrow-width fractures, the reaction time of acid is very short, perhaps a matter of seconds. With wider fractures or channels, the reaction time is prolonged. Likewise, in a wellbore (open hole), acid will have an extended reaction time because of the limited surface for reaction, as would be the case in an open-hole acid-soak treatment.

Matrix acidizing treatments result only in limited penetration of the acid because of the high area/volume ratio. In hydraulic fracturing, with its wider fractures, the restricted area of acid contact allows greater penetration of the acid into the reservoir before reaction is complete.

Formation Composition

The most important factor that governs the effectiveness of an acidizing treatment is the rock or formation composition—its chemical and physical makeup. The reaction time and, therefore, the possible penetration are the key factors in the acidizing process; they determine how and where the acid will efficiently react with and dissolve the rock.

There are relatively small differences between the acid reactions of limestones and dolomites, as mentioned above. The subtle heterogeneity of chemical and physical properties, however, promotes a selective type of acid reaction within the formation. Pore-size distribution and pore shapes contribute to wide differences in acid responses from one well treatment to the next, often within a given producing interval. Thus, it is this reaction factor (rock composition, texture, and structure) that renders it impractical to make acidizing a pure science in application. The knowledge gained in understanding all these relationships and taking them into account, however, has set them apart as guidelines in treatment design and techniques.

Additives in Acidizing

Since the first use of arsenic as a corrosion inhibitor, a host of auxiliary additives has been developed to modify acidizing solutions. These materials are employed to initiate some physical and chemical changes in the treating acids. Most acids used have one or more of these additives in addition to the necessary inhibitor. The following is a brief description of the most common types of acid additives and their general purposes.

Corrosion Inhibitors

An acid inhibitor temporarily slows down the reaction rate of acid with metal. It has very little effect, however, on the acid reaction with the formation rock. The length of time that the inhibitor is effective depends on several factors: (1) temperature and pressure, (2) type and concentration of acid, (3) type and concentration of inhibitor, (4) type of metal, and (5) other acid additives or contaminants.

With rising temperature, which is a major factor, corrosion rate increases and protective time decreases. (Protective time is that period for which a metal loss of less than 0.02 lb per sq ft of metal exposure is maintained. Uninhibited acids generally have corrosion rates in excess of 1.0 lb of metal per sq ft under the same conditions.) Pressure has very little effect; an increase in pressure, however, generally causes a reduction in corrosion rates.

Inhibited hydrochloric acid generally has higher corrosion rates than inhibited organic acids. Increasing the concentration of any type of acid, in turn, reduces effective protection times.

There are two fundamental types of inhibitors, inorganic and organic. Although the inorganic compound arsenic oxide is still an effective inhibitor, the organic types are more commonly used today. They consist of acetylenic alcohols, nitrogen-based compounds, or blends of both along with detergents. Protective times of several days can be obtained at temperatures below 200°F. Above this temperature, a limited time of only a few hours is possible for hydrochloric acid. Organic acids, however, are capable of being inhibited for at least 24 hr up to 400°F.

The common oil-field steels are protected with essentially equal ease. Those of high strengths are generally more difficult to protect, however. The amount of inhibitor used ranges from 0.2 to 2.0% by weight and is dependent on environmental conditions and desired protective time.

Surfactants

Surfactants are used primarily to lower surface and interfacial tensions, as dispersants and as wetting agents. Nonionic, anionic, and cationic formulations are employed. The properties they impart to the acid solutions provide improved reactions and cleanup and return of the treating solutions. Concentration ranges of 0.01–0.05% by weight of the surfactants are common. Specific surfactants are also added to suppress asphaltic sludge precipitation from certain crude oils. Others are used to make the formation oil-wet and thus reduce the acid reaction with it.

Gelling and Fluid-Loss Additives

Natural gums and synthetic polymers are added to acid to provide a viscous nature to the acid solutions. The natural gums most commonly used are guar and karaya. Synthetic polymers are essentially of the linear, nonionic type. From less than 0.1% to as much as 2% by weight is used. At the lower concentration, a friction-controlling property is provided, particularly at high pump rates. At higher concentrations, the viscous nature imparted to the acid prevents excessive fluid loss into the smaller pores and flow channels of the formation. Some reduction in acid reaction rate with the rock may be provided by the gelling agents but is not very significant.

Alcohols

Methyl or isopropyl alcohols are occasionally used as additives to acid. When used at 5–20% by volume, their primary advantage is improvement in rate and degree of cleanup following the acid treatment. This has been particularly helpful in dry gas wells. High vapor pressures and low surface tensions are the two properties of alcoholic solutions that provide these advantages. Unlike surfactants, these alcohols do not become adsorbed on the rock surfaces. Thus, low surface tensions are maintained throughout the volume of acid.

Other additives used include sequestering agents and gyp inhibitors. They help to prevent detrimental side effects of the acid treatment, such as secondary precipitation or reprecipitation of calcium sulfate and iron compounds from spent acids. The use of N_2 and CO_2 in acid has also been found helpful in some treatments. These gases act primarily to provide more gaseous energy to aid in returning the spent acids to the surface in putting the well back on production. They have very little effect on the reaction of the acid.

Retarded Acids

In an effort to achieve deeper penetration of the formation by the acid action, several types of acid formulations have been used. By extending the penetration of reactive acid, the effectiveness of the acidizing treatment is often greatly improved. A more extensive increase in flow capacity throughout the drainage area of the well is possible.

A generally accepted definition of a "retarded acid" is as follows: "an acid which, because of its physical or chemical properties or its concentration, has a longer comparative reaction time than 15% hydrochloric acid under the same conditions." When forced into the matrix of most carbonates, even retarded acids have limited penetration because of the large surface areas for reaction. The effect of retardation is most dramatic in vugular pore structure and in wide fractures where the area/volume ratio is low. Four types of acid retardation are:

1. Hydrochloric acid formulations that introduce a barrier on the rock surface to prevent its normal contact and reaction with the acid are "retarded." Emulsions of acid in oil, gelling agent, or fluid-loss films, and oil-wetting agents are examples of barriers that promote a retarded reaction of acid.
2. An increase in acid concentration serves to extend the reaction time of the acid. This longer spending time, and deeper penetration, occur in spite of the greater reaction rate produced by the higher initial concentration of acid and the presence of common ions in the partially spent acid. These ions suppress the reaction rate more as their concentration increases. The acid reaction, thus, is not truly one of first-order kinetics.
3. A weakly ionized acid such as formic or acetic acid inherently reacts more slowly than hydrochloric acid solutions.
4. As shown earlier, blends of hydrochloric and organic acids are also used to provide the relatively fast reaction of the hydrochloric acid followed by the slow, retarded reaction of the organic acid.

Although retarded acids have longer reaction times than hydrochloric acid under the same conditions, within the matrix of carbonate rocks, their comparative reaction times may be quite different from those expected at first. The flow of acid within heterogeneous rocks such as carbonate matrices provides a constantly changing environment of flow patterns and reaction of the acids.

Acetic acid, with its slow reaction rate per square foot of surface area, actually can spend faster and penetrate less than the faster-reacting hydrochloric acid. The hydrochloric acid enlarges a selected number of flow paths or channels rapidly, and subsequent volumes of acid, therefore, may have less area for reaction. Thus, a longer spending time and greater penetration may result. The acetic acid does not react and change the pattern of flow as rapidly. Therefore, the large area/volume ratio of the smaller channels is maintained, and reaction time and penetration are limited. This is shown schematically in Fig. 3.

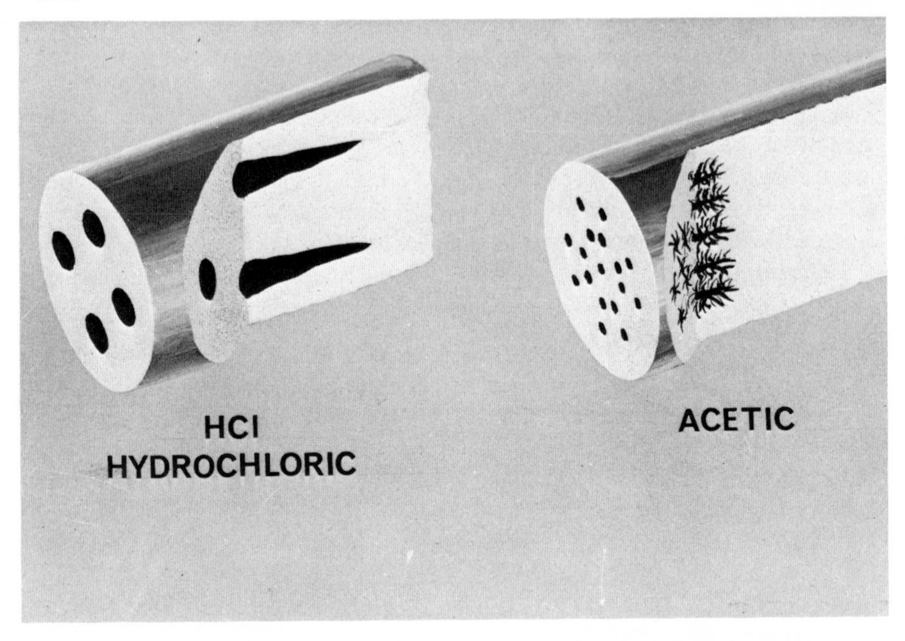

FIG. 3. Flow paths which might be expected from fast-reacting hydrochloric and slow-reacting acetic acids. (Courtesy of Dowell Division of the Dow Chemical Co.)

In a hydraulic fracture 0.05–0.50 in. in width or in a previously acidized channel system, the retarded acids will, in fact, penetrate more than hydrochloric acid. Acid-in-oil hydrochloric acid emulsions penetrate matrix and hydraulic flow paths to a greater degree than hydrochloric acid alone.

Testing Methods

No standards have been established in the study and evaluation of acids and their reactivities with formations. This nonconformity has led to some variations in interpretation and treatment designs. Expanded research programs and studies, however, have all contributed to a better understanding and greater usefulness of the acidizing processes.

The types of tests used in studying the acidizing reaction for treatment design can be placed in four categories: (1) static and dynamic methods, (2) high and low area/volume ratio conditions, (3) reaction rate, spending time, and spending distance, and (4) acid and formation compatibility testing.

Although many tests have been based on static (nonflow) or controlled-flow procedures in the past, recent trends are toward dynamic controlled-flow procedures. Recognition of the importance of fluid flow velocity on acid reactivity has apparently been responsible for this trend. In such a study realistic conditions can be used to duplicate more closely the actual occurrences in a treatment. Of course, complete duplication on a laboratory scale to account for all of the factors involved in acid treating is impractical. The overall purpose of testing is to establish comparative parameters and their importance in acidizing treatments. This correlation is used as a guide and should not be interpreted as providing absolute values.

Inasmuch as the area/volume ratio is so high in a rock matrix, actual core flow tests with acid more closely represent acidizing the rocks than tests on chunks or pieces of rock do. Of course, flow distribution in the rock itself is also of primary importance. Core flow–slot flow tests representing fracture environments have led to many new and improved concepts. The chemical and physical characteristics of the fluids, along with the hydraulic flow properties, play important roles in affecting the treatment results. A limit to the size of the test-model core or fracture lengths in the laboratory studies has prevented more exact duplication of reservoir conditions.

The relative reaction rate or spending-time tests and penetration calculations of acids are used as a basis for treatment recommendations. A spending-time test *per se* is quite limited in its value unless it is related in some way to effective penetration of the formation by the reactive acid solutions.

The following simplified equation can be used to approximate a theoretical acid spending time, t, in minutes:

$$t = 10{,}000\, C_x W/T \text{ or } 5{,}000\, C_x d/T \tag{5}$$

where C_x = relative spending time of the acid, with 1.0 as the basis for 15% HCl spent to 1.5% HCl; W = fracture widths into which acid is pumped (in.); d = channel or open-hole diameter into which acid is pumped (in.); and T = formation temperature (°F).

The distance the acid will penetrate before it is spent depends on the pump rate, the number of fractures or channels and their geometry, etc. Increased pump rates tend to raise the reaction rate and reduce the spending time, but result in greater penetration before the acid is spent. Thus, although spending may be shortened considerably at higher pump rates, the acid can penetrate effectively much deeper into the formation.

Often of equal importance are specific compatibility tests of an acid with the formation or well fluids. These considerations may lead to the use of additives necessary to minimize side effects and reactions that might detract from the acidizing process. Such aspects as emulsion tendencies and the secondary deposition of reaction products can be identified in designing a specific well treatment.

Regardless of the test methods and well data available, the data interpretation furnishes information for guidelines that is of only relative value. The final analysis should be made by careful control of treatments followed by a thorough evaluation of actual field results.

Acidizing Techniques and Treatment Design

In acidizing, as in any well treatment, success depends on correct analysis of the well problem so that proper materials and technique can be selected. Well problems may be classified into two categories: (1) damaged permeability, and (2) low natural permeability.

Wells with damaged permeability are candidates for matrix treatments. Those with low natural permeability may be improved by fracturing as well as, perhaps, a complementary matrix treatment. Treatment results may be carefully analyzed to determine how subsequent treatments can be improved or optimized with respect to cost versus results. The results should be evaluated over a long enough period of time to determine the true values and not just flush production results. These aspects of acidizing programs are a vital part of treatment design. They not only determine the type of treatment, materials, and technique, but also largely determine the overall results of the acidizing process.

The chemical and physical properties of the formation that must be considered include the following:

Chemical Properties	*Physical Properties*	
Composition	Permeability	Temperature
Acid solubility	Porosity	Hardness
Acid response	Thickness	Saturations
Mineral content	Pressure	Strength

The chemical composition of a given carbonate rock can require a specific type of acid or formulation. The physical properties determine other factors, such as type and size of treatment as well as technique to be used.

Solubility is the percentage of formation that is dissolved on contact with an acid. Solubility tests made on core chips or cuttings can be misleading when used as the only basis for selecting an acid. Acid responses in core flow tests with cores of the same solubility may vary considerably as a function of physical properties (porosity and permeability), pore-size distribution, and pore system geometry and type, such as natural fractures versus vugular porosity or more homogeneous, intergranular porosity. Thus, chemical and physical properties are closely related to the effect of an acidizing treatment.

In view of the infinite variation in formation composition, it is impractical to predict with certainty the precise response of a formation to an acid treatment. Most carbonate formations, however, can be successfully treated

—restoring, improving, or creating "new" permeability to improve the flow capacity of the well.

Acidizing Techniques

There are three fundamental techniques in acidizing: (1) *Wellbore Cleanup.* A fillup wash and soak of acid in the wellbore. Fluid movement is at a minimum unless some mechanical means of agitation or surging action is used. (2) *Matrix Acidizing.* Injection into the matrix structure of the formation below the hydraulic fracturing pressure. Flow occurs essentially through the natural fractures and pores. (3) *Acid Fracturing.* Injection into the formation above hydraulic fracturing pressure. Flow occurs essentially through hydraulic fractures. Much of the fluid, however, leaks off into the matrix along the fracture faces.

Wellbore Cleanup

The wellbore cleanup technique is used to remove carbonate and $CaSO_4$ deposits, corrosion products, mud, or other acid-soluble accumulations within the wellbore or perforations. Reference to this technique is often included when discussing matrix acidizing. Although the principal purposes are the same, major differences are evident. A wellbore treatment need only partially dissolve the damaging material; the remaining portion is sloughed or dispersed so that it may be washed from the well. The reaction time for the acid can be extensive (several hours) because of the limited area/volume ratio of the wellbore. In perforated intervals, only a very limited contact of acid with the perforation is available, as compared to open-hole completions.

Agitation of the acid, rather than soaking at static condition, speeds up the reaction and results in a more complete dissolution and/or dispersion of the damaging material. The volume requirements are normally restricted to one or just a few wellbore volume "fillups," depending on the extent of damage. For example, 50 gal of 28% hydrochloric acid will dissolve about 1 cu ft of carbonate scale. The displacement of multiple volumes into the formation is often made. It is best, however, to displace these up into the annulus, perhaps in a rocking action. This provides a form of agitation without forcing the reaction products or dispersed material into the "critical matrix," possibly leading to further damage within the formation itself.

Matrix Acidizing

Matrix acidizing (and wellbore cleanup treatments) may be selected as the proper technique for one or more of the following reasons: (1) to remove formation damage either natural or induced, (2) to achieve low-pressure breakdown of formation before fracturing, (3) to achieve uniform breakdown of all perforations, (4) to leave zone barriers intact, and (5) to reduce treating costs.

Sufficient natural permeability of the formation should be available to provide desired flow capacity in matrix acidizing. If the permeability is limited, increased productivity must be obtained through a fracturing treatment. The primary purpose of matrix acidizing treatments is to remove the plugging materials and to restore *natural* formation permeability. The type or sources of damage may include mud, saturation changes, cement, migration of fines, gyp ($CaSO_4$) deposits, and corrosion products.

An increase of three- to tenfold or more in flow capacity can be realized by removing the damage and restoring permeability. Creating an increased permeability above the natural level can provide stimulation to a limited degree. Although large increases in injectivity or productivity can be realized with matrix acidizing, these are a result of restoration, rather than extensive stimulation, of flow capacity. If sufficient natural permeability is not available from the drainage radius to the well, fracturing should be considered.

Acid Fracturing

The primary purpose of fracturing is to achieve injectivity or productivity beyond the natural reservoir capability. An effective fracture may create a new permeability path, interconnect existing permeability streaks, or break into an untapped portion of the reservoir. The success of any fracture treatment depends on two factors: fracture conductivity and extent of the effective penetration of the drainage area of the well.

Formations with low permeability require greater fracture penetration, but less fracture conductivity development, for a given level of stimulation. Formations with higher permeability require less penetration, but a high fracture conductivity. In some cases, a matrix treatment may be most economically efficient in formations with high natural permeability.

An optimum acidized fracture conductivity exists for an optimum penetration in any case. This, in turn, depends on and is interrelated with the well and reservoir properties, injection rate, and the type and volume of acid. The orientation and shape of the fracture are also important factors affecting treatment results. All these factors have been correlated by several companies into guides to fracture treatment design.

This brief description of techniques illustrates the theory of the relationship between various well problems and the possible designs of treatments to control them. It can be helpful as a comparative guide for the choice of these treatment techniques, along with correct interpretation of the well problem.

Inasmuch as the background information is never complete for a perfectly accurate analysis of a given well situation, acidizing is perhaps best described as a "scientific art." Thorough evaluation of experience factors and prior results may significantly contribute to successful results.

Acid Treatment Design

Matrix Acidizing Design

In restoring permeability, damage can be removed in either of two ways: by dissolving the damaging material itself, or by dissolving a portion of the formation in which the damage exists and thus releasing or dispersing the damage. In carbonates, the formation is usually dissolved, rather than just the material causing the damage. Where mud damage exists, dispersing the mud is more effective than dissolving it.

Hydrochloric acid is the normal type of acid used. Modifications or blends with other types to meet specific situations can be made. Hydrofluoric acid solutions can be used if extensive mud damage exists. Care should be taken, however, to prevent excessive reaction of the hydrofluoric acid with the carbonate formation, because this could cause the precipitation of calcium fluoride. As mentioned previously, acetic and formic acids are considered most suitable for wells with high temperatures (in excess of 250°F or 300°F) because of the relative ease of inhibition. These organic acids are also of value in lower-temperature wells when long contact with the metal parts of the well is anticipated.

In matrix acidizing, the volume of the acid is generally limited to less than 100–200 gal/ft of treating interval. As little as 10–25 gal/ft may be all that is required if good distribution of the acid is obtained and damage is shallow. If suspected damage is deeper than 5–10 ft, then larger volumes, a means of retarding the acid, or, perhaps, fracturing techniques may be required.

Actually, only a small amount of formation (perhaps less than 1%) needs to be dissolved to restore the initial condition by removing the damage or compensating for it. Only 50 gal of 28% hydrochloric acid per ft of interval height will dissolve 1% of carbonates within a 5-ft radius of a well. Yet, over 5,000 gal will be required to completely dissolve all of the carbonates within this radial distance. The penetration and the flow of acid in a carbonate rock are far from uniform. In almost every rock, especially carbonates, the pores are of different sizes and shapes. The pore structure may be in the form of vugs, hairline fissures, or tortuous capillary-like pores. Because of these heterogeneities, a "channeling" or "wormholing" occurs with almost any acid formulation. The resultant effect is the attainment of much greater acid penetration of the matrix than expected. Much less formation needs to be dissolved to achieve a given permeability change.

The following example illustrates the wide distribution of flow in a formation of varying pore diameters:

Diameter of Pore (μ)	Pore Volume (%)	Flow through Pores (% of total flow)
Less than 1	60	10
1 to 2	25	15
2 to 5	12	30
5 and above	3	45

Of course, variations in vugular and hairline fracture permeability also provide an even greater distribution of flow within a formation. The effect of acidizing is to further accentuate these flow distributions, which result in greater acid penetration before the acid is spent.

The extremely high area/volume ratios encountered in the matrix of carbonates limits the acid-spending times in spite of the selectivity of flow and channel enlargement. Extended penetration of the reservoir, therefore, is generally not achieved with reactive acid within matrix flow patterns.

Hydrochloric acid actually may penetrate the matrix to a greater degree than a retarded type of acid. The slow reaction of some retarded acids would not change the flow distribution rapidly. The result would be a more uniform acid reaction in the matrix for shorter distances, as opposed to an acid penetration into a smaller number of larger, longer channels within the rock matrix.

The pattern of acid flow and resulting enlarged channels or fracture enlargement assumes the shape of a tree root system. A few enlarged channels penetrate from the wellbore into the formation. Branching of these channels into a myriad of smaller-diameter ones occurs rapidly at increased distances. Because of this branching of radial flow patterns from the wellbore, exponentially greater amount of acid is required to increase the acid penetration that creates significant new formation flow capacity. Thus, only limited stimulation occurs with the matrix acidizing technique.

In summarizing the design of a matrix treatment for restoring flow capacity and providing limited stimulation, the following general conclusions can be reached:

Acid inhibitor selection must be based primarily on treating temperature and, to some extent, on the type of acid formulation.

Surfactant type and concentration should be selected to prevent emulsion tendencies and perhaps to aid in dispersing fine solids that are not dissolved. These may be mud or cement solids or natural material released from the formation. Suspension and removal of these materials can play an important part in the overall treatment results.

Diverting agents may be used to promote uniform penetration in long sections. Acid-swellable synthetic polymers, controlled-solubility particulate solids, balls, gel slugs, etc., have been applied successfully. Assuring the distribution of acid into the entire interval is a critical part of carrying out a matrix treatment. Without uniform distribution, large portions of the interval may get very little, if any, acid.

Preferably, injection rates should be controlled so that the formation is not fractured. The use of as high a rate as possible without exceeding the fracturing pressure, however, is recommended. In certain cases, it may be necessary to create a fracture to open perforations; pressure can then be reduced below the fracturing level to maintain essentially a matrix flow pattern.

Actually, injection pressure, and not rate, is the correct factor to consider. A control of bottom-hole pressures below hydraulic fracturing pressures may restrict injection rates to less than a barrel per minute, but may allow an increasing rate as the treatment progresses.

An overflush in the matrix acidizing treatment is recommended to assure placement of the acid efficiently into the matrix. A minimum shut-in time before returning the spent acid to the well is desirable. Inasmuch as the spending time of acid is short, a long shut-in time of several hours is not necessary, even for so-called retarded acids. Such practices may lead to an incomplete cleanup and a less effective treatment. The overflush fluid, which may be brine, water, oil, or a weak acid, should be used in sufficient volume to ensure maximum penetration of the last portion of the acid before it is spent.

Examples of situations in which various types of acids and formulations might be used beneficially in matrix acidizing include the following:

Organic Acids Such as Formic and Acetic

1. Formations with good natural productive capabilities where shallow damage has been incurred.
2. Formations with thick productive intervals.
3. Cleanup before fracturing, particularly in deep, high-temperature wells.

In these instances, the organic acids allow a prolonged inhibitor protective time and provide a more uniform permeability restoration over the entire pay section.

Hydrochloric Acid Solutions

1. Formations with good natural productive capabilities where damage is extensive.
2. Cases where some penetration is desired, but where weak zone boundaries might prevent fracture treatments.
3. Cleanup before the use of retarded acids such as emulsion formulations or hydrochloric-organic acid blends.
4. Extended penetration following organic acid treatment.

In these instances, deeper penetration may be accomplished with hydrochloric acid than with the organic acids. Penetration is not uniform over the entire face of the pay, and there is a pattern of deeper "channels" owing to the pore-size distribution. Such a pattern might provide enough productivity or might only provide clean channels for the more extensive use of emulsified acids or blends of different types of acids.

Acid-in-Oil Emulsions, Oil-Wetting Surfactant Acids, or Blends of Hydrochloric and Organic Acids

1. Formations with weak zone boundaries where maximum possible penetration of acid is necessary for stimulation.

In these instances, the acid formulations might be used to obtain maximum stimulation benefits with a matrix-type treatment. Such acids could be used as an initial treatment where little permeability damage had occurred, or as a second stage following one of the other acid types where extensive damage was suspected.

Other combinations of the various types of acids can be considered for other specific problems. In selecting an acid formulation for matrix acidizing, the solubility of the material to be removed and the reaction products must also be considered in conjunction with the factors previously discussed.

Matrix acidizing of carbonate formations, along with a thorough wellbore cleanup, is a treating technique that can be overlooked in the quest for large production increases. As a result, zone boundaries are frequently destroyed and high water/oil ratios encountered after fracture-type treatments. In many cases, with careful consideration of the reservoir problems in conjunction with a knowledge of the penetration patterns of various acids, a matrix treatment can be applied to achieve good productivity at a reduced cost and without the inherent danger of fracturing, or, in fact, to complement a program of fracturing.

Fracture Acidizing Design

When additional production capacity beyond the natural (undamaged) capacity is desired, a hydraulic fracturing treatment to provide deep penetration of the reservoir is required. If all the productive undamaged intervals of the formation are producing, the result of any fracture stimulation treatment depends on two primary factors: first, the degree to which the resultant fracture conductivity exceeds the natural formation flow capacity; and, second, the distance into the reservoir for which the increased conductivity is provided. Physical penetration without adequate fracture conductivity does not yield lasting results. Equally important, adequate fracture conductivity without effective penetration will provide only limited stimulation.

To obtain a two- to fourfold increase over natural production, extensive fracture penetration is required with a resulting fracture conductivity several times as great as that of the formation. The increase in well-flow capacity in reservoirs where formation damage exists can be many times greater than in formations without damage, as may be seen readily by comparison of Figs. 4 and 5. Partial restoration of the damaged permeability occurs in acid fracturing treatments, and/or a bypass of the damage is provided by the highly conductive fracture. This can often more than compensate for damage restrictions in the vicinity of the wellbore.

A correlation of the various factors that can be considered in designing optimum treatments takes the form of a guide. The physical properties of the formation, the reservoir characteristics, and the properties of the acid play an important role in design. Such a guide in design can consist of three fundamental considerations:

A. Achieve an effective hydraulic fracture area.
   Parameters include (1) stimulation desired, (2) fracture geometry, (3) fluid-loss efficiency, (4) pump rate, and (5) fluid volume.

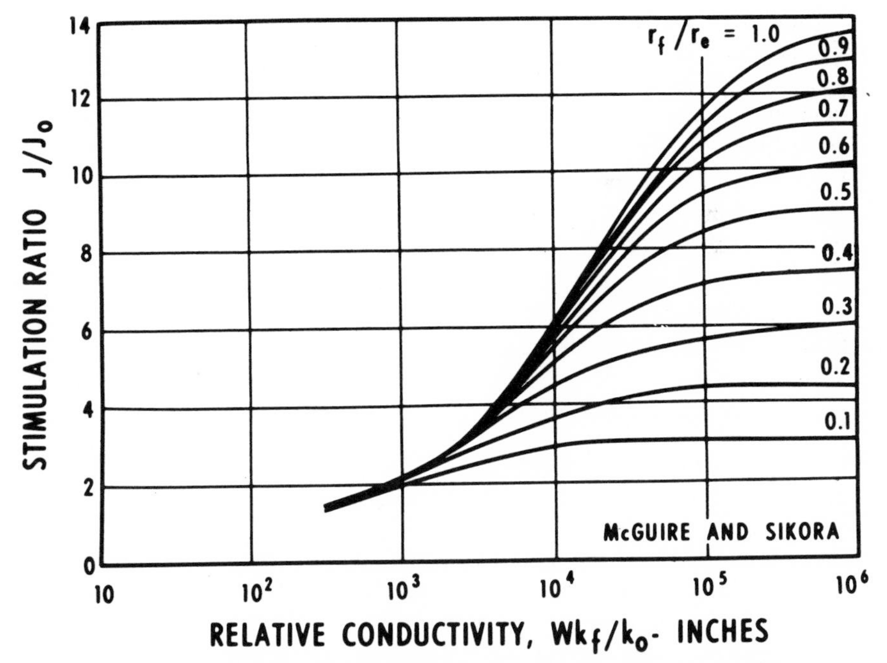

FIG. 4. Stimulation ratios for undamaged wells. J=productivity index of well after stimulation, J_o=productivity index before stimulation, r_f=radius of fracture (ft), r_e=drainage radius (ft), k_f=permeability of fracture (md), k_o=permeability of formation (md), and W=fracture width (in.). (After McGuire and Sikora,[16] courtesy of AIME.)

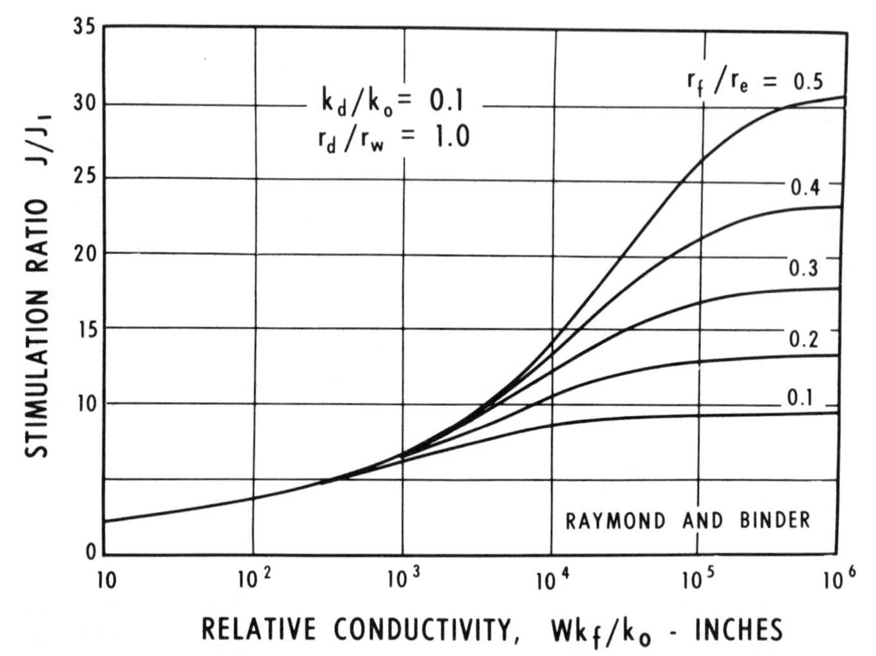

FIG. 5. Stimulation ratios for damaged wells. J_1 = productivity index before stimulation (B/D/psi-ft), k_d = permeability of damaged zone (md), k_o = permeability of formation (md), r_d = radius of damaged zone (ft), and r_w = radius of wellbore (ft). (After Raymond and Binder,[51] courtesy of AIME.)

B. Be sure the acid penetrates effectively.
Parameters include (1) fracture geometry, (2) effective pump rate, (3) formation properties, and (4) type of acid.

C. Create adequate fracture conductivity.
Parameters include (1) fracture geometry, (2) effective pump rate, (3) formation properties, (4) type of acid, and (5) volume of acid.

A. *Fracture area* is directly proportional to pump rate and inversely proportional to the acid fluid efficiency—the leakoff rate of fluid from the fracture faces into the matrix of the rock. This efficiency may vary from less than 10% to 30% or more, and is defined as "the percent of injected fluid that remains in the fracture." It is a function of the fluid properties, reservoir characteristics, and hydraulic pressures. The additives available to improve the efficiency restrict the rate of leakoff, increase the viscosity of the fracturing acids, or impart a fluid-loss film on the fracture faces. This film or cake acts to restrict leakoff.

"Effective fracture area" refers to the portion of the treating interval that the total fracture area covers. This coverage of the production interval depends on the orientation and shape of the fracture plane with respect to the thickness of the interval. Most of the hydraulic fractures are considered to be vertically oriented, intersecting the wellbore appproximately parallel to its axis. Hydraulic fracturing gradients of 1.0 psi/ft or less are commonly encountered, although some as high as 1.5 psi/ft occur in rocks with abnormal tectonic stresses. In any case, however, the plane of the fracture is believed to be parallel to the weakest stress and is often confined by rock barriers of greater strength or stress than that at the points of fracture initiation.

In horizontal fractures, more generally perpendicular to the wellbore axis, where overburden stresses are weakest, the fracture plane tends to take on a radial or pancake-like shape. With vertical fracture planes, the confinement of upper or lower formations can restrict fracture growth and it may take on a more linear pattern. It is thought, however, that all fractures have a somewhat radial shape initially.

B. The fundamental difference between fracturing treatments with acids and those with non-acid fluids, of course, involves the rock-dissolving aspects of the acid fluids. The acid not only increases the flow capacity of the fracture but also provides a matrix acidizing treatment of the rock along the fracture faces. In carbonates, this difference in reactivity accounts for the success of acid fracturing stimulation. The factors that govern the reaction rate of acid with the carbonate surfaces of the fracture determine the distance the acid will penetrate the reservoir before being spent. The *spending distance* is related to the reaction time, the pump rate, and the shape and width of the fracture as well as to the formation properties. An example of this relationship is given in the series of equations for single-phase hydraulic fracture systems. The radial system has slightly different flow and velocity effects from the vertical one, but the above-mentioned relationship is essentially the same.

In a symmetrical, linear fracture:

$$L_a = 33.5 q_i t / hW \tag{6}$$

where L_a = acid spending distance (ft), and

$$R_c = 100[1 - \exp\{-[1 + (q_i/h)^{0.8}/10W^{1.6}]1.92 \times 10^{-2} C_A t/W\}] \tag{7}$$

where R_c = percentage of the acid reaction completed.

In a symmetrical, radial fracture:

$$L_a = 4.6(q_i/W)^{1/2} \tag{8}$$

and

$$R_c = 100[1 - \exp\{-[1 + q_i^{0.4}/52W^{1.6}t^{0.4}]1.92 \times 10^{-2} C_A t/W\}]. \tag{9}$$

The acid coefficient, C_A, is defined as the relative reactivity of the acid:

$$C_A = S(T - 10)/70C_x \tag{10}$$

These reaction time equations are similar to Eq. 5, except for the effect of flow velocity on the reaction rate (Figs. 6 and 7). At low pump rates or wide fracture widths, this effect is relatively small. Because of the acid fluid loss, the efficiency of the treatment should be taken into account in determining the pump rate values to be used. For example, if the actual pump rate is 10 BPM and the efficiency is 25%, the q_i in these calculations should be only 2.5 BPM. This will account for the overall lower flow velocities of the acid owing to the leakoff of the acid from the fracture.

The penetration of acid before it is spent increases at higher injection rates and in wider fractures (Fig. 8). Yet, it is of interest to note that the spending times actually are much shorter at the higher injection rates than at the lower ones (Fig. 9). Spending time, therefore, is not a direct guide to acid treatment design. Some degree of retardation may be necessary to obtain the desired penetration of the acid before it is spent. To ensure the efficient use of acid, an overflush and a minimum shut-in time are often employed. The correct amount of overflush (nonacid fluid) depends on the spending time of the acid.

A set of equations similar to Eqs. 6–10 can be used for cylindrical channel

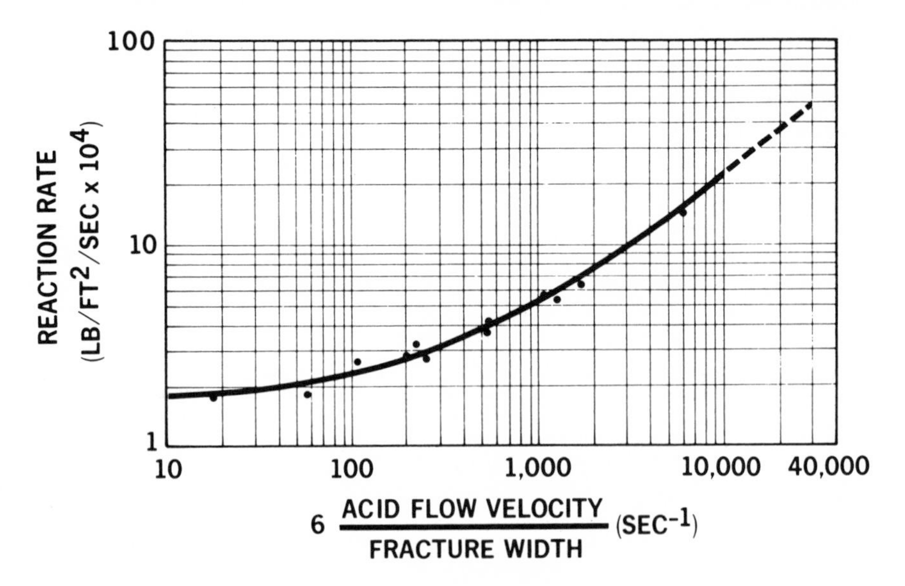

FIG. 6. Effect of acid flow velocity on reaction rate. Shear rate ($G = v/W$) plotted on abscissa is multiplied by 6. (Courtesy of Dowell Division of the Dow Chemical Co.)

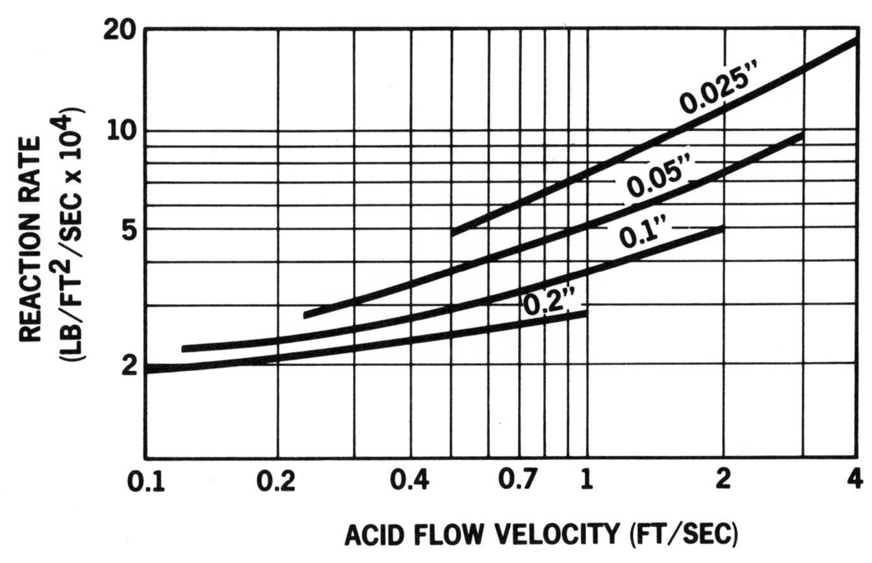

FIG. 7. Effect of acid flow velocity on reaction rate at constant fracture width. (Courtesy of Dowell Division of the Dow Chemical Co.)

flow. However, the number of channels involved in matrix acidizing and the wide variation in their diameters, along with the myriad of branching patterns, make these expressions of limited value as compared to those used in single hydraulic fracture calculations.

The actual fracture width is not a constant. It varies, depending on pump rate, fluid properties, and rock properties, as well as the amount of acid reaction. The use of an approximate average value for width, however, can provide an estimate of acid reaction time and acid penetration. With these calculations of the acid concentration and spending distance, the acid volume required to obtain the approximate fracture conductivity necessary can be determined.

C. Unlike hydraulic fracturing treatments with nonacid fluids, acid fracturing of carbonates does not necessarily require the use of propping agents. The acidizing fracture conductivity of a carbonate rock is provided by the uneven acid reaction pattern with the fracture faces. This *etching characteristic* will normally roughen the surfaces; this, in turn, prevents the fracture from fully closing after the pressures of the treatment have been released. The etching characteristics of a carbonate rock are not predictable in an absolute way. Even the polished surfaces of pure carbonates in laboratory flow models exhibit an uneven etching pattern, probably caused by subtle variations of the acid flow across the surfaces. Physical and chemical

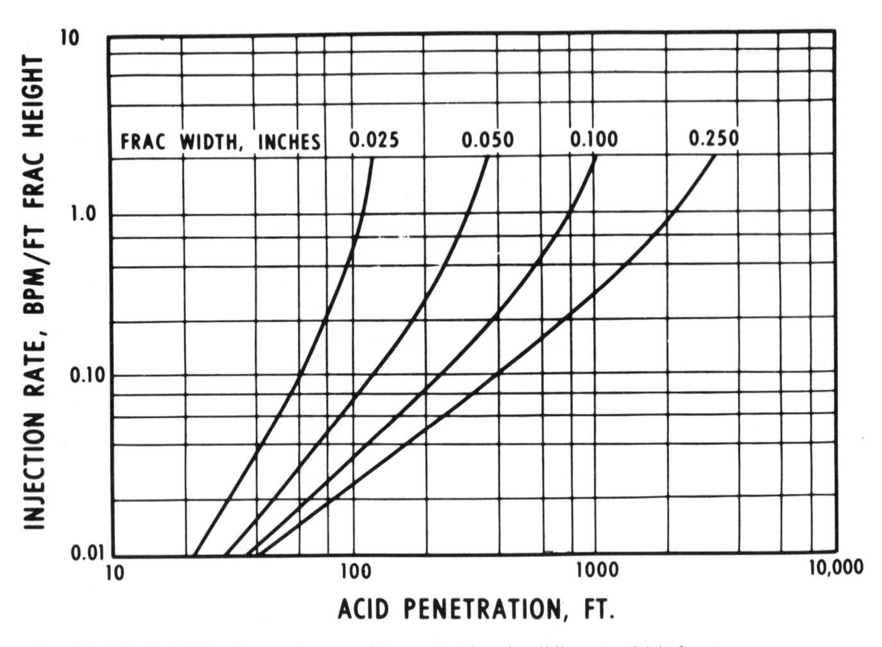

FIG. 8. Effect of injection rate on acid penetration in different width fractures. (Courtesy of Dowell Division of the Dow Chemical Co.)

heterogeneities of fracture faces in a treatment add to the variations of acid flow effects and result in a rougher etch or reaction pattern. Oil-wetting surfactants and oil blends with acids have been used in attempts to exaggerate this etching. If enough acid reaction takes place, a highly conductive fracture flow capacity is created. The portion of acid that has leaked off from the fracture undoubtedly can contribute indirectly to this flow capacity. Its actual contribution, however, cannot be calculated. The effect of acid reaction and equivalent fracture conductivity can be estimated by using the following equations:

In a symmetrical linear fracture for 1.5% hydrochloric acid or its equivalent:

$$e_r = [15.2(q_i/hW^2)^{0.8} + 156]C_A \times 10^{-6} \tag{11}$$

where e_r = etching rate (in./min).

In a symmetrical radial fracture for 1.5% hydrochloric acid or its equivalent:

$$e_r = [6.1(q_i/r_f W^2)^{0.8} + 156]C_A \times 10^{-6}. \tag{12}$$

Fracture conductivity (md-ft) is defined as:

$$k_f W = 4.53 \times 10^9 W^3 \tag{13}$$

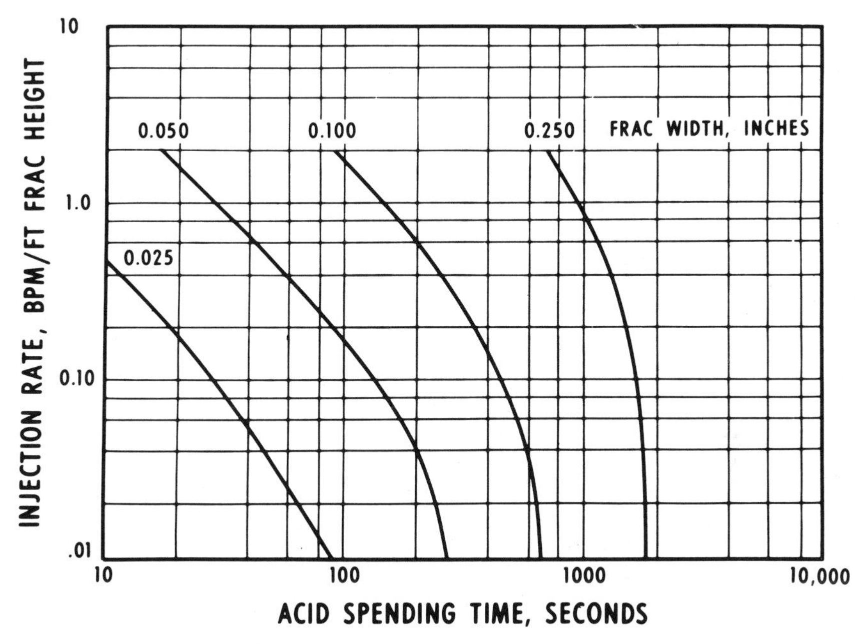

FIG. 9. Effect of injection rate on acid spending time in different width fractures. (Courtesy of Dowell Division of the Dow Chemical Co.)

Thus, when the etching rate of the acid solution in terms of inches of fracture face dissolved per minute and the desired width enlargement equivalent to the desired fracture conductivity in terms of inches are known, the required etching time can be calculated. Etching time is the time that the fracture should be exposed to the acid under the given conditions to dissolve the necessary volume of carbonate and to provide the calculated, equivalent fracture conductivity. The acid volume requirement is calculated by multiplying this time by the pump rate of the acid.

It is of interest to note that retarded acids may provide deeper penetration of unspent acid than unretarded acids. If there is too much retardation, however, the acid reaction may be too slow to develop a sufficient fracture conductivity.

Combination Fracture-Acid Treatments

Massive carbonate rocks form one type of reservoir that until recently had practically defied effective stimulation. These sections generally are deep and extremely thick, and are characterized by low permeability, high pressures, and high temperatures. Hydraulic fracturing of such sections is extremely hazardous, because a screen-out of propping agent could result in a junked hole and loss of a large investment. Acidizing is prohibitively expensive owing to the tremendous volumes of acid required. The solution to effective stimulation of massive carbonates has proved to be a combination fracture-acid treatment.

This combination treatment is basically a fracturing procedure. The primary difference is that acid is used to provide conductivity by etching of the fracture faces. Conventional propping agents are not employed. Essentially, a low-cost fracture fluid is used to achieve the desired fracture area and penetration of the fracture into the reservoir. This is followed by the volume of acid required to etch the fracture faces at the end of the fracture and thus provide the necessary conductivity. Overflush is applied to assure contact of all unspent acid with these fracture faces.

With the combination treatment, much less acid is required and its penetration before spending is greater because of reduced leakoff from the fracture. A greater volume of acid reaches more area. In addition, a greater amount of etching of the fracture faces will result with higher conductivity than would occur with greater leakoff.

In fracturing treatments, as the volume of fluid is raised, an increase in fracture area results. The rate at which the area increases, however, becomes less. Nevertheless, each succeeding increment of fluid becomes more efficient owing to the decreased leakoff in the area created by previously injected volumes. This, in effect, increases the effective injection rates, and the acid will penetrate deeper into the fracture plane before it is spent.

A further benefit of this combination-of-fluids technique is the cooling effect of the initial volume of nonacid fluids. Possible corrosion by acid is minimized and acid reaction with the formation is reduced, allowing deeper penetration of the live acid.

The combination treatment, which can provide predictable production increases of two- to fourfold, is applicable to all carbonate reservoirs. It has been applied mainly to the massive carbonates because it is the only method of obtaining such increases economically in these reservoirs.

Summary and Conclusions

Stimulation of carbonate reservoirs is achieved chiefly by acidizing treatments. Acids may be injected at matrix or hydraulic fracturing rates, depending on the well problem and the results desired. Regardless of the technique used, the purpose of the acid is to increase permeability by dissolving either the reservoir rock or materials causing damage within the rock.

Hydrochloric is the principal acid used, but several other types fulfill requirements for specific well or reservoir conditions. Many additives are available to change acid characteristics to meet specific requirements. Selection of acid type and volume must be based on a knowledge of reservoir rock and fluids characteristics, reservoir conditions, well conditions, desired results, and acid properties. Acid solutions can be tailor-made to meet almost any set of such conditions.

Whereas laboratory tests cannot duplicate down-hole conditions exactly, they have provided data of sufficient accuracy for engineered acidizing treatment design. Carefully planned and designed treatments, based on these guidelines, can provide predictable stimulation results in nearly all carbonate reservoirs.

Nomenclature

C = fracture fluid coefficient, ft/min$^{\frac{1}{2}}$.
C_A = acid reaction coefficient, dimensionless ratio.
C_x = relative spending time of an acid, compared to 1.0 as a basis for 15% HCl spent to 1.5% HCl under the same conditions, dimensionless.
d = channel or wellbore diameter, in.
e_r = etching rate, in./min.
h = fracture height, ft.
$k_f W$ = fracture conductivity, md-ft.
L_a = acid spending distance, ft.
q_i = injection or pump rate, BPM.
r_e = theoretical radial drainage of well, ft.
r_f = radial fracture radius, ft.
R_c = percentage of the acid reaction completed.
R_r = reaction rate, lb/sq ft/sec.
S = acid solubility of the formation rock, fractional.
T = formation temperature, °F.
t = spending time, min.
v = fluid flow velocity, ft/sec.
W = fracture width, in.

References and Bibliography

1. Love, F. H.: "Acid Treatment of Wells Is Finding Wide Application", *Pet. Eng.* (Oct., 1932) Vol. 32.
2. King, C. V.: "Reaction Rates at Solid-Liquid Interfaces", *J. Am. Chem. Soc.* (1935) Vol. 57, 828.
3. Chamberlain, L. C. and Boyer, R. F.: "Acid Solvents for Oil Wells", *Ind. Eng. Chem.* (Apr., 1939) Vol. 31, No. 4, 400.
4. Stone, J. B. and Hefley, D. G.: "Basic Principles in Acid Treating Lime and Dolomites", *Oil Weekly* (Nov. 11, 1940).
5. Morrison, P. W.: "Acidizing Fundamentals", *J. Pet. Tech.* (Sept., 1950) Vol. 1, 10.

6. Clason, C. E. and Hower, W. F.: "Removal of Flow Restrictions in Well Completion", *API* No. 851-27-C (Mar., 1953).
7. Clason, C. E.: "Acid Kinetics in Limestone Formations", *Pet. Eng.* (Mar., 1955) B-111.
8. Waldschmidt, W. A.: "Classification of Porosity and Fractures in Reservoir Rocks", *Bull.*, AAPG (May, 1956) Vol. 40, No. 5, 953.
9. Howard, G. C. and Fast, C. R.: "Optimum Fluid Characteristics for Fracture Extension", *Drill. and Prod. Prac.* (1957) 261–270.
10. van Poollen, H. K.: "Productivity vs Permeability Damage in Hydraulically Produced Fractures", *API* No. 906-2-G (1957) 103.
11. Hubbert, M. K. and Willis, D. G.: "Mechanics of Hydraulic Fracturing", *J. Pet. Tech.* (June, 1957) Vol. 9, 153–168; *Trans.*, AIME, Vol. 210, 153–168.
12. Rowan, G.: "Theory of Acid Treatments of Limestone Formations", *J. Inst. Pet.* (1959) Vol. 45, 431.
13. Pollard, P.: "Evaluation of Acid Treatments from Pressure Build-up Analysis", *Trans.*, AIME (1959) Vol. 216, 38–43.
14. Hendrickson, A. R., Hurst, R. E. and Wieland, D. R.: "Engineered Guide for Planning Acidizing Treatments Based on Specific Reservoir Characteristics", *Trans.*, AIME (1960) Vol. 219, 16–23.
15. Dunlap, P. M. and Hegwer, J. S.: "An Improved Acid for Calcium Sulfate-Bearing Formations", *Trans.*, AIME (1960) Vol. 219, 337–340.
16. McGuire, W. J. and Sikora, V. J.: "The Effect of Vertical Fractures on Well Productivity", *Trans.*, AIME (1960) Vol. 219, 401–403.
17. Hendrickson, A. R., Rosene, R. B. and Wieland, D. R.: "The Role of Acid Reaction Rates in Planning Acidizing Treatments", *Trans.*, AIME (1961) Vol. 222, 308.
18. Fast, C. R., Flickinger, D. H. and Howard, G. C.: "Effect of Fracture-Formation Flow Capacity Contrast on Well Productivity", *Drill. and Prod. Prac.* (1961) 145–151.
19. Perkins, T. K. and Kern, L. R.: "Width of Hydraulic Fractures", *J. Pet. Tech.* (Sept., 1961) Vol. 13, 937–949.
20. Barron, A. N., Hendrickson, A. R. and Wieland, D. R.: "The Effect of Flow on Acid Reactivity in a Carbonate Fracture", *J. Pet. Tech.* (Apr., 1962) Vol. 14, 409–415.
21. Lasater, R. M.: "Kinetic Studies of the HCl-$CaCO_3$ Reaction", *Am. Chem. Soc. Pet. Div.* (Dec., 1962).
22. Grubb, W. E. and Martin, F. G.: "A Guide to Chemical Well Treatment", *Pet. Eng. Reprint Series* (1963).
23. Crawford, H. R., Neill, G. H., Bucy, B. J. and Crawford, P. B.: "Carbon Dioxide—a Multipurpose Additive for Effective Well Stimulation", *J. Pet. Tech.* (Mar., 1963) Vol. 15, No. 3, 237–242.
24. Foshee, W. C. and Hurst, R. E.: "Improvement of Well Stimulation Fluids by Inclusion of a Gaseous Phase", *J. Pet. Tech.* (July, 1965) 768.
25. Knox, J. A., Lasater, R. M. and Dill, W. R.: "A New Concept in Acidizing Utilizing Chemical Retardation", *SPE* T.P. 975 (Oct., 1964).
26. Horton, H. L., Hendrickson, A. R. and Crowe, C. W.: "Matrix Acidizing of Limestone Reservoirs", *API* No. 906-10-F (Mar., 1965).
27. Broddus, E. C. and Knox, J. A.: "Influence of Acid Type and Quantity in Limestone Etching", *API* No. 851-39-1 (Mar., 1965).
28. Davis, J. J., Mancillas, G. and Melnyk, J. D.: "Improved Acid Treatments by Use of the Spearhead Film Techniques", *SPE* T.P. 1164 (June, 1965).
29. Anonymous: "New 'Beads' Help Acidizing, Fracturing", *Oil and Gas J.* (Aug. 30, 1965) 52–54.
30. Sewell, F. D.: "Hydrogen Embrittlement Challenges Tubular Goods Performance", *SPE* T.P. 1399 (Sept., 1965).
31. Moore, E. W., Crowe, C. W. and Hendrickson, A. R.: "Formation, Effect and Prevention of Asphaltene Sludges During Stimulation Treatments", *J. Pet. Tech.* (Sept., 1965) Vol. 17, 1023–1028.

32. Smith, J. E.: "Design of Hydraulic Fracture Treatments", *SPE* T.P. 1286 (Oct., 1965).
33. Anonymous: "Atomized Acid Treats Problem Wells", *Oil and Gas J.* (June 6, 1966) 134.
34. McLeod, H. O., McGinty, J. E. and Smith, C. F.: "Deep Well Stimulation with Alcoholic Acid", *SPE* T.P. 1558 (Oct., 1966).
35. Harris, O. E., Hendrickson, A. R. and Coulter, A. W.: "High-Concentration Hydrochloric Acid Aids Stimulation Results in Carbonate Formations", *J. Pet. Tech.* (Oct., 1966) Vol. 18, No. 10, 1291–1296.
36. Smith, C. L. and Ritter, J. E.: "Engineered Formation Cleaning", *API* No. 801-43-B (May, 1967).
37. van Poollen, H. K.: "How Acids Work in Stimulating Production and Injection Wells", *Oil and Gas J.* (Sept. 11, 1967).
38. van Poollen, H. K.: "How Acids Behave in Solution", *Oil and Gas J.* (Sept. 25, 1967).
39. Williams, G. F.: "Formation Cleaner Increases Stimulation Results", *Pet. Eng.* (Oct., 1967).
40. Parameter, C. B.: "Prescriptions Can Make Healthy Wells", *Can. Pet.* (Dec., 1967).
41. Scott, J.: "Fracture-Acid Shows Promise in Tight Carbonates", *Pet. Eng.* (Feb., 1968).
42. Crenshaw, P. L. and Flippen, F. F.: "Stimulation of the Deep Ellenburger in the Delaware Basin", *J. Pet. Tech.* (Dec., 1968) Vol. 20, No. 12, 1361–1370.
43. Hubbard, M. G.: "Atomization of Treating Fluids with Nitrogen", *API* No. 906-13-1 (Mar., 1968).
44. Hendrickson, A. R. and Cameron, R. C.: "New Fracture-Acid Technique Provides Efficient Stimulation of Massive Carbonate Sections", *CIM* T.P. (May, 1968).
45. Coulter, A. W. and Claiborne, T. S.: "Stress Corrosion Cracking of Oilfield Tubing in Aqueous Hydrochloric Acid", *Mater. Protect.* (June, 1968).
46. Gallus, J. P. and Dye, D. S.: "Deformable Diverting Agent Improves Well Stimulation", *SPE* T.P. 2161 (Sept., 1968).
47. Crenshaw, P. L., Flippen, F. F. and Pauley P. O.: "Stimulation Treatment Design for the Delaware Basin Ellenburger", *SPE* T.P. 2375 (Sept., 1968).
48. McLeod, H. O. and Coulter, A. W.: "The Stimulation Treatment Pressure Record—an Overlooked Formation Evaluation Tool", *SPE* T.P. 2287 (Sept., 1968).
49. van Poollen, H. K. and Jargon, J. R.: "How Conditions Effect Reaction Rate of Well-Treating Acid", *Oil and Gas J.* (Oct. 21, 1968) 84–91.
50. Smith, C. F., Crowe, C. W. and Nolan III, T. J.: "Secondary Deposition of Iron Compounds Following Acidizing Treatments", *SPE* T.P. 2358 (Nov., 1968).
51. Raymond, L. R. and Binder, G. G. Jr.: "Productivity of Wells in Vertically Fractured, Damaged Formations", *J. Pet. Tech.* (Jan., 1967) Vol. 19, No. 1, 120–130.

APPENDIX A

Application of Petrography and Statistics to the Study of Some Petrophysical Properties of Carbonate Reservoir Rocks

HERMAN H. RIEKE, III, GEORGE V. CHILINGAR, AND ROBERT W. MANNON

Appendix A describes several methods by which some of the petrophysical properties of porous carbonate rocks—porosity, permeability, mean pore diameter, etc.—can be determined. Because of the complex nature of the porous systems in most carbonates, statistical measures are employed to describe the size and distribution of pores in carbonate rocks.

Determination of Permeability from Pore-Space Characteristics of Carbonate Rocks

Teodorovich's petrographic method of relating the pore-space characteristics of carbonate rocks to permeability was investigated by Aschenbrenner and Chilingar in 1960.[1] This method is used to calculate permeabilities when they cannot be measured directly in the laboratory. According to Teodorovich,[2] the permeabilities of limestones and dolomites can be computed effectively by assigning numerical values to a series of empirical coefficients and then multiplying the coefficients together to obtain the permeability in millidarcys. The principal factors which determine the permeability of carbonate reservoir rocks are as follows: (A) pore-space type and pore space subtype (Table 1; also see p. 9); (B) effective porosity (Table 2); (C) average size of the pores (Table 3); (D_1) elongation of the pores (Table 4); (D_2) type and amount of cement; (D_3) degree of aggregation of the pores; and (E) smoothness of the wall surface in the pores and in the interconnections between the pores.

The influence of these seven factors on permeability was expressed by Teodorovich as follows:

$$k = A \times B \times C \times D_1 \times D_2 \times D_3 \times E. \quad (1)$$

Teodorovich has indicated, however, that Eq. 1 can be reduced to

$$k = A \times B \times C \times D_1 \quad (2)$$

where k is the permeability (md), and the variables A through D_1 are represented by empirically derived numbers. These empirical coefficients are

References p. 354.

TABLE 1. Empirical Coefficient A for Pore-Space Type

Pore-space type	Characteristic of subtype (as seen in thin sections)	Empirical coefficient A
I	With very narrow conveying canals (avg. diameter $\approx$ 0.01 mm), usually not visible in thin section under the petrographic microscope using normal range of magnification	2
	With rare relatively wide canals (avg. diameter $\approx$ 0.02 mm) visible in thin sections	4
	With few relatively wide canals, visible in thin section	8
	With many relatively wide canals, visible in thin section, or with few wide canals (avg. diameter $\geqslant$ 0.04 mm)	16
	With abundant wide canals, or few to many very wide conveying canals	32
II	With poor porosity, the pores being relatively homogeneous in size and distribution	8
	With good porosity and (or) porosity ranging from poor to good:	
	pores being of different size,	16–32
	pores being vuggy and irregular in outline	32–64
III	With very poor porosity inside the conveying canals	6
	With poor porosity inside the conveying canals	12
	With conveying canals finely porous	24
IV	With interconnected pore space between rhombohedral grains	10
	With interconnected pore space between subangular-subrounded rhombohedral grains	20
	With interconnected pore space between rounded to well rounded grains	30

TABLE 2. Empirical Coefficient B for Porosity

Effective porosity		Empirical coefficient B
Descriptive term	Limits (%)	
Very porous	>25	25–30
Porous	15–25	17
Moderately porous	10–15	10
Pores abundant	5–10	2–5
Pores present	2–5	0.5–1.0
Some pores present	<2	0

TABLE 3. Empirical Coefficient C for Pore Size

Descriptive term	Maximum size of pore (mm)	Empirical coefficient C
Large vugs	>2	16
Medium to large vugs	0.5 –2.0	4
Medium pores	0.25–1.0	2
Fine to medium pores	0.1 –0.5	1
Very fine to fine pores	0.05–0.25	0.5
Very fine pores	0.01–0.1	0.25
Pinpoint to very fine pores	<0.1 and in part <0.01	0.125
Mostly pinpoint porosity	<0.03 and in part <0.01	0.0625

TABLE 4. Empirical Coefficient D for Pore Shape

Descriptive term	Empirical coefficient D
More or less isometric pores	1
Elongate pores	2
Very elongate pores or pores arranged in bands (with emanating conveying canals)	4

obtained from Tables 1–4, whereas the petrographic criteria which these empirical values represent are determined from thin-section study. It should be remembered that the permeabilities calculated from thin-section analysis may be much too small, depending on the degree of fracturing in the rock mass.

Permeability can vary considerably for the same pore space owing to differences in the abundance and the width of the pore interconnections. The range of empirical coefficients for different types and subtypes is from 2 to 64. Carbonate rocks, the pore-space configuration of which is classified as Type II, are usually better reservoir rocks than those having other types of porosity (Table 1).

In limestones and dolomites the effective porosity normally ranges from 2% to 25% and the corresponding empirical values have very nearly the same range, that is, from 0 to 30 (Table 2). Table 3 shows that the diameters of the pores range from less than 0.01 mm to more than 2 mm, whereas the empirical coefficients range from 0.0625 to 16. The range for pore sizes involves 2 orders of magnitude, whereas the corresponding empirical coefficient C has a range involving 3 orders of magnitude.

The influence of pore elongation is expressed by empirical coefficients

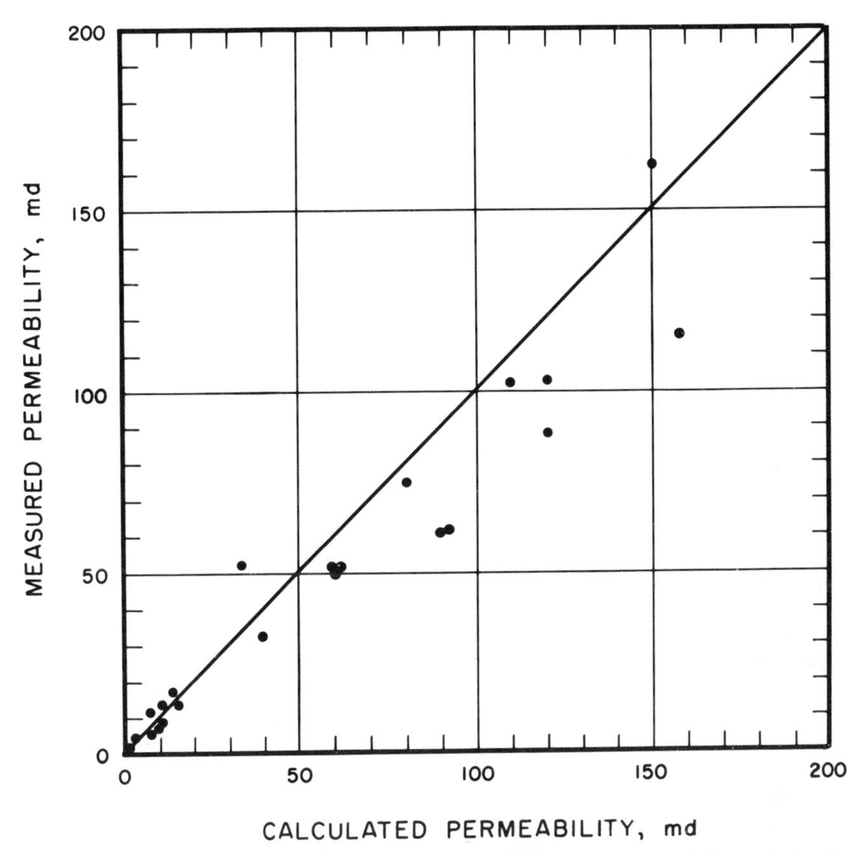

FIG. 1. Relation between laboratory-measured permeability and permeability calculated by Teodorovich's method (data from Teodorovich[2] and Aschenbrenner and Chilingar[1]). In the case of Teodorovich's data the arithmetic mean of the percent deviations is +13.8%, whereas Aschenbrenner and Chilingar found the arithmetic mean of all percent deviations to be −6.4%. The 45° straight line indicates a theoretically perfect correlation between measured and calculated permeabilities.

which vary from 1 to 4 (Table 4). Inasmuch as these empirical values stay within the same order of magnitude, factor D_1 is probably somewhat less significant than the other factors.

Independently of each other, Aschenbrenner and Chilingar[1] measured and calculated the permeabilities of 10 carbonate rock samples. Their results were in good agreement with those published by Teodorovich;[2] there tends to be an average discrepancy of approximately 10% between calculated and measured permeability values. In Fig. 1, 23 laboratory-measured permeabilities are plotted against permeabilities calculated from thin sections. The following observations were made by Aschenbrenner and Chilingar[1] after studying numerous thin sections:

1. The calculated permeability ranged from 2 to 16 md whenever the average diameter of the conveying canals was about 0.01 mm.
2. With an average diameter of the pore interconnections of approximately 0.02 mm, the permeability values varied between 30 and 75 md.
3. If a large number of the canals had an average diameter of about 0.03 mm, the permeability increased by an order of magnitude to a range from 600 to 800 md.
4. An abundance of wide interconnecting canals ($\approx$0.04 mm in diameter) was associated with permeabilities between 1200 and 1300 md.
5. There seems to be a similarity between carbonate rocks with Type I pore-space configuration and sandstones with less than 12% porosity.

Trebin[3] derived the following expression for sandstones with less than 12% porosity:

$$k = 2 \times e^{0.316\phi} \tag{3}$$

where k is permeability (md), ϕ is effective porosity (%), and e is 2.71828 (Napierian base). This relationship seems to be a fairly good approximation for low-porosity carbonates as well as sandstones.

Determination of the Basic Geometrical Characteristics of Porous Carbonates

In order to present a proper description of the internal geometry of carbonates, it is necessary to evaluate quantities that characterize the configuration of the pore space. Perez-Rosales[4] has proposed a simple but effective method of obtaining information about the internal geometry of porous media. The technique involves superimposing a type of grid over a photomicrograph of the rock surface, from which values for parameters of experimental interest, such as specific surface, pore width, and grain thickness, can be determined. In cases where the Kozeny equation is applicable, the absolute permeability can be obtained. In the procedure, the carbonate rock is impregnated with a colored material, such as plastic or synthetic resin, so that in a photomicrograph the void spaces can be easily distinguished from the solid rock.

The porosity of the rock is calculated from the equation:

$$\phi = l/L \tag{4}$$

where L is the total length of the lines of the grid, and l is the sum of the length of the line segments within the void spaces.

It can be shown that[5] the specific surface, which is the surface area of the pores per unit bulk volume, is equal to:

$$S_s = 2c/L \tag{5}$$

where c represents the number of intersections between the grid lines and the perimeter of the pores.

If the medium is homogeneous and isotropic, the mean pore width of the sample ($\bar{p}$) is given by:

$$\bar{p} = \frac{l}{c/2} \cdot \tag{6}$$

Inasmuch as each line segment within a solid space represents, by definition, a particular grain thickness, the mean grain thickness of the sample ($\bar{g}$) is:

$$\bar{g} = \frac{2(L-l)}{c}. \tag{7}$$

The process consists of obtaining a photomicrograph of the treated surface and then placing a grid, consisting of evenly spaced horizontal and vertical lines that intersect at points termed nodes, over an enlargement of the photomicrograph. From this arrangement the values of the following parameters are obtained:

c = number of intersections between the grid lines and the perimeter of the pores.
L = total length of the grid lines.
m = linear magnification of the photomicrograph.
n = number of grid nodes within pore areas.
N = total number of grid nodes.

The specific surface is then given by:

$$S_s = 2cm/L. \tag{8}$$

Inasmuch as

$$l = Ln/N \tag{9}$$

an equation for the mean pore width is derived by substituting the value for l from Eq. 9 in Eq. 6, or:

$$\bar{p} = 2Ln/Ncm \tag{10}$$

and similarly an expression for the mean grain thickness is obtained by using Eq. 6 in conjunction with Eq. 7:

$$\bar{g} = 2L(N-n)/Ncm. \tag{11}$$

Perez-Rosales[4] has shown that the Kozeny equation can be related in terms of the above parameters as follows:

$$k = CL^2n^3/4N^3c^2m^2 \qquad (12)$$

where k = absolute permeability, and C = constant of proportionality. He noted[4] that, although the method applies to homogeneous and isotropic materials only, it can be extended to include heterogeneous media through the use of statistical methods.

Statistical Concepts Used in Describing Porous Carbonate Media

Reliable predictions of carbonate pore-system characteristics, which are related to lithology and environment, depend on meaningful environmental and lithologic information. As illustrated by Jodry in Chapter 2, it is important to know how the injected volume of mercury is related to the frequency distribution of pore throat sizes, and how pore and pore-throat-size distributions obtained from thin-section measurements will aid in the development of a reservoir pore-space model. Pore size is defined as the diameter of the largest sphere that can be fitted inside a pore, whereas pore throat size is defined as the diameter of the largest sphere that will pass through the pore throat.[6] In Scheidegger's definitions[6] the pore throats are assumed to be circular in shape.

Washburn[7] first suggested the use of mercury injection under pressure to determine the pore-size distribution of a porous solid. Ritter and Drake[8] applied external pressure to force the mercury into porous material and then measured the volume of mercury intruded as a function of pressure. The mercury injection technique consists of placing a clean rock core in a mercury-filled pressure cell where the pressure is raised by predetermined steps. The amount of mercury that enters the core at each pressure increment is measured. The basis of this method is the concept that mercury is forced into successively smaller evacuated pores against capillary forces as the mercury pressure is increased. In this system liquid mercury represents the nonwetting phase (oil) in the pores, whereas the mercury vapor present is comparable to the water in the reservoir. Although the liquid mercury-mercury vapor system is only roughly comparable to the oil-water system in a carbonate reservoir, Ritter and Drake[8] reported that the mercury injection technique gives good results.

Some important considerations in using the mercury injection technique are as follows: (1) Deciding whether to make corrections for the compressibility of the rock, mercury, and residual air remaining in the carbonate core, or to assume these factors to be negligible. (2) Determining the effect of the breakdown of the rock on the readings. (3) Deciding whether or not injection equilibrium at a given pressure is reached in a reasonable time. (4) Making corrections for shape. The shape factor [$(V_p/(A_p \times r_p)$, where V_p is total pore volume, A_p is area, and r_p is the average radius of the pore] ranges from 0.5 (cylindrical pore) to 0.117 (hexagonal pore). (Most investigators assume a cylindrical pore-shape model.) (5) Choosing the value to use for the surface tension of mercury (a constant value at the ambient temperature of injection is normally employed). (6) Solving the problem of what to assume for contact angle (a value ranging from 130° to 140° is often used).

The reliability of measuring pore size in this manner was proved satisfactory when compared with the results obtained by using an independent method of nitrogen adsorption at liquid-nitrogen temperature (73.5°K). In physical adsorption, the amounts adsorbed at equilibrium depend on the pore sizes. Stevenson *et al.*[9] devised an automatic porosimeter employing a nitrogen-dosing method whereby the nitrogen dose is based on the measurement of rate of change of the pressure with time. The use of this porosimeter may result in more accurate pore-size determinations.

Grade size analysis of pores was developed specifically as a mathematical device to permit the application of conventional statistical practices to the prediction and description of the pore-size distribution in carbonate reservoir rocks. Pore- and pore-throat-size distributions represent an arrangement of values according to size and frequency of occurrence. In carbonate rocks, however, pores form an extremely complicated system of voids within the rock, and it is not easy to determine pore size in a visually satisfactory manner. Theoretically, the diameter of pores can be calculated by using Washburn's[7] equation:

$$d = \frac{4\sigma_{Hg} \cos \theta}{P} \tag{13}$$

where σ_{Hg} is surface tension of mercury (dynes/cm), θ is contact angle, P is injection pressure (dynes/cm^2 $\times$ 10^6), and d is diameter of pores (cm).

In Fig. 2, capillary pressure values are graphically presented as the mercury injection pressure (psia) on the left-hand vertical axis, and the amounts of the wetting phase and the nonwetting phase are indicated on the horizontal axes. Pore throat diameters can be shown on a micron (1/1,000 mm), millimeter, or phi (ϕ) grade scale; the last of these has been placed on the right-hand side of the graph in Fig. 2. The P_d is a pressure at which the nonwetting phase is present in the pore system as a continuous phase. The best manner of determining P_d is to extend the flat portion of the mercury injection curve; the point at which this line (tangent) intersects the vertical axis is P_d (Fig. 2). A low P_d indicates that a significant number of the pore throats are in the large size range.

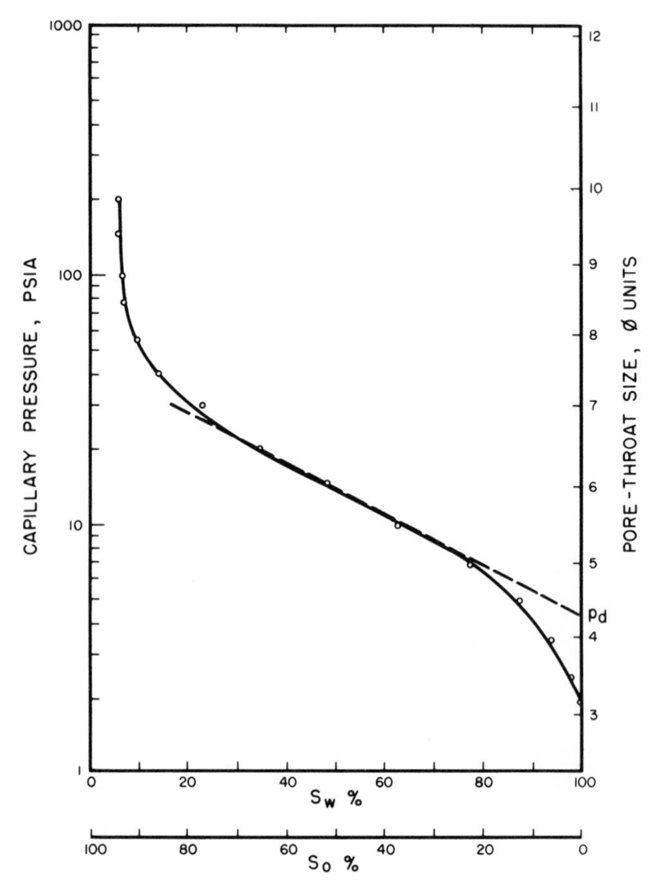

FIG. 2. Example of mercury injection curve. The approximate pore throat size that mercury will enter at a particular pressure is also shown.

Pore systems in carbonate rocks usually contain both pores and interconnections between these pores. Isolated vugs, however, are common in some carbonates. The pore system may have two extremes: (1) the size of the pores approaches that of the interconnecting pore throats, and (2) the size difference between the two is very large. In the mercury injection test, the size distribution between pores and pore throats is an artificial one. Mercury injection pressure is indicative of the pore throat sizes. At greater pressures, the mercury will enter pore throats of smaller size. It is usually assumed, therefore, that the pore throat sizes control the injection.

Phi values are equal to the negative logarithm to the base 2 of the millimeter values (Table 5). The phi units are related to the Wentworth size scale

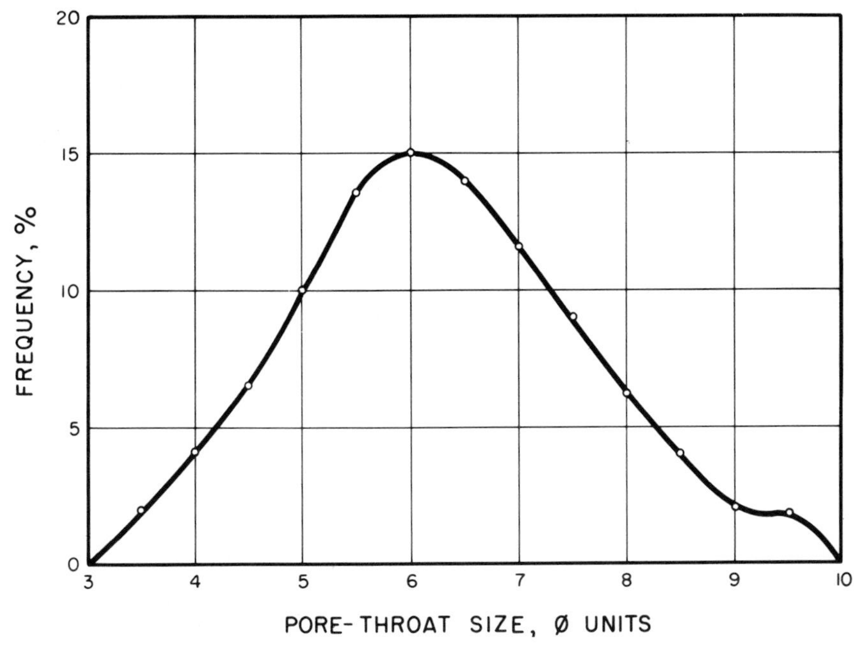

FIG. 3. Frequency distribution of pore throat sizes in a carbonate rock.

in that the limits of the size classes are whole phi numbers, and each Wentworth class is 1 phi unit (Griffiths[10]). Phi scale is generally preferred because of having integers for the class limits, thus making statistical calculations much easier. Inasmuch as phi intervals are equal, ordinary arithmetic paper can be used in plotting curves.

Figure 3 illustrates a simple frequency curve showing a normal distribution with approximately 68% of the pore throat diameters occurring between D_{16} and D_{84} [one standard deviation (1.3ϕ) on either side of the mean (6.1ϕ); $\sigma\phi = (D_{84} - D_{16})/2 = 1.3$ or use Eq. c, Table 6]. Figure 4 illustrates the data from Fig. 3 as a cumulative frequency curve, whereas in Fig. 5 the cumulative frequency of pore throat sizes is plotted on probability paper. The probability scale is designed in such a manner that a symmetrical cumulative pore-size frequency curve plots as a straight line on the graph.

The same statistical measures used to characterize a sediment can serve to describe the pore system of a carbonate rock. Table 6 summarizes these various measures of central tendency, dispersion, asymmetry, and peakedness. Pore-size distribution curves as determined by the mercury injection method are normally slanted toward the small pore sizes because of the lag effect caused by the "bottle neck"-shaped pores. The effect of the irregular

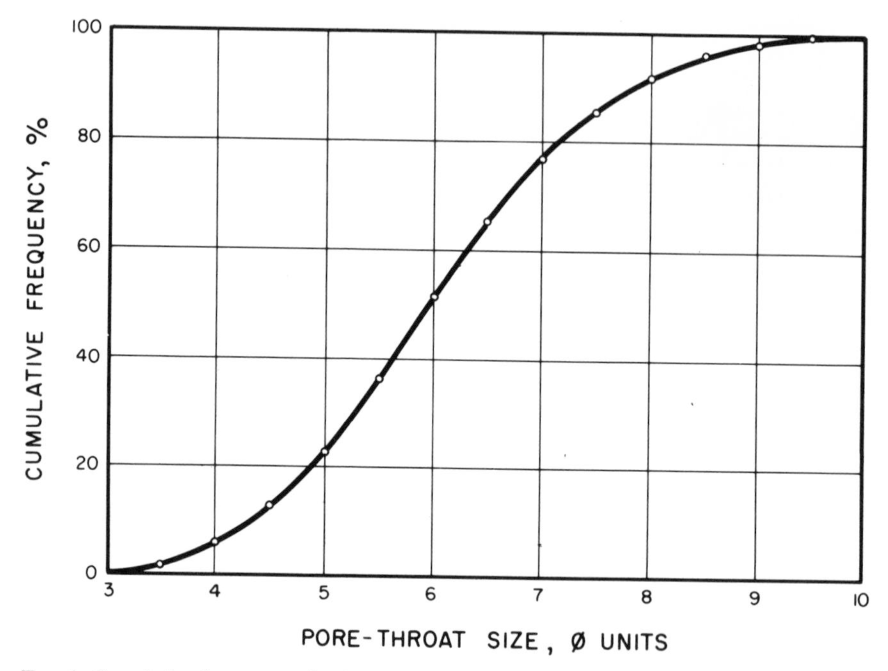

FIG. 4. Cumulative frequency distribution of pore throat sizes (data from Fig. 3).

pore shapes is to assign too small a portion of the pore space to the large pores and too large a part to the small pores. Meyer[11] employed probability theory to correct the mercury intrusion data so that reliable pore-size distribution data could be obtained. The correction is for the pores which could fill at a given pressure but which do not, because they are in communication with the mercury source through smaller pores only. The expression derived by Meyer was satisfied by Poisson's distribution. He ignored the fact that pores are mutually exclusive in deriving Poisson's distribution. This assumption, however, may introduce a serious error inasmuch as Meyer used the "zero" term in Poisson's distribution.

Aschenbrenner and Achauer[14] found that both pore and pore throat sizes were essentially log-normally distributed (straight-line relationship between logarithm of pore size or pore throat size and cumulative percentage on probability paper) in Paleozoic carbonates of the Williston Basin and in the Rocky Mountains. Inasmuch as most pore- and pore-throat-size distributions tend to be normal, this provides a method for estimating their size. Harbaugh[15] pointed out that pore- and pore-throat-size measurements based on two-dimensional thin sections may not always effectively describe the pore systems that exist in three-dimensional space. An attempt was made

TABLE 5. Pore-Size Diameter Conversion Table (millimeter to phi units)

mm	ϕ	mm	ϕ
4.00	−2.00	0.044	4.50
3.36	−1.75	0.037	4.75
2.83	−1.50	0.031	5.00
2.38	−1.25	0.026	5.25
2.00	−1.00	0.022	5.50
1.68	−0.75	0.019	5.75
1.41	−0.50	0.016	6.00
1.19	−0.25	0.013	6.25
		0.011	6.50
1.00	0.00	0.0093	6.75
		0.0078	7.00
0.841	0.25	0.0066	7.25
0.707	0.50	0.0055	7.50
0.500	1.00	0.0047	7.75
0.420	1.25	0.0039	8.00
0.351	1.50	0.0033	8.25
0.297	1.75	0.0028	8.50
0.250	2.00	0.0023	8.75
0.210	2.25	0.0020	9.00
0.177	2.50	0.0016	9.25
0.149	2.75	0.0014	9.50
0.125	3.00	0.0012	9.75
0.105	3.25	0.00098	10.00
0.088	3.50	0.0005	11.00
0.074	3.75	0.0003	12.00
0.062	4.00		
0.052	4.25		

by Fara and Scheidegger[16] to characterize a porous medium from data which can be read from a photomicrograph. Two types of harmonic analyses yield the most satisfactory descriptions of a porous medium. Mathematically, Fara and Scheidegger showed that spectral coefficients do indeed characterize a given porous medium if the medium is homogeneous in a statistical sense. Meyer[11] assumed randomness of pore distribution which is acceptable in the case of homogeneous oolitic limestones. Vuggy limestones would probably then be excluded from the statistical approaches of both Fara and Scheidegger[16] and Meyer.[11]

It has been found that both the P_d and the 50% saturation pressure (P_c^{50}), which is a pressure necessary to fill half the pore system with the nonwetting phase, are directly related to the mean pore throat size of the pore system. Mean pore throat sizes are essentially log-normally distributed

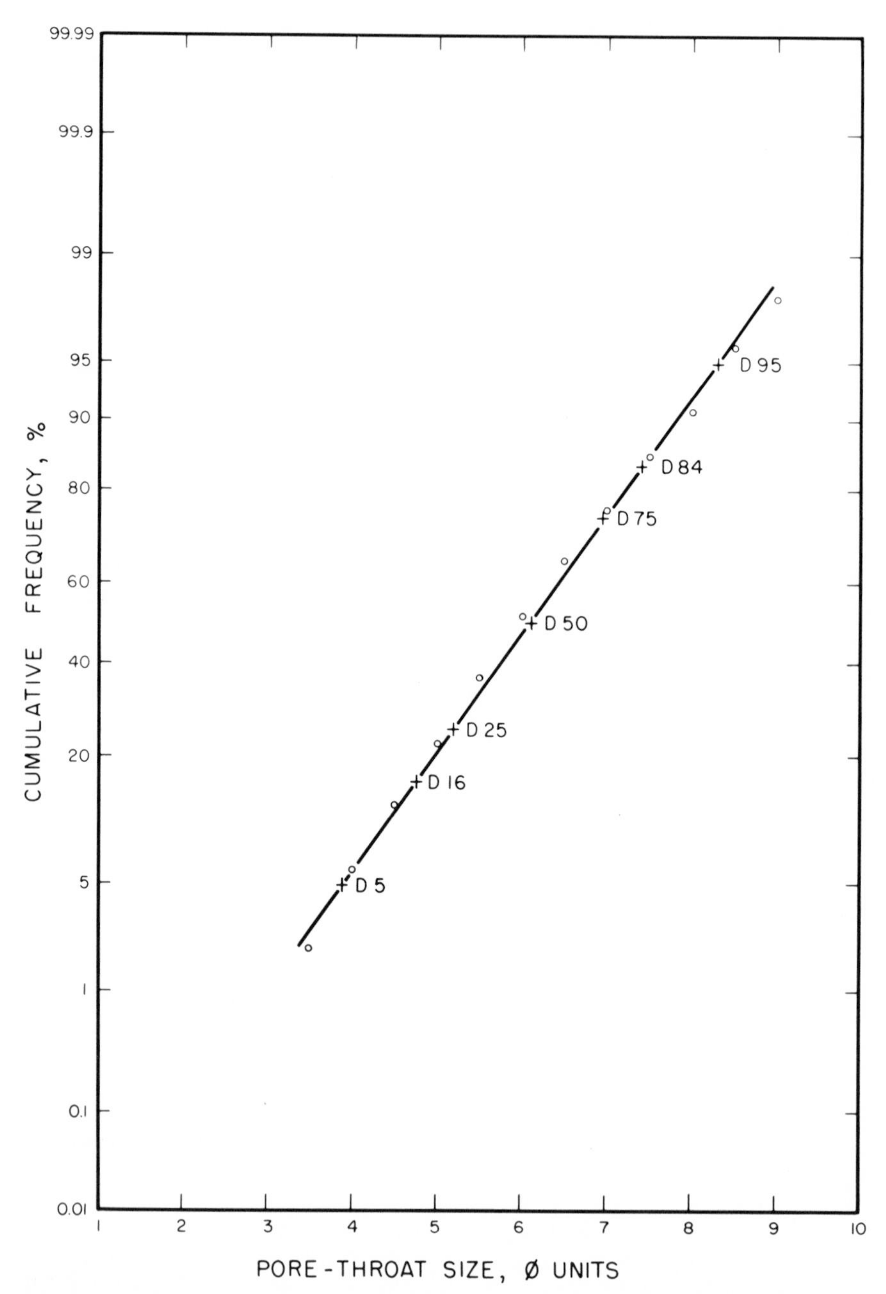

FIG. 5. Cumulative normal curve on probability paper, showing positions of various percentile ordinates (data are from Fig. 3).

TABLE 6. Summary of Pore- and Pore-Throat-Size Measures Used in Conjunction with Graphic Analysis

Measures of central tendency:

1. *Median* (D_{50}) is the diameter which is larger than half of the pores in the distribution and smaller than the other half (i.e., the middlemost member of the distribution). It reflects the overall pore size as influenced by the chemical or physical origin of the rock and any subsequent alteration. It may be a very misleading value, however.
2. *Mean* (D_M) is the measure of the overall average pore size:

$$D_M = \frac{D_5 + D_{15} + D_{25} \ldots D_{85} + D_{95}}{10} \quad \text{(a)}$$

or

$$D_M = \frac{D_{16} + D_{50} + D_{84}}{3}. \quad \text{(b)}$$

3. *Mode* (D_m) is the most frequently occurring pore diameter (peak of frequency curve). If two dominant pore sizes are present, which could result when there is a mixture of two or more different porosity types (vugy, oolitic, fracture, intergranular, etc.), then the frequency curve is bimodal.

Measure of dispersion:

Pore Sorting (S_p) is a standard deviation measure of the pore sizes in a sample:[12]

$$S_p = \frac{(D_{84} - D_{16})}{4} + \frac{(D_{95} - D_5)}{6.6}. \quad \text{(c)}$$

Measure of asymmetry:

Skewness (Sk_p) measures the nonnormality of a pore-size distribution:[12]

$$Sk_p = \frac{(D_{84} + D_{16} - 2D_{50})}{2(D_{84} - D_{16})} + \frac{(D_{95} + D_5 - 2D_{50})}{2(D_{95} - D_5)}. \quad \text{(d)}$$

A symmetrical curve has a Sk_p value of 0; limits in which Sk_p varies are as follows: $-1 \leqslant Sk_p \leqslant 1$. Positive values indicate that the curve has a tail in the small pores. Negative values indicate that the curve is skewed toward the larger pores.

Measure of peakedness:

Kurtosis (K_p) is a measure of the degree of peakedness, that is, the ratio between the spread of the pore diameters in the tails and the spread of the pore diameters in the central portion of the distribution:[12]

$$K_p = \frac{D_{95} - D_5}{2.44(D_{75} - D_{25})}. \quad \text{(e)}$$

Normal curves have a K_p of 1, whereas platykurtic (bimodal) distributions may have a K_p value as low as 0.6. A curve represented by a high narrow peak (very leptokurtic) may have K_p values ranging from 1.5 to 3.

D_n is the pore diameter in phi units at the *n*th percentile.

The methodology of pore- and pore-throat-size statistics has been borrowed from the measures used in grain-size analysis (see Folk and Ward[12] and McCammon[13]). Refer to Figs. 2, 3, 4, and 5.

in a logarithm of P_d or P_c^{50} versus mean pore-throat-size plots. Various other relationships employing sorting, skewness, kurtosis, etc., can be constructed to determine the nature of the pore-size distribution in a carbonate reservoir rock.[17]

An excellent treatment of the application of pore-space geometry and statistics to the study of reservoir rocks is given in the book by Khanin.[18]

References

1. Aschenbrenner, B. C. and Chilingar, G. V.: "Teodorovich's Method for Determining Permeability from Pore-Space Characteristics of Carbonate Rocks", *Bull.*, AAPG (1960) Vol. 44, 1421–1424.
2. Teodorovich, G. I.: "Structure of the Pore Space of Carbonate Oil Reservoir Rocks and Their Permeability as Illustrated by Paleozoic Reservoirs of Bashkiriya", *Dokl. Akad. Nauk SSSR* (1943) Vol. 39, No. 6, 231–234.
3. Trebin, F. A.: *Permeability to Oil of Sandstone Reservoirs*, Gostoptekhizdat, Moscow (1945).
4. Perez-Rosales, C.: "Simultaneous Determination of Basic Geometrical Characteristics of Porous Media", *Soc. Pet. Eng. J.* (Dec., 1969) Vol. 9, No. 4, 413–416.
5. Perez-Rosales, C.: "A Simplified Method for Determining Specific Surface", *J. Pet. Tech.* (Aug., 1967) 1081–1084.
6. Scheidegger, A. E.: *The Physics of Flow through Porous Media*, The Macmillan Co., New York (1957) 313 pp.
7. Washburn, E. W.: "Note on a Method of Determining the Distribution of Pore Sizes in a Porous Material", *Proc.*, Nat. Acad. Sci. (1921) Vol. 7, 115–116.
8. Ritter, H. L. and Drake, L. C.: "Pore-Size Distribution in Porous Materials: Pressure Porosimeter and Determinations of Complete Macropore-Size Distribution", *Ind. Eng. Chem. Anal. Ed.* (1945) Vol. 17, 782–786.
9. Stevenson, D. G., Hill, T. and Anderson, A.: "Automatic Control of Pore Size Distribution Measurement in Solids", *J. Sci. Instrum.* (1967) Vol. 44, 922–926.
10. Griffiths, J. C.: *Scientific Method of Analysis of Sediments*, McGraw-Hill Book Co., New York (1967) 508 pp.
11. Meyer, H. I.: "Pore Distribution in Porous Media", *J. Appl. Phys.* (1953) Vol. 24, No. 5, 510–512.
12. Folk, R. L. and Ward, W. C.: "Brazos River Bar, a Study in the Significance of Grain-Size Parameters", *J. Sediment. Petrol.* (1957) Vol. 27, 3–26.
13. McCammon, R. B.: "Efficiencies of Percentile Measures for Describing the Mean Size and Sorting of Sedimentary Particles", *J. Geol.* (1962) Vol. 70, No. 4, 453–465.
14. Aschenbrenner, B. C. and Achauer, C. W.: "Minimum Conditions for Migration of Oil in Water-Wet Carbonate Rocks", *Bull.* AAPG (1960) Vol. 44, No. 2, 235–243.
15. Harbaugh, J. W.: "Carbonate Oil Reservoir Rocks", in: G. V. Chilingar, H. J. Bissell, and R. W. Fairbridge (Editors), *Carbonate Rocks (A); Origin, Occurrence and Classification*, Elsevier Publ. Co., Amsterdam (1967) 349–397.
16. Fara, H. D. and Scheidegger, A. E.: "Statistical Geometry of Porous Media", *J. Geophys. Res.* (1961) Vol. 66, No. 10, 3279–3284.
17. Thomas, A., Alba, P. and Courand, G.: "Contribution aux Méthodes d'Étude du Milieu Poreux", *Rev. de l'Inst. Franc. du Pétrole* (1961) Vol. 16, No. 1, 36–47.
18. Khanin, A. A.: *Oil and Gas Reservoir Rocks and Their Study*, Izd. Nedra, Moscow (1969) 366 pp. (In Russian.)

APPENDIX B

Sample Problems and Questions

GEORGE V. CHILINGAR, ROBERT W. MANNON, AND HERMAN H. RIEKE, III

1. Given the following information, find the height of an oil column in feet, h, necessary to (1) saturate 50% of the sample with nonwetting phase (h_{50}), (2) cause displacement of water (h_d): mercury surface tension, $\sigma_{Hg} = 410$ dynes/cm; water-oil interfacial tension, $\sigma_{wo} = 30$ dynes/cm; mercury-rock contact angle, $\theta_{Hg} = 160°$; water-oil contact angle, $\theta_{wo} = 39°$; difference in density between the water and the oil (g/cc), $\Delta\rho = 0.4$; $P_c^{50} = 140$ psi; and $P_d = 30$ psi (mercury injection).

Solution:

$$P_c^{wo} = P_c^{Hg} \left(\frac{\sigma_{wo}}{\sigma_{Hg}}\right) \left(\frac{\cos \theta_{wo}}{\cos \theta_{Hg}}\right)$$

$$P_c^{50(wo)} = 140\left(\frac{30}{410}\right) \left(\frac{\cos 39°}{\cos 160°}\right) = 8.46 \text{ psi}$$

and

$$h_{50} = \frac{P_c^{wo}}{0.433\Delta\rho} = \frac{8.46}{0.433 \times 0.4} = \underline{49 \text{ ft}}$$

$$P_c^{d(wo)} = 30 \left(\frac{30}{410}\right) \left(\frac{\cos 39°}{\cos 160°}\right) = 1.82 \text{ psi}$$

and

$$h_d = \frac{1.82}{0.433 \times 0.4} = \underline{10.5 \text{ ft.}}$$

2. Calculate the height of oil column, h, in feet if $\sigma_{Hg} = 400$ dynes/cm; $\sigma_{wo} = 29$ dynes/cm; $\theta_{Hg} = 159°$; $\theta_{wo} = 40°$; $\rho_w = 1.1$ g/cc; $\rho_o = 0.85$ g/cc; and $P_d = 130$ psi (mercury injection).

Solution:

$$P_c^{wo} = P_c^{\mathrm{Hg}}\left(\frac{\sigma_{wo}}{\sigma_{\mathrm{Hg}}}\right)\left(\frac{\cos\theta_{wo}}{\cos\theta_{\mathrm{Hg}}}\right)$$

$$P_c^{wo} = 130\left(\frac{28}{400}\right)\left(\frac{\cos 40^\circ}{\cos 159^\circ}\right) = 7.46 \text{ psi}$$

$$h = \frac{P_c^{wo}}{0.433\Delta\rho} = \frac{7.46}{0.433\times 0.25} = \underline{68.9 \text{ ft.}}$$

3. Determine the factor to convert capillary pressure in pounds per square inch to height above the oil-water contact in feet. The following data are given: ρ_o at 267°F = 32.6 lb/cu ft; ρ_w at 267°F = 69.38 lb/cu ft; surface tension of water, σ_w = 72 dynes/cm; and water-oil interfacial tension, σ_{wo} = 30 dynes/cm.

Solution: As the oil moves upward through the water in a capillary, several forces act on this column of oil having height h: (1) weight of column of oil (lb) acting downward, $W_o = (\pi r^2 h)\rho_o$, where r is radius of capillary (ft) and ρ_o is density of oil (lb/cu ft); (2) upward buoyant force of water on oil (lb), $B = (\pi r^2 h)\rho_w$, where ρ_w is density of water (lb/cu ft); and (3) capillary force (lb), F_c, pulling downward $= 2\pi r(\sigma_{wo}\cos\theta_{wo})$, where σ_{wo} is water-oil interfacial tension, and θ_{wo} is water-oil-rock contact angle. Thus, $W + F_c = B$ and $(\pi r^2 h)\rho_o + 2\pi r(\sigma_{wo}\cos\theta_{wo}) = (\pi r^2 h)\rho_w$.
Solving for h, one obtains:

$$h = \frac{2\sigma_{wo}\cos\theta_{wo}}{r(\rho_w - \rho_o)} = \frac{2\sigma_{wo}\cos\theta_{wo}}{r\,\Delta\rho}\text{ ft}$$

and inasmuch as $(2\sigma_{wo}\cos\theta_{wo})/r = P_c^{wo}$, which is the capillary pressure in the water-oil system,

$$h = \frac{144\,P_c^{wo}}{\Delta\rho} = \frac{144\,P_c^{wo}}{36.78} = 3.92\,P_c^{wo}$$

where P_c^{wo} is in pounds per square inch, h is in feet and $\Delta\rho$ is in pounds per cubic foot. If $\Delta\rho$ is in grams per cubic centimeter,

$$h = \frac{P_c^{wo}}{0.433\,\Delta\rho}.$$

One can also calculate h in terms of capillary pressure in the water-air system: inasmuch as $\cos\theta_{wo}/\cos\theta_{wa} = 0.76604$ and $\sigma_{wa}/\sigma_{wo} = 72/30$,

$$h = \frac{144}{36.78}\left(\frac{P_c^{wa}\times 30\times 0.76604}{72}\right) = 1.25\,P_c^{wa}.$$

Note: the conversion factor should be calculated each time for a particular pore system and cannot be applied indiscriminately to all samples having wide ranges of lithology.

4. Given the following information, calculate the weight of dissolved pure limestone (or dolomite) and the radial distance acid will penetrate until it is spent: (*a*) *Matrix acidizing* of 40-ft-thick limestone producing section; (*b*) porosity = 0.16; (*c*) volume of acid = 600 gal of 15% hydrochloric acid; (*d*) spending time = 30 sec; (*e*) specific gravity of acid = 1.075; (*f*) pumping rate = 9 bbl/min; and (*g*) wellbore radius = 4 in. Given also:
Chemical equation for the reaction between HCl and calcite:

$$CaCO_3 + 2HCl \rightarrow CaCl_2 + H_2O + CO_2.$$

100 73 111 18 44 (relative weights)

One thousand gallons of 15% by weight HCl solution contains 1344.8 lb of hydrochloric acid (1000 × 8.34 × 1.075 × 0.15).
Chemical equation for the reaction between HCl and dolomite:

$$CaMg(CO_3)_2 + 4HCl \rightarrow CaCl_2 + MgCl_2 + 2H_2O + 2CO_2.$$

184.3 146 111 95.3 36 88 (relative weights)

Reference: Craft, B. C., Holden, W. R. and Graves, E. D., Jr.: *Well Design* (*Drilling and Production*), Prentice-Hall, Englewood Cliffs, N.J. (1962) 536–546.

5. For two different cores having the same P_c^{50} (pressure necessary to fill 50% of the pores with the nonwetting phase) draw a schematic diagram of pressure versus percentage wetting phase saturation curves if one core has better sorting than the other. Explain!

6. What relationship exists between P_d (pressure at which nonwetting phase is present in the pore system as a continuous phase), P_c^{50}, permeability, and mean pore throat size. Discuss!

7. What does a capillary pressure hysteresis loop show?

Reference: Pickell, J. J., Swanson, B. F. and Hickman, W. B.: "Application of Air-Mercury and Oil-Air Capillary Pressure Data in the Study of Pore Structure and Fluid Distribution", *J. Pet. Tech.* (Mar., 1966) 55–61.

8. What effect does dolomitization have on capillary pressure correlations?

Reference: Rockwood, S. H., Lair, G. H. and Langford, B. J.: "Reservoir Volumetric Parameters Defined by Capillary Pressure Studies", *J. Pet. Tech.* (Sept., 1957) 252–259.

9. Outline regularities in the relationship between capillary pressure curves (their shape, etc.) and pore configuration (size and shape of pores, type and width of intercommunicating channels, etc.). This question should be answered after examining thin sections and corresponding curves provided by the instructor.

10. Draw mercury capillary pressure curves (schematic diagrams) which are representative of a poor and a good carbonate reservoir rock. Label and explain all parts of the curves.

11. Draw approximate schematic diagrams of capillary pressure versus wetting phase saturation curves for:

(*a*) Dolomite: intercrystalline porosity = 20%, mean pore throat size = 0.0021 mm, size of pores (avg.) = 0.3 mm, sorting = 2.6 ϕ units, and permeability = 98 md.

(*b*) Oolitic limestone: intergranular porosity = 19%, mean pore throat size = 0.0035 mm, size of pores (avg.) = 0.19 mm, sorting = 6.6 ϕ units, and permeability = 40 md.

(*c*) Oolitic limestone: intergranular to vuggy porosity = 14%, mean pore throat size = 0.0004 mm, size of pores (avg.) = 0.3 mm, sorting = 3.5 ϕ units, and permeability = 9 md.

12. Name the following carbonate rocks: (*a*) containing 60% skeletal fragments and 40% micrite; (*b*) containing 91% coated grains (oopellets) and 9% micrite; and (*c*) GMR = 7.2/2.5, with detrital grains.

Reference: Chilingar, G. V., Bissell, H. J. and Fairbridge, R. W. (Editors): *Carbonate Rocks (Origin, Occurrence and Classification)*, 9A, Elsevier Publ. Co., Amsterdam (1967) 87–168.

13. Compare carbonates and shales as source rocks of petroleum, and list all differences.

Reference: Chilingar, G. V., Bissell, H. J. and Fairbridge, R. W. (Editors): *Carbonate Rocks (Physical and Chemical Aspects)*, 9B, Elsevier Publ. Co., Amsterdam (1967) 225–251.

14. What relationship exists between insoluble residue and organic matter in carbonate rocks?

Reference: as above.

15. Discuss in detail all geochemical techniques for recognizing carbonate source rocks.

Reference: as in Problem 13.

16. What relationship exists between porosity, insoluble residue, and Ca/Mg ratio in carbonate rocks? Explain!

Reference: Chilingar, G. V.: "Use of Ca/Mg Ratio in Porosity Studies", *Bull.*, AAPG (1956) Vol. 40, 2256–2266.

17. Discuss solution porosity and compare it with dolomitization porosity. What is the effect of solution and dolomitization on insoluble residue content and Ca/Mg ratio?

Reference: as above.

18. Why are porosity and permeability insensitive to percent mud-size matrix when a rock is 50–75% dolomite?

Reference: Ham, W. E. (Editor): *Classification of Carbonate Rocks (A Symposium)*, Am. Assoc. Pet. Geol., Tulsa, Okla. (1962).

19. Diagrammatically show the relationship between porosity, permeability, and percent dolomitization.

20. Define epigenetic dolomites and explain the Ca/Mg ratio distribution in such rocks.

Reference: Larsen, G. and Chilingar, G. V. (Editors): *Diagenesis in Sediments*, Elsevier Publ. Co., Amsterdam (1967).

21. Calculate the permeability of arenaceous dolomite, containing finely porous conveying canals, by using Teodorovich's method. The fine elongate pores are abundant, and the rock has an effective porosity of 10%.

22. Calculate permeability, using Teodorovich's method, if porosity = 13% and size of elongated pores = 0.25–1 mm (Type II good porosity, with pores of different size).

23. Would replacement of calcite by dolomite result in an increase or decrease in porosity? Show all calculations. The specific gravity of dolomite = 2.87 and of calcite = 2.71.

24. Is there an increase or decrease in porosity as aragonite is replaced by (*a*) calcite and (*b*) dolomite? Show all calculations. The specific gravity of aragonite = 2.95, of calcite = 2.71, and of dolomite = 2.87.

25. How are acid volumes and pumping rates determined for acidizing operations?

26. What effect does enlargement of pores have on acid velocity and on surface/volume ratio? Are these effects opposite in significance or not? Explain!

27. How much deeper would later increments of acid penetrate before being spent? Why?

28. On using stronger acid, does spending time decrease or increase? Why?

29. Is sludge formation more or less likely with stronger acid, and why? How can it be prevented?

30. In acidizing operations, what are the functions of (*a*) intensifier, (*b*) surfactant, and (*c*) iron retention additive?

31. How are pumping pressure and necessary horsepower determined in acidizing operations?

32. Is the spending time of acid lower or higher in the case of lower specific surface area?

33. Calculate the specific surface area of a carbonate rock with porosity = 15%, permeability = 8 md, and cementation factor, m = 2 (matrix acidizing).

References: (1) Chilingar, G. V., Main, R. and Sinnokrot, A.: "Relationship between Porosity, Permeability and Surface Areas of Sediments", *J. Sediment. Petrol.* (1963) Vol. 33, No. 3, 759–765. (2) As in Problem 4.

34. Calculate the productivity ratio for a horizontal fracture if fracture width = 0.1 in., net pay zone thickness = 60 ft, permeability of propping agent in place = 32,000 md, horizontal permeability = 0.6 md, r_e/r_w = 2,000, and fracture penetration, r_f/r_e = 0.3.

Reference: Craft, B. C., Holden, W. R. and Graves, E. D., Jr.: *Well Design* (*Drilling and Production*), Prentice-Hall, Englewood Cliffs, N.J. (1962) 483–546.

35. Discuss the major operational problems associated with the water-flooding of carbonate reservoirs.

36. Discuss the problems associated with gas injection in carbonate reservoirs.

37. Would relative permeabilities to oil and to water be higher or lower if sandstone contains considerable amounts of carbonate particles? Explain for both k_{rw} and k_{ro}!

Reference: Sinnokrot, A. A. and Chilingar, G. V. "Effect of Polarity and Presence of Carbonate Particles on Relative Permeability of Rocks", *Compass of Sigma Gamma Epsilon* (1961) Vol. 38, 115–120.

38. Do oil-wet reservoirs tend to have higher or lower recovery than water-wet reservoirs? Explain your answer!

39. Draw performance curves for closed and open combination-drive pools and discuss the differences.

40. Discuss the theoretical proposals of Jones-Parra and Reytor regarding the effect of withdrawal rates on recovery from reservoirs having the fracture-matrix type of porosity.

Reference: Jones-Parra, Juan and Reytor, R. S.: "Effect of Gas-Oil Rates on the Behavior of Fractured Limestone Reservoirs", *Trans.*, AIME (1959) Vol. 216, 395–397.

41. By using density logs, calculate S_w on assuming (*a*) limestone and (*b*) dolomite, when $R_w = 0.02$, $R_t = 20$, and $m = 2.2$. Explain the difference in the values obtained for S_w.

42. What is the porosity of a clastic limestone that shows a sonic transit time on the log of 90 μsec/ft?

43. When using Achie's formula ($F = \phi^{-m}$) for determining porosity from log analysis, what values of cementation factor, m, are appropriate for carbonate rocks?

Reference: Pirson, S. J.: *Handbook of Well Log Analysis*, Prentice-Hall, Englewood Cliffs, N.J. (1963) 23–24.

44. Estimate the initial oil- and gas-in-place for the "XYZ" pool from the following data. Can you explain the apparently anomalous GOR behavior?

Reservoir Data—XYZ Pool

Average porosity	16.8%
Average effective oil permeability	200 md
Interstitial water saturation	27%
Initial reservoir pressure	3,480 psia
Reservoir temperature	207°F
Formation volume factor of formation water	1.025 bbl/STB
Productive oil zone volume (net)	346,000 acre-ft
Productive gas zone volume (net)	73,700 acre-ft

Pressure-Production Data

Average reservoir pressure (psia)	Cumulative oil production (STB)	Cumulative GOR (SCF/B)	Cumulative water production (STB)
3,190	11,170,000	885	224,500
3,139	13,800,000	884	534,200
3,093	16,410,000	884	1,100,000
3,060	18,590,000	896	1,554,000

Flash Liberation Data

(Pertains to Production through One Separator at 100 psig and 75°F)

Pressure (psia)	B_o (bbl/STB)	R_s (SCF/STB)	z
3,480	1.476	857	0.925
3,200	1.448	792	0.905
2,800	1.407	700	0.888
2,400	1.367	607	0.880

APPENDIX C

Summary of Pool Characteristics and Primary Performance Data for Selected Carbonate Reservoirs

GERALD L. LANGNES, HERMAN H. RIEKE, III, ROBERT W. MANNON, AND GEORGE V. CHILINGAR

The following symbols are used in the table.

1. Exploration discovery methods:
   - G Surface geology
   - GS Surface geology and seismic survey
   - R Redrill
   - S Seismic survey
   - SG Subsurface geology
   - SGS Subsurface geology and seismic survey
   - W Wildcat
   - WS Wildcat and seismic survey
2. Geologic age:
   - T Tertiary (e = Eocene, p = Paleocene)
   - K Cretaceous
   - J Jurassic
   - P Permian
   - IP Pennsylvanian
   - M Mississippian
   - D Devonian
   - S Silurian
   - O Ordovician
   - C Cambrian
3. Trap mechanism:
   - 1 Anticline
   - 2 Fault
   - 3 Unconformity
   - 4 Reef
   - 5 Pinch out
   - 6 Monocline
   - 7 Local porosity
   - 8 Syncline
4. Reservoir rock type:
   - D Dolomite (dolostone)
   - L Limestone
   - D-L Dolomite and limestone; dolomite predominates
   - L-D Limestone and dolomite; limestone predominates
5. Porosity is normally given as an average value; some values are absolute maximums.
6. Most of the permeabilities reported represent an average value; some values are shown as ranges, and others are reported in relative descriptive terms.
7. Footnotes: [a] Matrix has no porosity. [b] Oil viscosity in centipoises. [c] Cumulative oil production to date of publication. [d] Condensate reservoir. [e] Ultimate recovery in barrels per acre-foot. [f] STOP (stock tank oil-in-place, initial). Ultimate recovery is primary unless otherwise noted.

	Field name	State, country	County	Exploration discovery method	Productive formation	Geologic age	Trap mechanism	Reservoir rock type	Productive acreage (acres)

A: Solution Gas Drive

AA: Intercrystalline-intergranular porosity—solution gas drive

	Field name	State, country	County	Exploration discovery method	Productive formation	Geologic age	Trap mechanism	Reservoir rock type	Productive acreage (acres)
1	Adell	Kansas	Sheridan	W	Lansing—Kansas City	ℙ	1	L	1,200
2	Anton-Irish	Texas	Lamb, Hale, Lubbock	SGS	Clear Fork	P	5	D	9,000
3	Block 31	Texas	Crane		Devonian	D	1	L	6,000
4	Eubank	Kansas	Haskell	W	Ste Genevieve	M	1–5	L	240
	Eubank	Kansas	Haskell	W	Cherokee	M	1–5	L	400
	Eubank	Kansas	Haskell	W	Lansing—Kansas City	ℙ	1–5	L	1,280
5	Ft. Chadborne	Texas	Coke		Odom	ℙ	5	L	16,000
6	Foster	Texas	Ector		Grayburg	P	5	D	15,500
7	Fullerton	Texas	Andrews		Clear Fork	P	5	L	16,642
8	Gard's Pt.*	Illinois	Wabash	W	Ohara	M	5	L	725
9	Harmattan-Elkton	Alberta, Canada			Turner Valley	M	5	D	23,962
10	Haynesville	Louisiana, Arkansas			Pettit	K	1–5	L	11,269
11	Hortonville	Kansas	Sheridan	SGS	Lansing—Kansas City	ℙ	1	L	60
12	Luther S.E.	Texas	Howard	S	Fusselman	S-D	5–6	D	3,600
13	McElroy	Texas	Crane, Upton		Grayburg	P	1–5	D	30,000
14	New Hope	Texas	Franklin		Bacon	K	1	L	4,609
15	Panhandle	Texas	Carson, Gray		Brown	ℙ	3	D	200,000
16	Parks	Texas	Midland		Bend	ℙ	1	L	6,400
17	Pickton	Texas	Hopkins		Bacon	K	1–5	L	7,900
18	Plainville	Indiana	Daviess	S	McClosky	M	1	D-L	400
19	Rocker A	Texas	Garza	SG	Glorieta	P	1	D	1,120
20	Slaughter	Texas	Cochran		San Andres	P	5	D-L	87,500
21	So. Cowden	Texas	Ector		Grayburg	P	5	D	3,354
22	Umm Farud	Libya			Dahra-B	T (p)	1	D	
23	Waddell	Texas	Crane		San Andres	P	1	D	8,000
24	West Lisbon	Louisiana	Claiborne		Pettit	K	5	L	4,341

* Composed of four small oil pools. [1] Total recovery with miscible displacement is 50^{+}%. [2] Total recovery with water flood is 54%. [3] Total with gas-cap cycling and water injection; in addition 75% of condensate will be recovered. [4] Including secondary recovery (water flooding); this field has some vugular porosity and gas-cap drive. [5] Plus 17.5 MMbbl secondary. [6] Total recovery with gas injection is 61.7%. [7] Includes No. 36; induced water drive.

Oil and gas column (ft)	Pay thickness (ft)	Crude gravity (°API)	ϕ (%)	k (md)	S_{wi} (%)	p_i (psi)	p_b (psi)	T (°F)	Solution GOR (cu ft/bbl)	Ultimate recovery (MMbbl)	(%)	Ref. No.
50–60	140	30–41.5								5[c]		1
	400	30	9.4	16.2		2,040				39[c]		2
350	185	46	15	1	37	4,145	2,764	139	1,300	69	25[1]	3
27	11	37										1
25	5	35.8				600						1
20	14	34.4			44	1,330						1
235	42	47	4–5	35	35–45	2,230	2,230	134	975	26	28[2]	4
	200	36.4	8	1	35	158–1,760	1,140	96	371		13.6	5
330	216		7.7–11.4	6.7–12.6	24	2,980	2,370		1,590	246	26.5	6
	6–12	38	8–12	100		800–1,200				0.9[c]		7
82		36	11.3	113		3,636	3,636	203	871	45–60	30–40[3]	8
	11.3	38–40	19–20	23	23	2,440	2,331	179	460	14.9 28.5[4]	18 34[4]	9
10	2–4	29–40										1
100–125	8–65	43.5	15	16		4,246				13[5]	42.5	2
550		32	13	20	30	1,300	755	86	315			10
	11	43	18.9	379	25.4	3,425	907	203	256	8.1	19	11
	60	40–46	12	25	30–45	440	440	85	275	1,200[f]		12–14
215	21	45.5	6.8	2.6	33	4,567	3,508	174	1,817	3.9	17	15
220	8	50	20	250	25	3,578	3,578	209	1,915	6.6[6]	19.4	16
60	20	38.5				375				4.0[7]		7
60	40	34	13.2	4.6								2
	40	29–32	12.2	10	11	1,710	1,710	108	465	275[c]	20	17–19
	40	34.8	7.5	24	31	1,760	325	96	160	10.4		20
80		0.7[b]	28.1	43	22	1,003	253	136	90–100	75[f]		21
280	150	34	11.8	12.3	35.5	1,650	1,134	88	484	33[c]	21	22
	8	35	9–21	1–49		2,550			395	1.4	14	23

	Field name	State, country	County	Exploration discovery method	Productive formation	Geologic age	Trap mechanism	Reservoir rock type	Productive acreage (acres)

AB: Vugular-solution porosity—solution gas drive

	Field name	State, country	County	Exploration discovery method	Productive formation	Geologic age	Trap mechanism	Reservoir rock type	Productive acreage (acres)
25	Aneth	Utah	San Juan		Paradox	ℙ	4–5	L	55,000
26	Bar-Mar	Texas	Crane	SG	Tubb	D	2–5	D	80
27	Bateman Ranch	Texas	King	G	Strawn B-Zone	ℙ	1	L	1,700
	Bateman Ranch	Texas	King	G	Strawn C-Zone	ℙ	1	L	1,450
28	Big Eddy	New Mexico	Eddy	GS	Strawn	ℙ	4	L	640
29	Fairway*	Texas	Henderson	SG	James Lime	K	4	L	23,000
30	GMK	Texas	Gaines	SGS	San Andres	P	1	D	1,280
31	Golden Spike South, D-3A	Alberta, Canada			Leduc	D	4	L	1,390
32	Kelly-Snyder	Texas	Scurry	SGS	Canyon Reef	ℙ	4	L	51,000
33	Morrow County	Ohio	Morrow		Trempealeau	C	1	D	
34	North Anderson Ranch	New Mexico	Lea	GS	Wolfcamp	P	1	L	920
35	Nunn	Kansas	Finney	S	Marmaton	ℙ	1	L	1,100
36	Plainville	Indiana	Daviess	S	Aux Vases	M	1	D	400
37	Pleasant Prairie	Kansas	Haskell	W	Morrow	ℙ	5	L	10,000
38	Pollnow	Kansas	Decatur	S	Lansing—Kansas City	ℙ	5	L	800
39	Sharon-Ridge	Texas	Scurry		Canyon Reef	ℙ	4	L	11,000
40	Sun City	Kansas	Barber	G	Marmaton	ℙ	5	L	300
41	Sycamore-Millstone	West Virginia	Calhoun	W	Greenbrier	M	8	D	6,000

* Considerable gas-cap drive. ** Includes No. 208. *** Includes No. 18. **** Plus 10.95 MMbbl owing to water injection; estimated ultimate recovery under water injection is over 50%.

AC: Fracture-matrix porosity—solution gas drive

	Field name	State, country	County	Exploration discovery method	Productive formation	Geologic age	Trap mechanism	Reservoir rock type	Productive acreage (acres)
42	Bitter Lake South and West	New Mexico	Chaves		San Andres	P	1–7	D	680
43	Brown	Texas	Gaines	SGS	Glorieta	P	6	D	960
44	Cottonwood Creek	Wyoming	Washakie	S	Phosphoria	P	5–6	D	14,200
45	Devil's Basin	Montana	Musselshell	G	Van Duzen	M	1	L	80
46	Elk Basin	Montana, Wyoming	Park-Carbon		Madison A Zone	M	1	D	8,960
47	Reeves	Texas	Yoakum	S	San Andres	P	5	D-L	5,480
48	Yellow House	Texas	Hockley	S	San Andres	P	6	D	3,230

Oil and gas column (ft)	Pay thickness (ft)	Crude gravity (°API)	ϕ (%)	k (md)	S_{wi} (%)	p_i (psi)	p_b (psi)	T (°F)	Solution GOR (cu ft/bbl)	Ultimate recovery (MMbbl)	(%)	Ref. No.
250		40–42	10.2	20	22	2,200	1,780	132	667	48	14.3–18.5	24
60	68	35.8	25			1,532				0.3	52[e]	2
28	28	34–38	11	103		2,265				9.1		2
14	14	36	7.7	172		2,390				2.0		2
	50	50.1	7.5	100	37	5,875		170		0.12[c]		25
	75	46–48	11			5,211	4,225		1,608	70	11	26
	86	32	9.6	3		2,040				5.1		2
624	477	37	8.8	320	11.7	2,093	1,393	154		320[f]		27
775	233	43	7	15	28	3,122	1,820	130	885	670	23.6	28
	100	42	7.8	49	25	1,080	1,045	88	307	90–140[e]	30	29
	35	41.7	9.6	124	20	3,569			1,500	3.6[c]		25
	8	33–35	17.8	38	50.8					2.5–3.0[c]**		1
75	20	38.5				300				4.0[c]***		7
	5–10	29–32.5				983						1
	4	38	20.5	176	32.8	1,050				0.3[c]		1
	98	44	8.8	44	28	3,135	1,900	128	1,153	51.9****		30
3–13	10–12	34.4				1,570				1.6[c]		31
	13	46	17	0.57	38	300						32
	10	20	13	2.5	38					0.07[c]		25
168	44	33.8	7	16	35	2,900	85	125	150	3.0*	62	2
55	21	30	10.4	16			1,126–1,810		313–452	182[f]		33
	5–10	28.2	17		64					0.03[c]	10	34
920	13	28	12	47–368		2,264	700	200		1,000[f]		35
		33	11.6	2.7		2,000	825	106	361	21		2
	20	32	12.1	0.5	20–51	1,540				25		2

	Field name	State, country	County	Exploration disovery method	Productive formation	Geologic age	Trap mechanism	Reservoir rock type	Productive acreage (acres)

AD: Undefined porosity—solution gas drive

	Field name	State, country	County	Exploration disovery method	Productive formation	Geologic age	Trap mechanism	Reservoir rock type	Productive acreage (acres)
49	Big Spring	Texas	Howard	SG	Fusselman	S	5	L	1,840
50	Cairo North	Kansas	Pratt	W	Viola	O	5	D	40
51	Camp Springs	Texas	Scurry	SG	Strawn	ℙ	4	L	80
52	Chaveroo	New Mexico	Chaves, Roosevelt	GS	San Andres	P	5	D	11,000
53	Davis Ranch	Kansas	Wabaunsee	G	Kansas City	ℙ	1	L	160
54	Gove	Kansas	Gove	SG	Lansing—Kansas City	ℙ	1	L	80
55	Greensburg	Kentucky	Green, Taylor	W	Laurel	S	3	D	14,953
56	Hanson	Texas	Crockett	SG	Grayburg	P	1–5	D	160
57	Huat	Texas	Gaines	SGS	Wolfcamp	P	1–5	L	550
58	Lamesa West	Texas	Dawson	S		M	1	L	800
59	Monarch	Montana	Fallon	S		S	1–2	D	1,600
60	Montgomery***	Indiana	Daviess	S	Salem	M	1	L	160
61	Ocho Jaun	Texas	Scurry, Fisher	SG	Canyon Reef and Strawn	ℙ	4	L	1,160
62	Pennel	Montana	Fallon	GS	Interlake	S	1	D	1,760
	Pennel	Montana	Fallon	GS	Stoney Mt.	O	1	D	1,760
	Pennel	Montana	Fallon	GS	Red River	O	1	D	2,589
63	Rhodes	Kansas	Barber	W		M	5	L	6,000
64	Rocker A	Texas	Garza	SG	San Andres	P	1	D	1,740
65	Rojo Caballos	Texas	Pecos	S	Morrow	ℙ	1–2	L	
66	Snyder North	Texas	Scurry	R	Strawn Upper B Zone	ℙ	5	L	1,000
	Snyder North	Texas	Scurry	R	Strawn Lower B Zone	ℙ	5	L	1,000
	Snyder North	Texas	Scurry	R	Strawn C Zone	ℙ	5	L	
67	Welch North	Texas	Dawson, Terry	S	San Andres	P	6	D	6,500
68	Westbrook	Texas	Mitchell		Upper Clear Fork	P	5	D	480
	Westbrook	Texas	Mitchell	R	Lower Clear Fork	P	5	D	20,010

* Plus 1.9 MMbbl secondary. ** Includes No. 144. *** Intergranular porosity.

Oil and gas column (ft)	Pay thickness (ft)	Crude gravity (°API)	ϕ (%)	k (md)	S_{wi} (%)	p_i (psi)	p_b (psi)	T (°F)	Solution GOR (cu ft/bbl)	Ultimate recovery (MMbbl)	Ultimate recovery (%)	Ref. No.
88	14	48.3	11.9	4.7		4,495				4.5		2
20	4	37										31
40	5	43.5				2,644				0.02	50[e]	2
	40	26	6	0.7	25	1,340		110	400–1,000	1.6[c] 1.3×10^9 Mcf		25
20	15	31	15	41.5	50	645				3.0[c]**		36
	4–6	36–39										1
15	30	41.8	12	560		45				18	60[e]	7
12	15	27	5							0.13	53[e]	2
	16	31.2	4.5	0.4		4,600			283	1.1		2
	20	35	9	3		4,800				0.96		2
	106	29	9.5	0.75		3,692				2.0		34
40	20	37.5								0.8		7
400	162	42.6	7.6	1.85	45	2,435				5.1		2
	22											
185	180	33	11	1		3,794						34
210	62	33	11	1		3,794				0.4[c]		34
270	67	31	12.6	1	30						24	34
	0–50	18–26								2.1[c]		31
250		34	23.6	3.6						3.0		2
		53	5.5	0.1		13,221						2
66	40	40	16			3,337				5.6	140[e]	2
65	40	40	16			3,337						2
35		40	16			3,401						2
95	40	34	10	0.3		2,100				3.6[c]		2
	450	25–26	5.75	3.5		420			270			2
255	600	24	6.34	4.9		1,100			133	51		2

B: Water Drive

BA: Intercrystalline-intergranular porosity—water drive

	Field name	State, country	County	Exploration discovery method	Productive formation	Geologic age	Trap mechanism	Reservoir rock type	Productive acreage (acres)
69	Beaver Creek	Wyoming	Fremont		Madison	M	2–6	L	1,260
70	Buckner	Arkansas	Union		Smackover	J	1	L	1,610
71	Coldwater	Michigan	Isabella		Rogers City	D	1	D	3,200
72	Damme	Kansas	Finney	SG	St. Louis	M	1	L	
73	Dorcheat	Arkansas	Union		Smackover	J	1	L	2,130
74	Gila	Illinois	Jasper	SG	McClosky	M	5	L	540
75	Grayson	Texas	Reagan	W	San Andres	P	1	D	320
76	Lerado S.W.	Kansas	Reno	SG	Lansing	ℙ	1	L	200
	Lerado S.W.	Kansas	Reno	SG	Viola	O	1	D	200
77	Little Beaver	Montana	Fallon	GS	Red River	O	1	L	1,100
78	Little Beaver East	Montana	Fallon	GS	Red River	O	1	L	500
79	Llanos	Kansas	Sherman	SGS	Marmaton	ℙ	1	L	160
	Llanos	Kansas	Sherman	SGS	Cherokee	ℙ	1	L	40
80	Magnolia	Arkansas	Columbia	W	Smackover	J	1	L	4,494
81	Mt. Holly	Arkansas	Union		Smackover	J	1	L	360
82	New Richland	Texas	Navarro	W	Rodessa	K	1–2	L	10
83	Ross Ranch	Texas	King	G	Strawn	ℙ	1	L	650
84	Schuler	Arkansas	Union	W	Smackover	J	1	L	1,200

BB: Vugular-solution porosity—water drive

	Field name	State, country	County	Exploration discovery method	Productive formation	Geologic age	Trap mechanism	Reservoir rock type	Productive acreage (acres)
85	Acheson	Alberta, Canada			Leduc	D	4	L	3,640
86	Amrow	Texas	Gaines	S	Devonian	D	1	D	1,460
87	Ashburn	Kansas	Wabaunsee	SG	Viola	O	1	D-L	40
88	Bannatyne	Montana	Teton	G	Sun River	M	1	D	100
89	Bateman Ranch	Texas	King	G	Bunger	ℙ	4	L	680
90	Bonnie Glen	Alberta, Canada			Leduc	D	4	L	8,800
91	Bough Devonian	New Mexico	Lea	S	Devonian	D	1	D	240
92	Breedlove	Texas	Martin	S	Devonian	D	1	D	3,520
93	Bronco	Texas	Yoakum	S	Devonian	D	1–2	D	1,080
94	Comiskey	Kansas	Morris	G	Viola	O	1	D-L	0
95	Comiskey North-east	Kansas	Morris	SGS	Viola	O	1	D-L	240
96	Corning	Missouri	Holt	W	Kimmswick	O	1	D	
97	Delphia	Montana	Mussel-shell	S	Amsden	ℙ	1	D	480
98	Glen Park	Alberta, Canada			Leduc	D	4	L	433
99	Homeglen-Rimbey	Alberta, Canada			Leduc	D	4	L	14,053
100	Lea	Texas	Crane	S	Ellenburger	O	1–2	D	2,066
101	Livengood	Kansas	Brown	W	Hunton	S-D	1–5	D	240
102	Llanos	Kansas	Sherman	SGS	Lansing—Kansas City	ℙ	1	L-D	440
103	Lundgren	Kansas	Gove	GSG	Spergen	M	1	D	
104	Lundgren South	Kansas	Gove	GSG	Spergen	M	1	D	185

Oil and gas column (ft)	Pay thickness (ft)	Crude gravity (°API)	ϕ (%)	k (md)	S_{wi} (%)	p_i (psi)	p_b (psi)	T (°F)	Solution GOR (cu ft/bbl)	Ultimate recovery (MMbbl)	(%)	Ref. No.
500	250	42	8.7	10	10	5,300	673	232	288	6		37
			20	50		3,250				20	49	38
	30–35	49	7.1	0–1,368	15	1,453	1,190		512	20.1[c]	70	39
	10–18	30–32										1
		43.5	14	155		4,250	4,250			21.8[r]		38
	8	38.6				270				1.27	293[e]	7
15	14	32								1.2	117[e]	2
3	3	41										31
10	11	41								0.14	61[e]	31
	95	29	12	1.3	35	3,828				1.64	12	34
	65	30	13.6	4.5	35	3,828				0.85	5	34
		35.9–41.4				1,373						1
		20.2–23.5				1,115						1
308	170	39	18.5	1,500	20	3,465	3,465		700	220	50	38, 40
			20	1,130		3,405	3,405			10.5[r]	35	38
	20	23								0.01[f]		41
62	40	37	10	19		2,360				4.7		2
26		36–38	16.7	1,200		3,530	3,530	200	930	13	40	38, 42
234			9.1	3,100	10	2,530				149[f]*		43
159	40	35.4	4	2–280 (20)		5,455				10.1	203[e]	2
26		22–25	4.8–25.5	0.1–3,890	26–86	1,085				0.03[c]		36
25	5	27	10–20	2–25	43	285				0.1	40	34
20	20	33	16.5	115		1,631				2.7	198[e]	2
711			9.4	350	6	2,560				625[f]**		43
	78	43	6	400	26	4,580		164		0.09[c]		25
145	35	41.3	9			5,600				31.2	253[e]	2
265	140	43.4	6	150		4,789				25	166[e]	2
	6	19				1,015				0		36
	4	22				1,030				0.2[c]		36
	200–250	31.5				1,195						36
245		33.6				2,800				0.7		34
421			9.6	1,604	7.6	2,560				28[f]		43
553			7.6	250	10	2,570				110[f]***		43
430	75	43	1.8	v.h.		3,730	750		367	20.9	135[e]	2
18	15–20	25–27								0.1[c]		36
		33.8–39.5				1,320				0.03[c]		1
	21	35				1,820				0.005[c]		1
17	16	35	13	15	54	1,120				0.1[c]		1

	Field name	State, country	County	Exploration discovery method	Productive formation	Geologic age	Trap mechanism	Reservoir rock type	Productive acreage (acres)
105	Magutex	Texas	Andrews	S	Devonian	D	1	D-L	5,760
106	McFarland	Texas	Andrews	S	Ellenburger	O	1	D	400
107	Mill Creek	Kansas	Wabaunsee	W	Viola	O	1	D-L	80
108	Mound Lake	Texas	Terry	SG	Fusselman	S	5	D	880
109	Newbury	Kansas	Wabaunsee	W	Viola	O	1–3	D-L	160
110	Outlook and South Outlook	Montana	Sheridan	S	Winnipegosis	D	1	D	2,240
	Outlook and South Outlook	Montana	Sheridan	S	Interlake	S	1	D	2,240
	Outlook and South Outlook	Montana	Sheridan	S	Red River	O	1	D	2,240
111	Rocker A	Texas	Garza	SG	Clear Fork	P	1	D	280
112	Rosedale	Kansas	Kingman	S	Lansing—Kansas City	ℙ	1	L	320
113	Sabetha	Kansas	Nemaha	S	Hunton	D	1	D	40
	Sabetha	Kansas	Nemaha		Viola	O	1	D	30
114	Strahm	Kansas	Nemaha	S	Hunton	D	1	D	200
	Strahm	Kansas	Nemaha	S	Viola	O	1	D	80
115	Strahm East	Kansas	Nemaha		Hunton	D	1	D	80
116	Sumatra, North-west	Montana	Rosebud	S	Amsden	ℙ	1	D	200
117	Terre-Haute East	Indiana	Vigo	SG		D	1	D-L	210
118	Tex-Hamon	Texas	Dawson	S	Fusselman	S	1	D	1,200
119	Unger	Kansas	Marion	SGS	Hunton	D	1–3	D	2,700
120	Vealmoor East	Texas	Howard, Borden	S	Cisco-Wolf-camp	P	4	L	3,600
121	Wapella East	Illinois	De Witt	W	Niagaran	S	1–4	D	400
122	Wellman	Texas	Terry		Wolfcamp	P	4	L	1,200
123	Wilmington	Kansas	Wabaunsee	W	Viola	O	1	D	320
124	Wizard Lake	Alberta, Canada			Leduc	D	4	L	3,250

* Plus 10 Bcf gas-in-place (GiP). ** Plus 430 Bcf GiP. *** Plus 1.3 Tcf GiP. (B = billion $= 10^9$; T = trillion $= 10^{12}$.) [r] Original reservoir volume, oil zone.

BC: Fracture-matrix porosity—water drive

	Field name	State, country	County	Exploration discovery method	Productive formation	Geologic age	Trap mechanism	Reservoir rock type	Productive acreage (acres)
125	Woman's Pocket Anticline	Montana	Golden Valley	G	Amsden	ℙ	1	D	10
126	Big Wall	Montana	Mussel-shell	WS	Amsden[a]	ℙ	1	D	600
127	Black Leaf	Montana	Teton	S	Madison	M	1	D	
128	Cabin Creek	Montana	Fallon	S	Madison	M	1–2	D	6,700
129	Crossroads South	New Mexico	Lea	S		S-D	1–5	D-L	600
130	Deer Creek	Montana	Dawson	S	Interlake	S	1–2	D	160
	Deer Creek	Montana	Dawson	S	Red River	O	1–2	D	480
131	Dupo	Illinois	St. Clair	G	Kimmswick	O	1	L	1,020
132	Elk Basin	Wyoming	Park		Madison B Zone	M	1	D	8,960
133	Richey	Montana	Dawson, McCone	S	Madison	M	1	D	1,700
134	Sumatra, North-west	Montana	Rosebud	S	Piper	J	1	L	3,260
135	Wolf Springs	Montana	Yellow-stone	S	Amsden	ℙ	1	D	2,560

Oil and gas column (ft)	Pay thickness (ft)	Crude gravity (°API)	ϕ (%)	k (md)	S_{wi} (%)	p_i (psi)	p_b (psi)	T (°F)	Solution GOR (cu ft/bbl)	Ultimate recovery (MMbbl)	(%)	Ref. No.
150	40	46	6–7	41		5,350			635	46	200[e]	2
78	2 (?)	47.5	3.9–6	612–670		5,702				2.0		2
	4–8	24				1,000				0.3[c]		36
120	17–25	38	6–15	50–5,000	30	4,680				1.9		2
	3–5	24				1,178				0.3[c]		36
	60	40	1.5–8							}		34
125										} 3.8		
	18	40	1.5–8			4,186				}		34
20	20	33	5			4,486				0.3		34
70	15	32–34	14	165		1,229				0.3[c]		2
6	19	43				1,540						31
31		26.7	2.56–23.40	0–843	37.6	1,000				0.05[c]		36
12			7.61–14.30	0–6.06	62.9	1,180						36
None	65	23.2–26.9				910				0.2[c]		36
25	20	27.4–28.6	14.1	2.66–130.4		1,340						36
	65	24				640				0.003[c]		36
48	5–10	27.3	2–25	0–350 (2)	35	1,981				228[e]	25	34
58	12	44.1	10.5–15.8	3.6–45		630				1.8		7
167	15	40.2	7	25.3	35	5,066		154		1.1[c]	35	2
45	0–50	39–41	6.5–19.8	0.1–110	51.3–91.4	620				4.1[c]		36
610	100	46.1	10.2	267		3,362			900–1,000	37.8	105[e]	2
48	22	30.5	15	950		360				1.5		7
738	280	41–44				4,150	1,290	150	350	13.4[c]		2
17	33	20.5	18.3	1,018	48	985			50			36
646			9.4	700	7	2,510				380[f]		43
11		19				1,500						34
60	17	19	16		35	1,250				0.4[c]	25	34
	80 (?)		7.5			800						34
	25	33	12	3.25	20	4,180					30	34
	25	52	10		25	4,867		168	500			25
	38	43	7		35					135[e]	25	34
	90	41	6.7		35	4,500				83[e]	24	34
100	100	32.7	14	7.7						2.9		7
920	24	28	10–11.5	3–11		1,400			350	1,000[f]		35
	53	39	4		30	3,539				4	25	34
10	40	27.3				1,973						34
	5–30	31	1.6–4.8	0.1–1,000	25	2,758				5.4	50	34

	Field name	State, country	County	Exploration discovery method	Productive formation	Geologic age	Trap mechanism	Reservoir rock type	Productive acreage (acres)

BD: Undefined porosity—water drive

	Field name	State, country	County	Exploration discovery method	Productive formation	Geologic age	Trap mechanism	Reservoir rock type	Productive acreage (acres)
136	Ash Grove	Kansas	Dickinson	SG		M	1–5	D	480
137	Bantam	Nebraska	Harlan		Kansas City	IP		L	
138	Barada	Nebraska	Richardson		Hunton	D	1	D	
139	Bredette	Montana	Roosevelt	S	Charles	M	1	D-L	200
140	Bredette North	Montana	Daniels, Roosevelt	S	Madison	M	1	D-L	320
141	Brown	Texas	Gaines	SG	Wichita-Albany	P	4	D	680
142	Cabin Creek	Montana	Fallon	S		S-O	1–2	D	6,700
143	Cary	Mississippi	Sharkey	S	Selma		1	L	120
144	Davis Ranch	Kansas	Wabaunsee	G	Hunton	D	1	D	40
	Davis Ranch	Kansas	Wabaunsee	G	Viola	O	1	D	680
145	Dawson	Nebraska	Richardson		Hunton	D	1	D	
	Dawson	Nebraska	Richardson		Viola	O	1	L	
146	Dollarhide East	Texas	Andrews	S	Fusselman	S	1–2	L	800
	Dollarhide East	Texas	Andrews	S	Ellenburger	O	1–2	D	1,680
147	Eagle Springs	Nevada	Nye	W	Sheep Pass	T (e)	2	L	
148	Fairplay	Kansas	Marion	G	Hunton	S-D	1	D	120
149	Falls City	Nebraska	Richardson	GS	Hunton	S-D	1–2	D	1,080
150	Fradean	Texas	Upton	SG	Ellenburger	O	1	D	800
151	Gingrass	Kansas	Harvey	SG	Hunton	S-D	1–5	D	30
152	Goodrich	Kansas	Sedgwick	W	Lansing—Kansas City	IP	1	L	600
	Goodrich	Kansas	Sedgwick	W		M	1	L	600
	Goodrich	Kansas	Sedgwick	W	Hunton	S-D	1	D	600
153	Greenwich	Kansas	Sedgwick	W	Hunton	S-D	1–5	D	400
	Greenwich	Kansas	Sedgwick	W	Arbuckle	O	1–5	D	80
154	Hausserman	Nebraska	Harlan		Kansas City	IP	1	L	
155	Howard Glasscock	Texas	Howard, Glasscock, Sterling, Mitchell	G	Glorieta—Clear Fork	P	1	D	7,120
156	Huat Canyon	Texas	Gaines	SGS		IP-P	4	L	850
157	Hugoton	Kansas	Morton, Stevens,	W	Herington	P	5	L-D	2,560,000
	Hugoton	Kansas	Seward, Stanton,	W	Krider	P	5	L-D	2,560,000
158	Hugoton	Kansas	Grant,	W	Winfield	P	5	L-D	2,560,000
	Hugoton	Kansas	Haskell, Hamilton,	W	Towanda	P	5	L	2,560,000
	Hugoton	Kansas	Kearney, Finney	W	Fort Riley	P	5	L	2,560,000
159	Hutex	Texas	Andrews	S	Devonian	D	1	D-L	4,480
160	John Creek	Kansas	Morris	G	Viola	O	1	D	1,960
161	Lerado	Kansas	Reno	W	Lansing	IP	1	L	750
	Lerado	Kansas	Reno	W		M	1	L	750
	Lerado	Kansas	Reno	W	Viola	O	1	D	750
162	Magutex	Texas	Andrews	S	Ellenburger	O	1	D	2,160
163	Pine	Montana	Fallon	GS	Interlake	S	1	D	8,380
	Pine	Montana	Prairie	GS	Stoney Mt.	O	1	D	8,380
	Pine	Montana	Wibaux	GS	Red River	O	1	D-L	8,380
164	Shubert	Nebraska	Richardson		Hunton	D	1	D	

Oil and gas column (ft)	Pay thickness (ft)	Crude gravity (°API)	ϕ (%)	k (md)	S_{wi} (%)	p_i (psi)	p_b (psi)	T (°F)	Solution GOR (cu ft/bbl)	Ultimate recovery (MMbbl)	(%)	Ref. No.
20	3	33.3				880				0.2[c]		36
	12	27										44
	30	29								2.6[c]		44
15	19	40	6		56	3,110				0.2	50	34
47	18	38	6		53	3,125				0.5	50	34
146	95	36.8	4.5	79	30	3,043	361	131	254	3.0		2
	695	33	12	3.25	20	4,180				51	30	34
	2	24.8	30	0–1,124	45					0.2[c]		45
7	7	19.3				992						36
61	20	29.5	12	188	64	1,120						36
	20	23								} 0.5[c]		44
	13	23										44
100	62	41	7		28	4,564				0.7[c]		2
250	107	45	5		30	4,609				1.9[c]		2
		26–29						200		2.0[c]		46
14	7	36–37	8			820				0.3[c]		36
93	23	30								5.0[c]	197[e]	36
370	35	50	5			4,200				6.0		2
11		34				975				0.04[c]		36
	52	34.2										36
	30									} 5.2[c]		36
	18	39										36
35	35	41.5								13[c]*		36
	2–3	36.5										36
	12	36										44
		27.4	10	4	40							2
500	115	37.6	6.7	61		4,539			361	{ 3.9 × 10⁶ Mcf		2
	5–10	Gas	7.7–15	0.3–0.8	50–65	435	350			{ 9.6		1
	25–35	Gas	8–15.6	0.6–4.5	24.8–38					}		1
	20–35	Gas	10.5–18.1	0.8–5.4	31.6–42.5					} 4.4 × 10⁹ Mcf[c]		1
	25–40	Gas	14.6–18.7	1.5–22	22.8–39.6					}		1
	15–20	Gas	14–18	3.8–5.4	27					}		1
270	60	44	4–7	12–76	34	5,400				25.4		2
75	18	25.5	9	135	63.5	1,050				1,3[c]		36
3		43.2								} 2.4		31
40		Gas								} 2.2 × 10⁶ Mcf		31
10		43.2										31
70	19	45.1	2–3	23		6,013			322	10[c]		2
	140	34	12	3.78	30	4,160				}		34
	80	34	12	3.78	30	4,160				} 55	30	34
	225	34	12	3.78	30	4,160				}		34
	7	23										44

No.	Field name	State, country	County	Exploration discovery method	Productive formation	Geologic age	Trap mechanism	Reservoir rock type	Productive acreage (acres)
165	Snethen	Nebraska	Richardson		Viola	O	1	L	
166	Tex-Hamon	Texas	Dawson	S	Montoya	O	1	D	400
167	Valley Center	Kansas	Sedgwick	W		M	1–5	L	
168	Wells Devonian	Texas	Dawson	SGS		S-D	1	D-L	1,200
169	Willowdale	Kansas	Kingman	S	Viola	O	1	D	840

* Includes No. 228.

C: Gas-Cap Drive

CA: Intercrystalline-intergranular porosity—gas-cap drive

No.	Field name	State, country	County	Exploration discovery method	Productive formation	Geologic age	Trap mechanism	Reservoir rock type	Productive acreage (acres)
170	Eubank	Kansas	Haskell	W	Lansing—Kansas City—E Zone	ℙ	1–5	L	960
171	Maydelle	Texas	Cherokee	G	Rodessa	K	1	L	1,200
172	Novinger	Kansas	Meade	SG	Marmaton	ℙ	4	L	2,400
173	Opelika	Texas	Henderson		Rodessa	K	1–2	L	11,200
174	Plainville	Indiana	Daviess	S	Salem	M	1	L	200
175	Wilde	Kansas	Morris		Lansing	ℙ	1	L	1,080
176	Wilsey	Kansas	Morris		Lansing	ℙ	1	L	640

CB: Vugular-solution porosity—gas-cap drive

No.	Field name	State, country	County	Exploration discovery method	Productive formation	Geologic age	Trap mechanism	Reservoir rock type	Productive acreage (acres)
177	Coyanosa	Texas	Pecos	SGS	Devonian	D	1–2	D	13,300
178	Fishhook	Illinois	Adams, Pike	W	Edgewood	S	1	D	5,000
179	Wichita	Kansas	Sedgwick	W	Viola	O	1	D	1,100

CC: Fracture-matrix porosity—gas-cap drive

No.	Field name	State, country	County	Exploration discovery method	Productive formation	Geologic age	Trap mechanism	Reservoir rock type	Productive acreage (acres)
180	Brown-Bassett	Texas	Terrell, Crockett	S	Ellenburger	O	1–2	D	16,000
181	Coyanosa	Texas	Pecos	SGS	Ellenburger	O	1–2	D	8,960

CD: Undefined porosity—gas-cap drive

No.	Field name	State, country	County	Exploration discovery method	Productive formation	Geologic age	Trap mechanism	Reservoir rock type	Productive acreage (acres)
182	Fradean	Texas	Upton	SG	Devonian	D	1	L	800
183	Greenwood	Kansas, Colorado	Morton, Baca	W	Greenwood	ℙ	5–6	L	160,000
184	Levelland Northeast	Texas	Hockley	SGS	Strawn	ℙ	1–5	D-L	1,440
185	Loring	Mississippi	Madison	S	Smackover		1	D-L	1,280
186	Puckett	Texas	Pecos	SG	Devonian	D	1–2	L-D	12,160
187	Puckett	Texas	Pecos	SG	Ellenburger	O	1–2	D	23,680

Oil and gas column (ft)	Pay thickness (ft)	Crude gravity (°API)	ϕ (%)	k (md)	S_{wi} (%)	p_i (psi)	p_b (psi)	T (°F)	Solution GOR (cu ft/bbl)	Ultimate recovery (MMbbl)	(%)	Ref. No.
		27				1,055				0.12[c]		44
67	113	39	7	23.1	20	5,051		157			35	2
	0–113	34–35										36
175	15	37	7–11			5,180				3.36		2
30	6	41.1										31
36	14	Gas	16	61		1,105						1
	20	52				4,035						41
	15.3	42								2.0[c]		1
40	20	44	11–17	59		3,400	3,400		1,400		20–30	41, 47
100	28	Gas				550				3.4×10^6 Mcf		7
		Gas										36
	12	Gas										36
1,068	162	53.4	8.1	10	20	5,937				5.6×10^6 Mcf[c]		2
	14	Gas	20	54		119				3 Bcf		7
20	4	38	7.8–14.6	0.8–4,240	44.3–75.9	1,280				0.8[c]		36
500	135	Gas	6.7	1–100	35	6,648				1.5×10^8 Mcf[c]		2
1,715	790	51.2[d]	1.33	0.56	10	6,964				4.3×10^6 Mcf[c]		2
580	140	56[d]	12	1		3,000				$1.0 + 3 \times 10^7$ Mcf		2
	65	Gas				435				10^9 Mcf		1
	40	44	10		27	3,250						2
	361	49.4	14	0–27.9	25	5,890			21,000	0.8[c]		45
181	180	49.5[d]								1.9×10^8 Mcf[c]		2
1,600	1,030	53[d]								7.9×10^8 Mcf[c]		2

	Field name	State, country	County	Exploration discovery method	Productive formation	Geologic age	Trap mechanism	Reservoir rock type	Productive acreage (acres)

D: Open Combination Drive

DA: Intercrystalline-intergranular porosity—open combination drive

	Field name	State, country	County	Exploration discovery method	Productive formation	Geologic age	Trap mechanism	Reservoir rock type	Productive acreage (acres)
188	Abqaiq	Saudi Arabia			Arab D	J	1	L	89,300
189	Adell Northwest	Kansas	Decatur	W	Lansing—Kansas City	IP	1	L	320
190	Aden Consolidated and Aden South	Illinois	Wayne, Hamilton	S	Ohara	M	1	L	2,700
	Aden Consolidated and Aden South	Illinois	Wayne, Hamilton	S	Rosiclare	M	1	L	2,700
	Aden Consolidated and Aden South	Illinois	Wayne, Hamilton	S	McClosky	M	1	L	2,700
	Aden Consolidated and Aden South	Illinois	Wayne, Hamilton	S	Salem	M	1	L	2,700
	Aden Consolidated and Aden South	Illinois	Wayne, Hamilton	S	Warsaw	M	1	L	2,700
191	Bahrain	Bahrain			Second Pay	K	1	L	
192	Dale Consolidated	Illinois	Hamilton, Saline	S	Ohara	M	1	L	20,000
	Dale Consolidated	Illinois	Hamilton, Saline	S	Rosiclare	M	1	L	20,000
	Dale Consolidated	Illinois	Hamilton Saline	S	McClosky	M	1	L	20,000
193	Damme	Kansas	Finney	SG	Marmaton	IP	1	L	
194	Feely	Kansas	Decatur	W	Lansing	IP	5	L	160
195	Ghwar-Ain-Dar	Saudi Arabia			Arab D	J	1	L-D	4,000
196	Ghwar-Fazran	Saudi Arabia			Arab D	J	1	L-D	136,500
197	Ghwar-Harah	Saudi Arabia			Arab D	J	1	L-D	110,100
198	Ghwar-Hawiyan	Saudi Arabia			Arab D	J	1	L-D	298,600
199	Ghwar-Shedgum	Saudi Arabia			Arab D	J	1	L-D	4,100
200	Ghwar-Uthmaniyan	Saudi Arabia			Arab D	J	1	L-D	47,400
201	Howard Glasscock	Texas	Howard, Glasscock, Sterling, Mitchell	G	San Andres Grayburg	P	1	D	15,620
202	Jingo	Kentucky	Ohio	SG	Salem	M	5	L	90
203	Pleasant Prairie	Kansas	Finney, Kearney, Haskell	W	Lansing—Kansas City	IP	5	L	
	Pleasant Prairie	Kansas	Finney, Kearney, Haskell	W	St. Louis	M	1–5	L	10,000

Oil and gas column (ft)	Pay thickness (ft)	Crude gravity (°API)	ϕ (%)	k (md)	S_{wi} (%)	p_i (psi)	p_b (psi)	T (°F)	Solution GOR (cu ft/bbl)	Ultimate recovery (MMbbl)	Ultimate recovery (%)	Ref. No.
	200	29–38	22	500	7–8	3,395	2,520		950	8,700		48
50	95	31–39					1,170			0.8[c]		1
		38–41.5								15.5		7
	5–12											
62						700						
16	8–27		7.5			600–1,040						
12	8–20		9			1,750						
	110	2.2[b]	20–30	40		1,236	278	140	142			49
			14			800						7
	5–15									125*		
		34.8–37.2				800						
	5	30–31								0.6[c]		1
20	15	34–39				1,150				0.3[c]		1
124	96	39	23	340			1,632	222	510	149		50, 51
1,140	143	35	22	290			1,920	215	570	7,515		50, 51
987	150	35	23	270			1,910	215	570	6,443		50, 51
1,369	103	33	21	220			1,830	215	520	10,989		50, 51
1,094	61	33	19	68			1,715	215	450	83		50, 51
1,126	60	33	16	52			1,595	215	420	429		50, 51
	80	31.7	10.5	5	40							2
82	8–27	41.5	8–12.5	0.1–125		280 DST						7
	12	33				1,025						1
										1.2[c]**		
	5–38	32–36	14.3	64.5	46.1	1,045						1

	Field name	State, country	County	Exploration disovery method	Productive formation	Geologic age	Trap mechanism	Reservoir rock type	Productive acreage (acres)
DB: Vugular-solution porosity—open combination drive									
204	Enlow	Kansas	Edwards	S	Lansing	ℙ	1	L	160
205	Gage	Montana	Mussel-shell	G	Amsden	ℙ	1–5	D	320
206	Leduc-Woodbend	Alberta, Canada			Leduc	D	4	L	21,640
207	New Baden, East	Illinois	Clinton	SG		S	4	L	320
208	Nunn	Kansas	Finney	S		M	5	L-D	1,100
209	Ozona, East	Texas	Crocket	SG	Ellenburger	O	2	D	
210	Pegasus	Texas	Midland, Upton		Ellenburger	O	1–2	D	10,000
211	Redwater	Alberta, Canada			Leduc	D	4	L	37,833
212	Ropes and South Ropes	Texas	Hockley	S	Canyon	ℙ	4	L	1,320
213	Shallow Water	Kansas	Scott	S	St. Louis	M	1–5	L	500
214	Westerose	Alberta, Canada			Leduc	D	4	L	1,757
DC: Fracture-matrix porosity—open combination drive									
215	Blackfoot	Montana	Glacier	S	Sun River	M	1–7	D-L	440
216	Gypsy Basin	Montana	Pondera, Teton	W	Madison	M	2	L-D	
217	Waterloo	Illinois	Monroe	GW	Kimmswick	O	1	L	230
218	West Brady	Montana	Pondera	SG	Sun River	M	1–5	D	10
219	West Edmond	Oklahoma	Canadian	W	Bois d'Arc	M	3–6	L	29,240

Oil and gas column (ft)	Pay thickness (ft)	Crude gravity (°API)	ϕ (%)	k (md)	S_{wi} (%)	p_i (psi)	p_b (psi)	T (°F)	Solution GOR (cu ft/bbl)	Ultimate recovery (MMbbl)	Ultimate recovery (%)	Ref. No.
	8	34								0.09[c]		31
	18	33.9	10		48					0.5	25	34
232	35.2	39	8.0	1,000	15	1,894	1,894	150	553	167	55	43, 52
										420 Bcf		
	161	39.4				655 DST				0.2		7
	15	26–28	9.5–11.9	22–166	23							1
	15	54–72[d]	7.4		45	3,391				275 Mcf/ac-ft		2
800		53	2–8		30	5,668	3,386		1,556		40	53
101	101	2.7[b]	6.5	500	25	1,050	485	94	195	817	64	54
100		42	8.5	66	26.9	3,754			368	59		2
	100	26–28								2.0[c]		1
	589	42	9.3	1,934	7	2,570	2,570	178	650	121	77***	43
										117 Bcf		
6–15		25	14.2	264		955						34
133	40	34.5	14	305		1,038						34
50	30	30.2								0.24		7
10	15	28.1	7.4	40						0.005		34
600	70	41	5.3	0–20	15	3,145	2,770	145–150	1,010	105	25–35	55, 56
										961 Bcf		

DD: Undefined porosity—open combination drive

No.	Field name	State, country	County	Exploration discovery method	Productive formation	Geologic age	Trap mechanism	Reservoir rock type	Productive acreage (acres)
220	Crosset South—El Cinco	Texas	Crockett	SG	Devonian	D	2–3	L	5,420
221	Dale Consolidated	Illinois	Hamilton, Saline	S	St. Louis	M	1	L-D	20,000
222	Deerhead	Kansas	Barber	G	Viola	O	2–5	D	880
223	Dwyer	Montana	Sheridan	S	Madison	M	1	L	
224	Fanska	Kansas	Marion	W	Warsaw	M	1	D	240
225	Fertile Prairie	Montana	Fallon	S	Red River	O	1	L-D	240
226	Gas City	Montana	Dawson	S	Red River	O	1–2	D	1,080
227	Glendive	Montana	Dawson	S	Red River	O	1–2	L	1,040
	Glendive	Montana	Dawson	S	Stony Mountain	O	1–2	D	1,040
228	Greenwich	Kansas	Sedgwick	W		M	1–5	L	1,400
229	Hardesty	Kansas	Decatur	W	Lansing—Kansas City	ℙ	1	L	160
230	Neva West	Texas	Schleicher	S	Strawn	ℙ	4	L	2,052
231	Pondera	Montana	Pondera, Teton	W	Madison	M	3–6	D	6,000
232	Valley Center	Kansas	Sedgwick	W	Lansing—Kansas City	ℙ	1	L	160
	Valley Center	Kansas	Sedgwick	W	Viola	O	5	D	1,800
233	Warner	Kansas	Decatur	SGS	Lansing—Kansas City	ℙ	1–2	L	360

* Includes No. 221. ** Includes No. 37. *** Efficient gravity segregation drive; with some gas injection. **** Plus 1.4×10^3 Mcf gas.

E: Closed Combination Drive

EA: Intercrystalline-intergranular porosity—closed combination drive

No.	Field name	State, country	County	Exploration discovery method	Productive formation	Geologic age	Trap mechanism	Reservoir rock type	Productive acreage (acres)
234	Turner Valley	Alberta, Canada		G	Turner Valley	M	1–2	D	16,275

Oil and gas column (ft)	Pay thickness (ft)	Crude gravity (°API)	ϕ (%)	k (md)	S_{wi} (%)	p_i (psi)	p_b (psi)	T (°F)	Solution GOR (cu ft/bbl)	Ultimate recovery (MMbbl)	Ultimate recovery (%)	Ref. No.
500	0–100	42	20	3.5	40	2,500			750	17		
	5–15					200						7
65	15 (?)	32								0.8[c]****		31
	50	32.9	10									34
	13	33				750				0.3[c]		36
	49	33.4	5–6	24	30	3,520				1.1	24	34
	31	38.2	11		30	160				4.0	25	34
320	150	38	6.5		35					10.4	24	34
60		38	6.5		35							34
	60	44								13.0[c]		36
16	4–15	29–39				1,170				0.3[c]		1
166		45	7	2.5	28	2,551			849	14.0		2
58	30	27–38	14	82	45					19.0	35	34
12	2–14	34–35								23.0[c]		36
76	20–30	34–35										36
	4	34.6				930						1

Oil and gas column (ft)	Pay thickness (ft)	Crude gravity (°API)	ϕ (%)	k (md)	S_{wi} (%)	p_i (psi)	p_b (psi)	T (°F)	Solution GOR (cu ft/bbl)	Ultimate recovery (MMbbl)	Ultimate recovery (%)	Ref. No.
	156	39	9.8–12.8	6.84	10	2,775	758	149	880	1,050[f]		57

No.	Field name	State, country	County	Exploration discovery method	Productive formation	Geologic age	Trap mechanism	Reservoir rock type	Productive acreage (acres)

EB: Vugular-solution porosity—closed combination drive

235	Allison Northwest	New Mexico	Lea, Roosevelt	S	Cisco	ℙ		L	12,175
236	Rainbow Area (typical)	Alberta, Canada		S	Keg River	D	4	L-D	978
237	Zama Area (typical)	Alberta, Canada		S	Keg River	D	4	L	

F: Unknown Drive

FA: Vugular-solution porosity—unknown drive

238	Tex Hamon	Texas	Dawson	S	Strawn	ℙ	1	L	80

FB: Fracture-matrix porosity—unknown drive

239	Indian Basin	New Mexico	Eddy	SGS		ℙ	1	D	34,000

FC: Undefined porosity—unknown drive

240	Bear's Den	Montana	Liberty	SG	Madison	M	1–5	L	250
241	Otto	Texas	Schleicher	SG	Strawn	ℙ	2–5	L	480
242	Tex-Hamon	Texas	Dawson	S		M	1	L	240
243	Valley Center	Kansas	Sedgwick	W	Hunton	D	5		

Field name	State, country	County	Exploration discovery method	Productive formation	Geologic age	Trap mechanism	Reservoir rock type	Productive acreage (acres)

Oil and gas column (ft)	Pay thickness (ft)	Crude gravity (°API)	ϕ (%)	k (md)	S_{wi} (%)	p_i (psi)	p_b (psi)	T (°F)	Solution GOR (cu ft/bbl)	Ultimate recovery (MMbbl)	Ultimate recovery (%)	Ref. No.
	10.5	47–49	7.07	200	26	3,363		154		17	35	25
500	180–302	36.1–40	7–11	159–460	5.6–8.2	2,501–2,623		187–195	355–765	483	32	58
800	172	32	7		19							58
70	32	49.1	6.5–7									2
	207		4.3	0.1–1,780	25	2,921		146		0.13[c]		25
	18	39	15		35	470				0.1	15	34
30	6	38–41							137	0.03[c]		2
		37.5	6–18							0.09[c]		2
												36

Oil and gas column (ft)	Pay thickness (ft)	Crude gravity (°API)	ϕ (%)	k (md)	S_{wi} (%)	p_i (psi)	p_b (psi)	T (°F)	Solution GOR (cu ft/bbl)	Ultimate recovery (MMbbl)	Ultimate recovery (%)	Ref. No.

References to Appendix C

1. Kansas Geological Society: *Kansas Oil and Gas Fields*, Vol. II, *Western Kansas*, Kansas Geol. Soc. (1959) 207.
2. West Texas Geological Society: *Oil and Gas Fields in West Texas*, West Texas Geol. Soc., Midland, Tex. (1966) 398.
3. Herbeck, E. F. and Blanton, J. R.: "Ten Years of Miscible Displacement in Block 31 Field", *J. Pet. Tech.* (June, 1961) 543–549.
4. Goss, L. E. and Vague, J. R.: "Pressure Maintenance Operations—Fort Chadbourne Field—Odom Lime Reservoir", paper SPE 406, 37th *Annual Fall Meeting, Los Angeles, Calif.* (Oct. 7–10, 1962).
5. Gealy, Jr., F. D.: "North Foster Unit—Evaluation and Control of a Grayburg-San Andres Waterflood Based on Primary Oil Production and Waterflood Response", paper SPE 1474 presented at SPE Permian Basin Oil Recovery Conference, Midland, Tex. (May 8–9, 1967).
6. Hoss, R. L.: "Calculated Effect of Pressure Maintenance on Oil Recovery", *J. Pet. Tech.* (Sept., 1947) 121–130.
7. Miller, Jr., D. N.: *Geology and Petroleum Production of the Illinois Basin*, Illinois and Indiana-Kentucky Geol. Socs. (1968) 301.
8. Donohoe, C. W. and Bohannan, D. L.: "Harmattan-Elkton Field—A Case for Engineered Conservation and Management", *J. Pet. Tech.* (Oct., 1965) 1171–1178.
9. Akins, Jr., D. W.: "Primary High Pressure Water Flooding in the Pettit Lime Haynesville Field", *Pet. Trans.*, AIME (1951) Vol. 192, 239–248.
10. Goolsby, J. L. and Anderson, R. C.: "Pilot Water Flooding in a Dolomite Reservoir, the McElroy Field", *J. Pet. Tech.* (Dec., 1964) 1345–1350.
11. Trube, Jr., A. S.: "High-Pressure Water Injection for Maintaining Reservoir Pressures, New Hope Field, Franklin County, Texas", *Pet. Trans.*, AIME (1950) Vol. 189, 325–334.
12. Gray, R. and Kenworthy, J. D.: "Early Results Show Wide Range of Recoveries in Two Texas Panhandle Water Floods", *J. Pet. Tech.* (Dec., 1962) 1323–1326.
13. Henry, J. C. and Moring, J. D.: "Flood Evaluation Yields Vital Guidelines", *Pet. Eng.* (Aug., 1968) 57–59.
14. Neslage, F. J.: "Gas Injection in Dolomite Reservoir, West Pampa Repressuring Association Project as of January 1, 1951", *Proc.*, Second Oil Recovery Conference, Texas Petroleum Research Committee (1951) 119–142.
15. Marrs, D. G.: "Field Results of Miscible-Displacement Program Using Liquid Propane Driven by Gas, Parks Field Unit, Midland County, Texas", *J. Pet. Tech.* (Apr., 1961) 327–332.
16. Barton, H. B. and Dykes, Jr., F. R.: "Performance of the Pickton Field", *Pet. Trans.*, AIME, *Reprint Series* 4, 83–88.
17. Sessions, R. E.: "How Atlantic Operates the Slaughter Flood", *Oil and Gas J.* (July 4, 1960) 91–98.
18. Hiltz, R. G., Huzarevich, J. V. and Leibrock, R. M.: "Performance Characteristics of the Slaughter Field Reservoir", *Proc.*, Second Oil Recovery Conference, Texas Petroleum Research Committee (1951) 146–157.
19. Sessions, R. E.: "Small Propane Slug Proving Success in Slaughter Field Lease", *J. Pet. Tech.* (Jan., 1963) 31–36.
20. Fickert, W. E.: "Economics of Water Flooding the Grayburg Dolomite in South Cowden Field", *Proc.*, Twelfth Annual Meeting of the Southwestern Petroleum Short Course, Texas Technological College, Lubbock, Tex. (Apr. 22–23, 1965) 21–31.
21. Allen, W. W., Herriot, H. P. and Stiehler, R. D.: "History and Performance Prediction of Umm Farud Field, Libya", *J. Pet. Tech.* (May, 1969) 570–578.
22. Borgan, R. L., Frank, J. R. and Talkington, G. E.: "Pressure Maintenance by Bottom-Water Injection in a Massive San Andres Dolomite Reservoir", *J. Pet. Tech.* (Aug., 1965) 883–888.

23. Miller, F. H. and Perkins, A.: "Feasibility of Flooding Thin, Tight Limestones", *Pet. Eng.* (Apr., 1960) B55–B75.
24. Burchell, P. W. and Coonts, H. L.: "Review of Secondary Recovery Operations in the Greater Aneth Area, San Juan County, Utah", *Producers Monthly* (July, 1964) 10–14.
25. Kinney, E. E. and Schatz, F. L.: *The Oil and Gas Fields of Southeastern New Mexico*, Roswell Geol. Soc., Roswell, N.M. (1967) 185.
26. Latimer, Jr., J. R. and Oliver, F. L.: "The Fairway Field of East Texas: Its Development and Efforts Toward Unitization", paper SPE 703 presented at SPE 38th Annual Fall Meeting, New Orleans, La. (Oct. 6–9, 1963).
27. Larson, V. C., Peterson, R. B. and Lacey, J. W.: "Technology's Role in Alberta's Golden Spike Miscible Project", paper P.D. 12(9) presented at the Seventh World Pet. Cong., Mexico City (1967).
28. Allen, H. H. and Thomas, J. B.: "Pressure Maintenance in SACROC Unit Operations, January 1, 1959", *J. Pet. Tech.* (Nov., 1959) 42–48.
29. Sutton, E.: "Trempealeau Reservoir Performance, Morrow County Field, Ohio", *J. Pet. Tech.* (Dec., 1965) 1391–1395.
30. Lacik, H. A. and Black, Jr., J. L.: "Pressure Maintenance Operations in the Sharon Ridge Canyon Unit, Scurry County, Texas", *J. Pet. Tech.* (July, 1961) 645–648.
31. Kansas Geological Society: *Kansas Oil and Gas Pools*, Vol. I, *South Central Kansas*, Kansas Geol. Soc. (1956) 97.
32. Wasson, J. A.: "Secondary Oil-Recovery Possibilities in the Basal Greenbrier Dolomite Zone, Sycamore-Millstone Field, Sherman District, Calhoun County, W. Va.", *USBM Rept. Invest.* 7049 (1967) 20.
33. Willingham, R. W. and McCaleb, J. A.: "The Influence of Geologic Heterogeneities on Secondary Recovery from the Permian Phosphoria Reservoir, Cottonwood Creek, Wyoming", paper SPE 1770 presented at SPE Rocky Mountain Regional Meeting, Casper, Wyo. (May 22–23, 1967).
34. Abrassart, C. P., Nordquist, J. W. and Johnson, M. C.: *Montana Oil and Gas Fields Symposium*, Billings Geol. Soc. (1958) 247.
35. Wayhan, D. A. and McCaleb, J. A.: "Elk Basin Madison Heterogeneity—Its Influence on Performance", *J. Pet. Tech.* (Feb., 1969) 153–159.
36. Curtis, G. R.: *Kansas Oil and Gas Fields*, Vol. III, *Northeastern Kansas*, Kansas Geol. Soc. (1960) 220.
37. Pollock, C. B: "Beaver Creek Madison, Wyoming's Deepest Water Injection Project", *J. Pet. Tech.* (Jan., 1960) 39–41.
38. Bruce, W. A.: "A Study of the Smackover Limestone Formation and the Reservoir Behavior of Its Oil and Condensate Pools", *Pet. Trans.*, AIME (1944) Vol. 155, 88–119.
39. Criss, C. R. and McCormick, R. L.: "History and Performance of the Coldwater Oil Field, Michigan", *Pet. Trans.*, AIME, *Reprint Series* 4, 55–63.
40. Winham, H. F.: "An Engineering Study of the Magnolia Field in Arkansas", *Pet. Trans.*, AIME (1943) Vol. 151, 15–34.
41. Herald, F. A.: *Occurrence of Oil and Gas in Northeast Texas*, Univ. of Texas, Bureau of Economic Geology, Austin, Tex. (1951) 449.
42. Kaveler, H. H.: "Engineering Features of the Schuler Field and Unit Operations", *Pet. Trans.*, AIME (1944) Vol. 155, 58–87.
43. Hnatiuk, J. and Martinelli, J. W.: "The Relationships of the Westerose D-3 Pool to Other Pools of the Common Aquifer", *J. Can. Pet. Tech.* (Apr.–June, 1967) 43–49.
44. Finch, W. C., Cullen, A. W., Sandberg, G. W., Harris, J. D. and McMahon, B. E.: *The Oil and Gas Fields of Nebraska*, Rocky Mtn. Assoc. Geologists, Denver, Colo. (1955) 264.
45. Frascogna, X. M.: *Mesozoic-Paleozoic Producing Areas of Mississippi and Alabama*, Mississippi Geol. Soc., Jackson, Miss. (1957) 139.

46. Garfield, R. F.: *International Oil and Gas Development*, Intern. Oil Scouts Assoc., Austin, Tex. (1969) Part 1, Vol. XXXIX, 496.
47. Clay, T. W.: "Pressure Maintenance by Gas Injection . . . in Opelika Field of Henderson County, Texas", *Oil and Gas J., Reprint Series on Secondary Recovery*, 54–57.
48. Stanley, T. L.: "Approximation of Gas-Drive Recovery and Front Movement in the Abqaiq Field, Saudi Arabia", *Pet. Trans.*, AIME (1960) Vol. 219, 274.
49. Cotter, W. H.: "Twenty-Three Years of Gas Injection into a Highly Undersaturated Crude Reservoir", *J. Pet. Tech.* (Apr., 1962) 361–365.
50. Bramkamp, R. A. and Powers, R. W.: "Classification of Arabian Carbonate Rocks", *Bull.*, GSA (Oct., 1958) Vol. 69, No. 10, 1305–18.
51. Steineke, M., Bramkamp, R. A. and Sander, N. J.: "Stratigraphic Relations of Arabian Jurassic Oil", *Habitat of Oil*, Am. Assoc. Pet. Geol. (1958) 1294–1329.
52. Horsfield, R.: "Performance of the Leduc D-3 Reservoir", *Pet. Trans.*, AIME, *Reprint Series* 4, 65–72.
53. Cargile, L. L.: "A Case History of the Pegasus Ellenburger Reservoir", *J. Pet. Tech.* (Oct., 1969) 1330–1336.
54. Willmon, G. J.: "A Study of Displacement Efficiency in the Redwater Field", *J. Pet. Tech.* (Apr., 1967) 449–456.
55. Elkins, L. F.: "Internal Anatomy of a Tight, Fractured Hunton Lime Reservoir Revealed by Performance—West Edmond Field", *J. Pet. Tech.* (Feb., 1969) 221–232.
56. Littlefield, M., Gray, L. L. and Godbold, A. C.: "A Reservoir Study of the West Edmond Hunton Pool, Oklahoma", *Pet. Trans.*, AIME, *Reprint Series* 4, 89–107.
57. White, R. J.: *Oil Fields of Alberta*, Alberta Soc. Pet. Geol. (1960) 272.
58. Robertson, J. W.: "New Life Begins for Alberta's Keg River Pinnacle Reefs", *World Oil* (Sept., 1969) 87–90.

Author Index*

Numbers set in *italics* designate the page numbers on which the complete literature citation is given.

* Prepared by Abdulrahman El-Nassir, John O. Robertson, Jr., Lili Chiang and George V. Chilingar.

Subject Index*

*Prepared by John O. Robertson, Jr., Abdulrahman El-Nassir and George V. Chilingar. Appendix C, which is presented in a tabular form, is not indexed.

The Editors

DR. GEORGE V. CHILINGAR is a Professor of Petroleum Engineering at the University of Southern California. He was awarded the degrees of B.E. and M.S. in petroleum engineering (1949 and 1950) and Ph.D. in geology and petroleum engineering in 1956 from the same university. Professor Chilingar has acted as consultant for many companies and held the position of senior petroleum engineering advisor for the United Nations. He was president and vice-president of two engineering firms and, in 1966, was awarded membership in Executive and Professional Hall of Fame. Dr. Chilingar, who is an author of more than one hundred research articles and ten books, is a recipient of numerous awards from various organizations and governments. Several of his books, including three books on carbonate rocks, were translated into foreign languages. Professor Chilingar is a member of the SPE of AIME, AAPG, AGU, GSA, ASEE, SEPM, Geochemical Society, American Institute of Chemists, and the Academies of Science of New York and California.

ROBERT W. MANNON, a consulting petroleum engineer since 1960, has worked for various major and independent oil companies in the United States, Canada, and the Middle East. From 1963 to 1965, he was Assistant Professor of petroleum engineering at the Montana State College of Mineral Science and Technology. Since 1965, he has been a lecturer in petroleum engineering at the University of Southern California. Awarded a B.S. in petroleum engineering (1950) by Stanford University and an M.S. (1961) and Ph.D. (1971) in the same subject by the University of Southern California, Dr. Mannon is a member of the Society of Petroleum Engineers and of the American Association of Petroleum Geologists.

HERMAN H. RIEKE, III is, at present, a research scientist of the Continental Oil Company in Ponca City, Oklahoma. His professional training includes a B.S. in geology from the University of Kentucky (1959) and, from the University of Southern California, an M.S. in petroleum engineering (1964), an M.S. in engineering geology (1965), and a Ph.D. in petroleum engineering (1970). Dr. Rieke has been a member of the petroleum engineering department of the University of Southern California, and a staff engineer and geologist for Electro-osmotics, Inc., and International Reosurces Consultants, Inc. He is a member of SPE of AIME, GSA, and SEPM.